Tropospheric Ozone and Food Security

Ashutosh K. Pandey · Elina Oksanen
Editors

# Tropospheric Ozone and Food Security

## Interactions of Ozone with Other Major Abiotic Stresses and its Impact on Food Security

Springer

*Editors*
Ashutosh K. Pandey
Department of Life Science, Sharda School of Bio-Science and Technology
Sharda University
Greater Noida, Uttar Pradesh, India

Elina Oksanen
Department of Environmental and Biological Sciences
University of Eastern Finland
Kuopio, Finland

ISBN 978-3-032-03607-0 ISBN 978-3-032-03608-7 (eBook)
https://doi.org/10.1007/978-3-032-03608-7



This Springer imprint is published by the registered company Springer Nature Switzerland AG
The registered company address is: Gewerbestrasse 11, 6330 Cham, Switzerland

# Contents

# Chapter 1
# Tropospheric Ozone and Crop Production: Special Reference to Horticultural Crops and Its Associated Damages

**Chongtham Allaylay Devi and Ashutosh K. Pandey**

**Abstract** In the twenty-first century, global warming and climate change are the two outmost challenges for the mankind and nature. Production of horticultural crops, especially the established commercial kinds of fruits, vegetables, and flowers, will perform unpredictably poorly as a result of the climate anomaly. Many horticultural crops, including apples, saffron, rhododendrons, and orchids, require a freezing temperature to flower. Since the industrial era, the concentration of tropospheric ozone ($O_3$) has been rising and is anticipated increase future. $O_3$ is a greenhouse gas and a secondary air pollutant. $O_3$ molecules can cause serious health issues when inhaled by both humans and animals; it can also enter plant leaves majorly through stomata and cause considerable phytotoxic damages. Plants exposed to $O_3$ became more sensitive to abiotic and biotic stresses. $O_3$-induced changes in plants result in significant production and quality losses in commercial crops at the genetic, metabolic, and physiological levels. Most of the studies related to $O_3$-induced damage are confined to the crops, for example, soybean, wheat, rice, and mustard, but the in-depth understanding related to commercial horticultural crops are far from complete. This eventually causes crops' productivity and its quality to decline and can severely jeopardises the world's food security. This chapter aims to create understanding on the present commercial cultivars tolerant to $O_3$-induced damages, with an especial emphasis on the horticultural crops.

**Keywords** Climate change · Horticultural crops · Plants · Tropospheric ozone

C. A. Devi (✉)
Amity Institute of Horticultural Studies and Research, Amity University, Noida, Uttar Pradesh, India
e-mail: chongthamallaylaydevi@gmail.com

A. K. Pandey (✉)
Department of Life Sciences, Sharda School of Bio-Science & Technology, Sharda University, Greater Noida, Uttar Pradesh, India
e-mail: Aashu.20@gmail.com; Ashutosh.pandey1@sharda.ac.in

 
A. K. Pandey and E. Oksanen (eds.), *Tropospheric Ozone and Food Security*,
https://doi.org/10.1007/978-3-032-03608-7_1

## 1.1 Introduction

With its diverse soil and climate, Indian soil has been divided into several agro-ecological regions, which offers favourable growth conditions for the cultivation not only the staple food crops but also to a wide range of horticultural crops. These crops include fruits, vegetables, root and tuber crops, ornamentals, medicinal and aromatic plants, spices, condiments, plantation crops, and mushrooms, and they account for a sizeable portion of the nation's total agricultural output. 90% of the $O_3$ gas that makes up the Earth's atmosphere is located in the stratosphere. The remaining 10%, also referred to as natural background and is present in the troposphere but at a lower concentration Vingarzan (2004). However, since the start of the twenty-first century, background $O_3$ concentrations in the Northern Hemisphere have been rising by 0.5–2% year. Wagg et al. (2012) reported that the upward trend is anticipated to continue, with mean $O_3$ concentrations rising by 20–25% by 2050 and by 40–60% by 2100. Wheat and beans are one of the most $O_3$-sensitive crops, and it has been investigated that the losses associated with these crops may range from 10 to 48 and 27 to 52% respectively. It is commonly acknowledged that the legume family contains some of the evaluated species with the highest $O_3$ sensitivity reported by Mills et al. (2011), Oksanen et al. (2013); Hayes et al. (2007). Numerous modelling studies indicate that $O_3$ levels will continue to rise especially, in South and East Asia, exacerbating a serious threat on agriculture (Brauer et al. 2013).

Plants growing in the real field condition are exposed more to $O_3$-induced damages, therefore, the physiological problems associated with the horticultural crops will become more obvious, such as, mango's spongy tissue, the litchi's fruit shattering, flowers and fruits abscission, etc. Additionally, $O_3$-induced impacts on fruit quality and changes in its nutritional content of beans have been reported to decline significantly and their market value in nations for example, UK (Iriti et al. 2009; Mills et al. 2011). Several horticulture crops' yields were also severely reduced as a result of air pollution, therefore, in order to ensure their viability in under the present climate change scenario, it is necessary to conserve these valuable crops. Use of sustainable agricultural practice, use of renewable energy, water and forest resource conservation and reforestation, etc. and application of horticultural techniques such as, greenhouse technology more frequently are two strategies to maintain productivity while reducing the impact of climate change.

The two primary strategies to address the impact of O3-induced damages will involve (1) development of new horticultural crop cultivars that are tolerant of high temperatures, disease- and pest-resistant, short-lived, and capable of producing a good yield under stress conditions (2) legislation regarding the release of $O_3$ precursors. In addition, other strategies such as, the adoption of high-tech horticulture and prudent management of natural resources can be a potential way out to reduce the future food insecurity to meet the increasing food demand (Dutta 2013). Detailed studies, forward planning, and improved management are required to address these problems caused by climate change. Plants are exposed to numerous stress factors in the field conditions adversely affecting agricultural output and potential yields.

According to climate change predictions (Houghton et al. 2001), average air temperature will rise by 1.4–5.8 °C, atmospheric $CO_2$ concentration will rise, and rainfall patterns will vary significantly on global scale.

Comparing horticultural crops to annual food crops, it is more difficult to address climate change issues. And there is limited information available on how $O_3$ may affect horticultural crops, unlike the annual crops. Instead of direct $O_3$ deposition on the plant surface, stomata absorb $O_3$ into the interior of the leaf, which is what mostly causes $O_3$-induced plant damages. The degree of stomatal opening, which is often measured as stomatal conductance and highly reliant on external factors including the availability of soil water, can significantly affect stomatal intake of $O_3$. $O_3$, one of the most potent air pollutants, and has a negative effect on natural vegetation and crops (Feng et al. 2015). $O_3$ pollution is expected to worsen in the future as a result of a warming climate and, depending on the region, increased anthropogenic emissions of $O_3$ precursors like nitrogen oxides ($NOx = NO_2 + NO$), methane ($CH_4$), and carbon monoxide. As a result of population expansion and a global shift towards a more meat-intensive diet from basic foods, future food production is expected to increase by at least 50% by the year 2050 in order to meet the rapidly expanding global food demand (Alexandratos and Bruinsma 2012).

## 1.2 Major Causes for Climate Change

### *1.2.1 Variation in the Ocean*

The ocean plays a crucial role in the climate system, and its variability is well-known. Since it is hundreds of times more massive and has much higher thermal inertia than the atmosphere, some changes occur there on a longer time period. One example of this is how the ocean depths are still catching up to the atmosphere in terms of temperature adjustment from the little ice Age.

### *1.2.2 Orbital Variation*

Small variations in the Earth's orbit have a significant impact on the amount of sunlight that reaches the planet's surface and how it is distributed globally. The area has not undergone many changes there can be considerable fluctuations in the amount of sunshine per year, however both seasonally and regionally.

### 1.2.3 Changes in the Sun

The majority of the energy that the Earth receives comes from the sun. It is recognised that both long- and short-term oscillations in solar intensity affect the world's climate. The sun's output was just 70% as strong three to four billion years ago as it is today. Liquid water shouldn't have existed on Earth if the atmosphere's makeup had been the same as it is now.

### 1.2.4 Volcanic Eruptions

Gas and particle emissions from volcanic eruptions are released into the atmosphere. The eruptions are powerful enough to cool the climate for a few years and have an impact on it multiple times every century.

### 1.2.5 Tectonic Plate Movement

The movement of tectonic plates alters the topography of the earth's surface over the course of millions of years. This may have an impact on regional and local climate and ocean circulation patterns. The positioning of the continents affects patterns of ocean circulation by determining the geometry of the oceans. The positions of the seas play a significant role in regulating the movement of heat and moisture around the world and, consequently, in shaping the global climate.

### 1.2.6 Human Activities

The majority of climate scientists agree that there is a change in the climate, that this change is mostly due to human activity, and that this change is irreversible. The increase in $CO_2$ levels brought on by the burning of fossil fuels is the anthropogenic element that is most concerning, followed by cement production and aerosols (particulate materials in the atmosphere). The climate, microclimate, and measures of climatic variables are influenced by additional factors such as land use, ozone depletion, agriculture, and deforestation, both independently and in combination with other causes. Power plants account for 40% of carbon emissions, autos for 33%, deforestation for 20–25%, buildings for 12%, and aeroplanes for 3.5% of global warming.

### 1.2.7 Greenhouse Effect

One of the key causes of climatic change is the change of the greenhouse effect. Because of its special environmental circumstances and the presence of water, the Earth is the only planet in our solar system that can sustains life. An environment with plenty of oxygen and an appropriate surface temperature. There is an atmosphere with the right depth and of 30% or so of the incoming solar energy is reflected back into space, while the remaining energy, plays a part in warming the atmosphere, oceans, and land, preserving a 15 °C surface temperature on an average. The quantity of nitrogen and 21% of the atmosphere is made up of oxygen, which all species depend on to stay alive on which 0.036%, or a very small percentage, is made up of carbon dioxide which is essential for the process of photosynthesis. Solar energy is absorbed and infrared waves are simultaneously released from the land, sea, mountains, etc. Instead of being lost of this heat to space, all of this emitted heat is partially absorbed by certain gases, or GHGs, which are present in the atmosphere in very small amounts, comprising ozone, carbon dioxide, methane, nitrous oxide, water vapour, which causes the greenhouse effect (Kricksen 2008). Some of the horticultural crops which are sensitive to high temperature and advanced line tolerant to abiotic stress are illustrated in Tables 1.1 and 1.2, respectively.

## 1.3 Formation of Tropospheric Ozone

One oxygen atom per molecule makes up $O_3$, a gas. It is created in the atmosphere when sunlight interacts with volatile organic chemicals and nitrogen oxides. A major contributor to smog is ground-level $O_3$. In addition to damaging lung tissue, $O_3$

**Table 1.1** Horticultural crops and varieties susceptible to abiotic stress

| S. No. | Abiotic stress | Crops |
|---|---|---|
| 1 | High temperature | Peas, tomato, potato, beans, capsicum, banana, papaya, litchi, citrus, bael, rose |
| 2 | Low temperature | Tomato, brinjal, onion, drumstick, Indian gooseberry, ber, senna, phalsa, gonad, rose, jasmine, tropical orchids, carnation |
| 3 | Drought | Chilli, turnip, tomato, onion, pomegranate, custard apple, fig, grape, mango, banana, guava, black pepper, cardamom |
| 4 | Salinity | Onion, radish, potato, beans, melons, peas, mango, fig, citrus, grape, guava, custard apple, apple, pear, strawberry |
| 5 | Flooding/ excess moisture | Chilli, onion, tomato, papaya, early cauliflower, banana |

*Source* Singh (2010), Muthukumar and Selvakumar (2013)

**Table 1.2** List of some varieties and advanced lines tolerant to abiotic stresses

| S. No. | Tolerant | Crop | Varieties/rootstocks | Advanced line |
|---|---|---|---|---|
| 1 | Drought/rainfed | Tomato | Arka Vikas, Arka Meghali | RF-4A |
| | | Chilli | Arka Lohit | IIHR sel.-132 |
| | | Aonla | Goma Aishwarya | Clonal sel. NA-7 |
| 2 | Photo insensitive | Grape | Dogridge (Rootstock) | |
| | | Lablab Bean | Arka Jay and Arka Vijay | IIHR Sel.-16–2 |
| 3 | Frost | French Bean | Arka Anoop, Arka Bold | IIHR-19–1 |
| 4 | High Temperature | Cauliflower | Pusa Meghna | IIHR-316–1 IIHR-371–1 |
| | | Cucumber | - | Pusa Barkha |

*Source* Rai and Yadav (2005), Yadav et al. (2012), Muthukumar and Selvakumar (2013)

aggravates pre-existing respiratory conditions like asthma in humans. It damages leaf tissue, lowering the plant's ability to photosynthesise and harming delicate plants.

The stratosphere is home to the 90% layer of atmospheric $O_3$. Only 10% of the total $O_3$ column comes from the troposphere, but its concentrations have been continuously increasing during the past century. Carbon monoxide (CO), methane ($CH_4$), and other hydrocarbons undergo photochemical reactions with NOx ($NO + NO_2$) to produce tropospheric $O_3$. $O_3$ can also be destroyed through photochemical processes involving NO, $HO_2$, or OH. NOx is predominantly produced by the combustion of fossil fuels (63%) but is also a byproduct of the burning of biomass (14%), lightning (10%), soils (11%), and other minor sources. In addition to the direct evaporation of fuel, the use of solvents, the production of chemicals, and natural vegetation, hydrocarbons are also produced through the burning of fossil fuels.

## 1.4 Uptake of Ozone by Plants

The complicated process of plant uptake of $O_3$, which introduces $O_3$ into the plant canopy, involves micrometeorology. The turbulent atmosphere carries $O_3$ to the surface of plants. Once in the canopy, $O_3$ mainly enters tissue via stomata and can also be absorbed by surfaces such as stems, leaves, and soil. Upon exposure to $O_3$, reduced stomatal conductance has been frequently reported. The flow of oxygen from the troposphere into the leaves is influenced by a range of resistances at different levels. High levels of $O_3$ concentrations have been reported typically in the spring and summer seasons (Oksanen et al. 2013). Additionally, high boundary layer and stomatal resistances impede $O_3$ uptake under conditions such as, high radiation, high temperature, and low wind speed.

Hence, the width of stomatal pores determines the amount of $O_3$ absorbed (at a given atmospheric condition), with high uptake occurring under circumstances that favour stomatal opening and low uptake occurring under circumstances that favour stomatal closure. The levels of ambient $O_3$ and stomatal conductance both affect how much $O_3$ is absorbed by leaves. The regulation of stomatal opening ensures that the plant gets enough carbon dioxide ($CO_2$) and does not experience leaf water scarcity. $O_3$ is considered phytotoxic because of its high oxidant capacity (redox potential = +2.07 V) and ability to produce other toxic species including OH–, $O_2$-, and $H_2O_2$. The main consequences of $O_3$ take over when these harmful molecules damage the composition, structure, and function of the plasma membrane. Transportable across membranes, $H_2O_2$ can trigger the production of more harmful species and initiate a signalling cascade involving messenger molecules. Second messengers for the plant defence response have been suggested to include $H_2O_2$, salicylic acid, jasmonate acid, and calcium.

## 1.5 Antioxidant Response and Photosynthesis Induced by Ozone

Because of its oxidative properties, $O_3$ diffuses into the aqueous phase of the sub-stomatal cavity of plant leaves, where it produces reactive oxygen species (ROS) like superoxide anions ($O_2{}^{\circ}-$), hydrogen peroxide ($H_2O_2$), hydroxyl radicals ($OH^{\circ}$), and singlet oxygen ($1O_2$) (Fiscus et al. 2005). According to Ishida et al. (1999), these ROS have the potential to harm proteins, lipids, and nucleic acids. They can also harm leaves, induce early senescence, increase leaf abscission, decrease stomatal conductance, and diminish net $CO_2$ absorption, all of which can lower crop output. The intracellular balance between ROS production and removal must be skilfully maintained to prevent any potential harm to the cellular components produced by $O_3$-induced ROS. Antioxidants that are enzymatic and non-enzymatic effectively maintain this balance (Caregnto et al. 2013).

Sharma et al. (2012) suggested that the enzymatic components consist of various antioxidant enzymes, including glutathione peroxidase (GPX), catalase (CAT), and superoxide dismutase (SOD), as well as enzymes of the ascorbate–glutathione (AsA-GSH) cycle, including glutathione reductase (GR) and ascorbate peroxidase (APX). Carotenoids and phenolic substances are examples of non-enzymatic components, together with the two main cellular redox buffers, glutathione (GSH) and ascorbate (AsA). The activities of SOD, APX, CAT, and GR are raised in response to ambient and elevated $O_3$ levels (Liu et al. 2015; Ueda et al. 2013). Additionally, the production of ascorbate and glutathione is accelerated (Yendrek and Ainsworth 2015; Dumont et al. 2014). Plants adopted various strategies involving upregulation of antioxidative defence system against the $O_3$-induced oxidative stress. Similar results, i.e. increased

SOD, APX and GR enzymes activities have been reported in mustard cultivars treated with EDU and grown in area experiencing high-$O_3$ concentrations in the ambient field condition in India (Pandey et al. 2014). Enzyme activities and the antioxidant pools may even vary among the cultivars of the same crop indicating that the differential strategies adopted by the cultivars to combat against $O_3$-induced damages.

### 1.5.1 *Effect of Ozone on Horticultural Crops*

The past 40–50 years have seen an alarming increase in air pollution in developing nations, which could pose a serious threat to crop productivity. $O_3$, hydrofluoride, nitrogen oxide, and sulphur-dioxide and acid rain are the main air pollutants. Vegetable production is negatively impacted by ozone, which lowers growth, yield, and quality. The risk of air pollution increases when vegetable crops are cultivated near places with high population density. According to a recent study, ambient air pollution dramatically reduced the production of *Brassica oleracia, Lactuca sativa* and *Raphanus sativus* by up to more than 50%. The vegetable crops, such as, water melon, potatoes, tomatoes, squash, cantaloupe, peas, carrots, beets, turnips, etc., have been reported to be more sensitive to air pollution. When daily $O_3$ concentrations above 50 parts per billion, vegetable yields can be decreased by 5–15% Raj Narayan (2009 reported that if the daily $O_3$ concentration are above 50 parts per billion, the vegetable yield can be decreased by 5–15%. Numerous growth, physiological, biochemical, yield, and other related parameters have been investigated in relation to the effects of $O_3$. $O_3$ causes apparent damage on plants, concomitant with increased enzymes activities, SOD, catalase, reactive oxygen species and peroxidase leading to early senescence. On the other hand, $O_3$ has detrimental effects on the following: plant height; leaf area and number; biomass; photosynthetic rate; stomatal conductance; rubisco enzyme activity; ascorbic acid; C-allocation; methane and NOx emissions; soil microbial community; grain yield; straw yield; harvest index; grain protein; and glucose and starch in grains.

## 1.6 Effect on Microbial Content

Soil microbial biomass is related to plant roots and the breakdown of plant leftovers. As organic carbon availability limits microbial growth and activity, changes in soil N levels due to modified decomposition and a decrease in C imports may result in changes to the size and makeup of soil microbial biomass, which in turn may impact microbial metabolic activities. In peatlands, microbial biomass and organic acid concentrations are raised by long-term exposure to $O_3$. The organisation of the soil microbial community has changed or not changed as a result of $O_3$ (Agathokleous et al. 2020). $O_3$ appears to affect soil bacteria more so than soil fungi, but it appears to have very little influence on soil nematodes. Increased *Trichoderma* colonisation was observed in strawberry fields treated with high rates of $O_3$. This suggests that either an early microbial knockback occurred, or that nutrients that are favourable for *Trichoderma* sp. growth were released.

## 1.7 Effect on Germination

Seeds from oilseed rape plants that were exposed to 100 nl $l^{-1}$ $O_3$ for six hours during blooming showed a minor delay in germination when compared to control plants, according to Bosac (1992). Stewart (1998) discovered that when the terminal raceme of *Brassica compestris* plants were exposed to 100 nl $l^{-1}$ $O_3$ (6 h $d^{-1}$) once or more times during blooming, germination got enhanced. All raceme positions showed this response, indicating a widespread impact on seed development. On the other hand, while the ultimate germination percentage remained unchanged, the final germination percentage was delayed in plant seed whose vegetative and reproductive structures were both exposed to 70 nl $l^{-1}$ $O_3$ for 10 days prior to first blooming. *B. compestris* pods exposed to $O_3$ showed a premature germination of certain seeds, especially in older, more developed pods. Both *B. napus cv.* and *B. compestris* cv. Macro showed premature germination.

## 1.8 Effect on Phenology of Flowering and Pollinators

$O_3$ exposure triggered plants to flower more quickly, especially those growing in the autumn when temperature and light cues which normally encourage flowering, were weaker. Plants exposed to $O_3$ intended to underinvest in vegetative growth at the same time. Exposure of plants to $O_3$ did not significantly alter pollinator preferences; nevertheless, bumblebees visited more flowers on $O_3$-exposed plants, primarily because these plants showed more open flowers, which indicated that a greater attractiveness signal. In comparison to control plants, honeybees spent more time per bloom on $O_3$-exposed plants.

Despite the detrimental effects of $O_3$ on plant development, plants pollinated by bumblebees, honeybees, and mason bees kept their reproductive fitness due to the acceleration of flower formation and the behavioural reactions of pollinators to $O_3$-exposed plants.

The reproductive fitness of plants pollinated by hoverflies was reduced in response to $O_3$. For plants that flower for short periods of time or that, in contrast to wild mustard, do not have compensating mechanisms to withstand the lack of pollinator activity early in the flowering season, this can have a significant effect (Fig. 1.1). When compared to pollen from plants raised in clean air (the control) and incubated in an $O_3$-enriched atmosphere, the pollen from plants grown in the $O_3$ environment germinated much more slowly and the pollen tubes were shorter. Pollen collected from clean-air-grown plants that germinated in $O_3$-enriched air also had negative

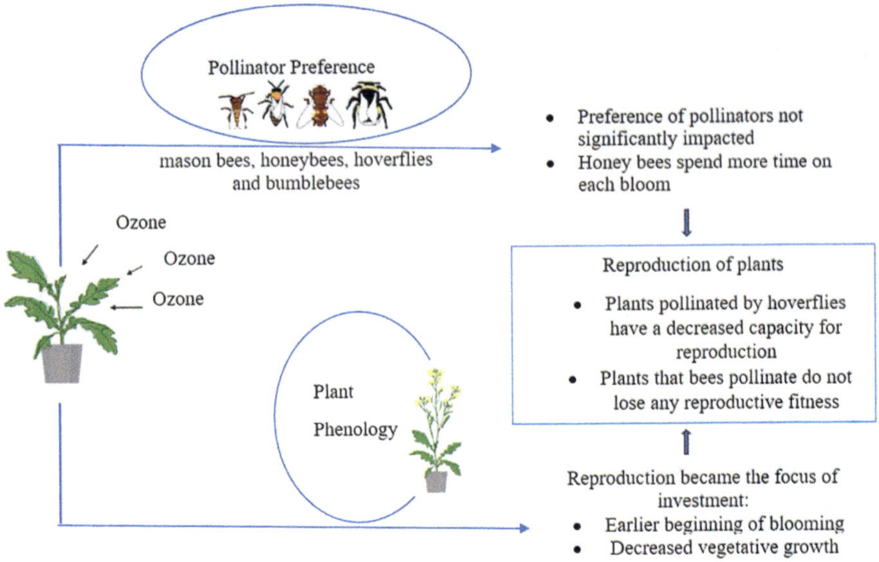


**Fig. 1.1** Impact of ozone stress on flowering phenology, different plant–pollinator interactions and plant reproductive success (*Modified Source* Duque et al. 2021)

impact. In comparison to the control, it also discovered that when pollen and plant had grown in $O_3$-enriched settings, fertilisation rates had decreased by 25% and also observed 26% lower percentage of viable seeds (Gillespie et al. 2015). The effects of $O_3$ on pollen may potentially be utilised in crop breeding initiatives to identify plant varieties that are tolerant of $O_3$ to provide food security in areas experiencing high-$O_3$ concentrations.

## 1.9 Effect on Plant Growth

Elevated levels of oxygen in the atmosphere have the power to impede various processes such as photosynthesis, carbon (sugar) production, and the allocation of carbon to roots and stems. They can also lessen the formation of carbohydrates in mycorrhizae, a symbiotic relationship between fungi and roots, as well as the uptake of vital minerals like N, P, K, and S, as well as root and stem growth. A steady rise in $O_3$ inhibits photosynthesis, speeds up leaf senescence, enhances leaf burn, irregularity necrosis and chlorosis (Fig. 1.2), slows down plant growth, and degrades yield characteristics, all of which have a substantial effect on agricultural productivity.

Crop leaves exposed to $O_3$ for an extended period of time experience an advanced beginning of senescence, and their rapid loss of protein and chlorophyll is caused by accelerated catalysis. Phloem loading is hampered by $O_3$, which also affects assimilation to roots and frequently reduces grain content while retaining carbs in

**Fig. 1.2** Examples of primary cotton leaves exposed to varying degrees of $O_3$. **a** No exposure; **b** 1200 nl $O_3$ exposure for 15 min; **c** 2400 nl $O_3$ exposure for 15 min. The arrows in **b** show "burn" on the leaf margin, whereas the arrow in **c** shows no necrosis close to the vein. Displaying the necrosis in **c** and the irregularity of the chlorosis in **b** 2018 (*Source* Grulke and Heath)

leaves. Thus, this leads to: (i) increased shoot/root biomass ratio; (ii) decreased harvest index (together with the consequence of a shorter grain filling period); and (iii) modified leaf chemistry. $O_3$ effects on development and allocation, rather than on photosynthesis and biomass buildup, may be more significant in reducing final yield. For a large variety of plant species, $O_3$ inhibits root growth more than shoot growth.

## 1.10 Effect on Yield Parameters and Crop Yields

Economic crop loss assessments have benefited from a plethora of data and exposure yield relationships that have been developed by research conducted in Europe and North America utilising the Open Top Chamber (OTC) approach with charcoal-filtered air, non-filtered ambient air, and $O_3$-enriched air. Horticultural crop such as tobacco, sugar beet, potato, oilseed rape, rice, maize, grape, and broccoli; barley, plum, and strawberry were categorised as $O_3$-tolerant; while the most $O_3$-sensitive crops were wheat, watermelon, pulses, cotton, turnip, tomato, onion, soybean, and lettuce. Because of the probable co-occurrence of $O_3$ peak levels with these crops' growing seasons, wheat, soybean and maize were found to be particularly sensitive to $O_3$. 1990 levels of $O_3$ resulted in a 23–27% drop in soybean crop load; by 2020, losses could surpass 30%.

The negative effects of ambient $O_3$-containing air pollution in a mixture with other contaminants on yield and seed quality of mung bean (*Vigna radiata*) throughout summer and of wheat (*Triticum aestivum*) and in the winter months, mustard (*B. campestris*). Grain number, grain mass, photosynthesis, biomass, and leaf area index (LAI) are all impacted by $O_3$. Raising cereals, pulses, vegetables, and oilseeds in various Asian nations has resulted in yield reduction owing to surface $O_3$ levels which is illustrated in Table 1.3. The influence of $O_3$ on distinct yield components

**Table 1.3** Yield loss in agricultural crops due to surface ozone in India

| Crops | Average $O_3$ Conc., h ppb | Yield reduction (%) |
|---|---|---|
| Wheat | 40–48 | 11–21 |
| Rice | 56 | 10–15 |
| Maize | 56 | 13–39 |
| Tomato | 37 | 24 |
| Mustard | 62 | 20 |
| Mungbean | 10–59 | 18–79 |

*Source* Bhatia et al. (2012)

seems to differ based on several aspects such as $O_3$ content, exposure circumstances, exposure time, crop species, and cultivars.

Many different types of agricultural production reductions are significantly impacted by air pollution. Between close to 0 and 39% of the total yield was lost, with high $O_3$ concentration showing the greatest decline in yield. Overabundance of $O_3$ uptake not only affects crop quality and development and production, but over time it may also modify the species and genetic makeup of semi-natural plant communities. The concentration of $O_3$ in the atmosphere is predicted to rise in the future in the absence of an efficient emission control system for $O_3$ precursors. Although it is also true that $O_3$ may limit the potential $CO_2$ aerial fertilisation impact in some plants, rising atmospheric $CO_2$ levels will probably lessen the harmful effects of $O_3$ on vegetation. Environmental $O_3$ pollution will continue to harm crop yield and quality in many parts of the world, necessitating more scientific analysis of the extent, distribution, and mechanism of the damage. Quantifying the effects of ambient $O_3$ in open field circumstances on newly developed crop cultivars, which are being produced without particular consideration of their sensitivity to $O_3$ requires understanding. To sustain and improve agricultural yields and nutritional quality, crop breeding programmes must include trait selection for increased plant tolerance to ambient $O_3$.

## 1.11 Effect on Postharvest Technology

$O_3$ has long been studied for its potential to prevent postharvest diseases and be used for various storage purposes. The assessment of $O_3$ for a variety of water treatment and air treatment applications in postharvest quality management is gaining popularity and generating a lot of empirical work. Apples, cherries, carrots, garlic, kiwis, onions, peaches, plums, potatoes, and table grapes are among the produce that has seen some commercial usage. Many infections that can cause harm can be inhibited by $O_3$, such as *Giardia* spp., *Cryptosporidium* spp., and *E. coli*. Pesticides and chemical residues can also be broken down by $O_3$. Food deterioration, food shrinkage, and cross-odour contamination can all be decreased by using $O_3$e, which is especially good at stopping the formation of mould.

In addition, $O_3$ prevents the growth of fungi, their sporulation, and the generation of mycotoxin. This is especially true for *Penicillium expansum*, which has been shown in multiple studies to suppress the production of patulin on a variety of fruits, including citrus, apples, and cherries. $O_3$ has no effect on the flavour, colour, or residue of food or drinkable water. It also doesn't leave behind any byproducts or residues. But in the presence of bromide ions, $O_3$ can create halogenated organic compounds. Extended shelf life of certain ethylene-sensitive goods has been associated with ozone destruction of ethylene in air filtration systems. Furthermore, it has been observed that ozone treatment induces natural plant defence response chemicals that are assumed to be involved in resistance to postharvest deterioration. Additional study is required to enhance the application on commercial needs.

## 1.12 Ozone Injury Symptoms on Plants

Plants affected by $O_3$-induced damages may exhibit interveinal chlorosis and necrosis, bleaching, stippling (little, darkly pigmented regions 2–4 mm in diameter), flecking (tiny, irregular, light-tan spots less than 1 mm in diameter), bronzing, and reddening. First to be impacted were often older and middle-aged leaves. Usually, the damaged area is limited to the top leaf surface. Affected leaf surfaces can be on the top or lower part of some plants. Similarities can be observed between the chlorosis and necrosis and typical leaf ageing and dying. A crop's susceptibility and particular symptoms vary from variety to variety. The degree of harm detected is also dependent on environmental factors and crop growth stage (Table 1.4). Cultivating tomatoes in $O_3$-rich had less weight, smaller size, and fewer seeds.

It was discovered through additional testing that $O_3$ also adversely damaged fruit quality by speeding up the ripening process, which resulted in 15% less vitamin C and less starch in the tomato fruit. These findings demonstrate the detrimental effects

**Table 1.4** Ozone affected on different stages of horticultural crops (*Modified source* Maynard 2020)

| | 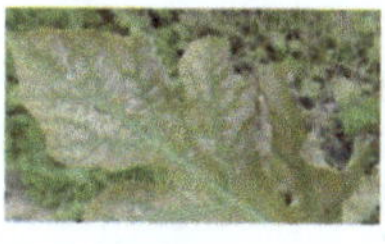 |  |  | 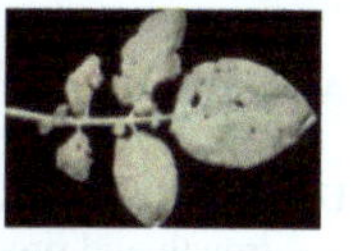 |
|---|---|---|---|---|
| Crops | Watermelon | Squash and Pumpkin | Snap beans | Potato |
| Occurs | Older leaves | Older leaves | Younger leaves | Younger leaves |
| Symptoms | Interveinal chlorosis, eventually turn dark and then white | Interveinal chlorosis and later appearance of bleaching | Bronzing appears on upper leaves and turn yellow and eventually die | Dark stippling appearance both in upper and lower leaves |

of $O_3$ pollution on tomato reproductive development, which eventually lowers fruit yield. In risk evaluations, the impact of ozone and perhaps other pollutants on pollen may serve as a quick warning of pollution-induced stress on crop plants.

## 1.13 Conclusion

Although with advancement in understanding different plants physiological mechanisms against $O_3$-stress, to assess agricultural productivity losses still remains challenging. Various factors, such as, crop, cultivar specific sensitivity to $O_3$, phenological age (vegetative or flowering), exposure time, different $O_3$-indices (AOT40, SUM60, CSRI, and POD), irrigation, growing season and other environmental factors, etc. influence the estimation of productivity losses due to $O_3$-induced damages.

Realising the fact, the most of the agricultural crops are still grown in the ambient field conditions, especially in the areas already experiencing high-$O_3$ risk, where multiple factors such as, other gaseous pollutants ($CO_2$, NOx, and $H_2S$), UV, heavy metal, nutrients, temperature, drought etc. may have interactive effects with $O_3$ in the real filed conditions. Therefore, studies on interactive effects of $O_3$ with other biotic and abiotic stresses are in the real field condition is warranted to meet the food security and nutritional demands.

## References

Agathokleous E, Feng Z, Oksanen E, Sicard P, Wang Q, Saitanis CJ, Araminiene V, Blande JD, Hayes F, Calatayud V, Domingos M (2020) Ozone affects plant, insect, and soil microbial communities: a threat to terrestrial ecosystems and biodiversity. Sci Adv 6(33)

Agrawal M, Singh B, Agrawal SB, Bell JNB, Marshall F (2006) The effect of air pollution on yield and quality of mung bean grown in peri-urban areas of Varanasi. Water Air Soil Poll 169:239–254

Agrawal M, Singh B, Rajput M, Marshall F, Bell JNB (2003) Effect of air pollution on peri-urban agriculture: a case study. Environ Poll 126:323–329

Alexandratos N, Bruinsma J (2012) Food and agriculture organization, world agriculture towards 2030/2050: the 2012 revision. FAO, Rome

Amundson RG, Kohut RJ, Schoettle AW, Raba RM, Reich PB (1987) Correlative reductions in whole-plant photosynthesis and yield of winter wheat caused by ozone. Phytopathol 77:75–79

Bhatia A, Tomer R, Kumar V, Singh SD, Pathak H (2012) Impact of tropospheric ozone on crop growth and productivity—a review. J Sci Ind Res 71:97–112

Brauer M, Freedman G, Frostad J, Van Donkelaar A, Martin RV, Dingenen RV, Estep K, Amini H, Apte JS, Balakrishnan K (2016) Ambient air pollution exposure estimation for the global burden of disease 2013. Environ Sci Technol 50(1):79–88

Bosac C (1992) The impact of ozone and sulphur dioxide on reproductive development in oilseed rape, Ph D thesis, University of Nottingham, UK

Caregnato FF, Bortolin RF, Divan JAM, Moreira JCF (2013) Exposure to elevated ozone levels differentially affects the antioxidant capacity and the redox homeostasis of two subtropical Phaseolus vulgaris L. varieties. Chemosphere 93(2):320–330

Datta S (2013) Impact of climate change in Indian horticulture—a review. Int J Sci Env Tech 2(4):661–671
Dohrmann AB, Tebbe CC (2005) Effect of elevated tropospheric ozone on the structure of bacterial communities inhabiting the rhizosphere of herbaceous plants native to Germany. Appl Environ Microbiol 71:750–7758
Dumont J, Keski-Saari S, Keinänen M, Cohen D, Ningre N, KontunenSoppela S, Baldet P, Gibon Y, Dizengremel P, Vaultier MN, Jolivet Y, Oksanen E, Le TD (2014) Ozone affects ascorbate and glutathione biosynthesis as well as amino acid contents in three Euramerican poplar genotypes. Tree Physiol 34(3):253–266. https://doi.org/10.1093/treephys/tpu004
Duque L, Poelman EH, Steffan-Dewenter I (2021) Effects of ozone stress on flowering phenology, plant-pollinator interactions and plant reproductive success. Environ Pollut 272
Feng ZZ, Hu EZ, Wang XK, Jiang LJ, Liu XJ (2015) Ground-level O-3 pollution and its impacts on food crops in China: a review. Environ Pollut 199:42–48. https://doi.org/10.1016/j.envpol.2015.01.016
Fiscus EL, Booker FL, Burkey KO (2005) Crop responses to ozone: Uptake, modes of action, carbon assimilation and partitioning. Plant, Cell Environ 28:997–1011
Gillespie C, Stabler D, Tallentire E, Goumenaki E, Barnes J (2015) Exposure to environmentally-relevant levels of ozone negatively influence pollen and fruit development. Environ Pollut 206:494–550
Gora JS, Verma AK, Singh J, Choudhary DR (2019) Climate change and production of horticultural crops from. Agric Impacts Clim Change 45–61
Grulke NE, Health RL (2018) Ozone effects on plants in natural ecosystems. Plant Biol 1–26
Hayes F, Jones M, Mills G, Ashmore M (2007) Meta-analysis of the relative sensitivity of semi-natural vegetation species to ozone. Environ Pollut 146:754–762
Horvitz S, Cantalejo MJ (2014) Application of ozone for the postharvest treatment of fruits and vegetables. Crit Rev Food Sci Nutr 54:312–339
Houghton J, Ding Y, Griggs D, Noguer M and Van der Linden P (2001) Climate Change 2001: the scientific basis. Published for the Intergovernmental panel on climate change. Cambridge University Press, Cambridge, 881p
Iriti M, Di Maro A, Bernasconi S, Burlini N, Simonetti P, Picchi V, Panigada C, Gerosa G, Parente A, Faoro F (2009) Nutritional traits of bean (*Phaseolus vulgaris*) seeds from plants chronically exposed to ozone pollution. J Agric Food Chem 57:201–208
Ishida H, Makino A, Mae T (1999) Fragmentation of the large subunit of ribulose-1,5-bisphosphate carboxylase by reactive oxygen species occurs near gly-329. J Biol Chem 274:5222–5522
Kricksen PJ (2008) Conceptualizing food systems for global environmental change research. Glob Environ Chang 18(1):234–245
Liu X, Sui L, Huang Y, Geng C, Yin B (2015) Physiological and visible injury responses in different growth stages of winter wheat to ozone stress and the protection of spermidine. Atmos Pollut Res 6:596–604
Maynard L (2020) Ozone injury on vegetable plants, vegetables crop hotline, Newsletter Perdeu University, p 678. Online source: https://vegcropshotline.org/article/ozone-injury-on-vegetable-plants/
McKee IF, Long SP (2001) Plant growth regulators control ozone damage to wheat yield. New Phytol 152:41–51
Mills G, Buse A, Gimeno B, Bermejo V, Holland M, Emberson L, Pleijel H (2007) A synthesis of AOT40-based response functions and critical levels of ozone for agricultural and horticultural crops. Atmos Environ 41:2630–2643
Mills G, Harmens H (2011) Ozone pollution: a hidden threat to food security. NERC/Centre for Ecology & Hydrology, Lancaster, UK
Mills G, Hayes F, Norris D, Hall J, Coyle M, Cambridge H, Cinderby S, Abott J, Cooke S, Murrells T (2011) Impacts of ozone pollution on food security in the UK: a case study for two contrasting years, 2006 and 2008; Report for Defra Contract

AQ08610. Available online: https://icpvegetation.ceh.ac.uk/impacts-ozone-pollution-food-security-uk-case-study-twocontrasting-years-2006-and-2008. Accessed on 13 July 2021
Muthukumar P, Selvakumar R (2013) Glaustas horticulture. New Vishal Publications, New Delhi, p 544
Narayan R (2009) Air pollution–a threat in vegetable production. In: Sulladmath UV, Swamy KRM (2009) International conference on horticulture (ICH-2009) Horticulture for livelihood security and economic growth, pp 158–159
Oksanen E, Pandey V, Pandey AK, Keski-Saari S, Kontunen-Soppela S, Sharma C (2013) Impacts of increasing ozone on Indian plants. Environ Pollut 177:189–200
Pandey AK, Majumder B, Keski-Saari S, Kontunen-Soppela S, Pandey V, Oksanen E (2014) Differences in responses of two mustard cultivars to ethylenediurea (EDU) at high ambient ozone concentrations in India. Agr Ecosyst Environ 196:158–166
Phillips RL, Zak DR, Holmes WE, White DC (2002) Microbial community composition and function beneath temperate trees exposed to elevated atmospheric CO2 and O3. Oecologia 131:236–244
Pryor A (2001) Field trials for the combined use of ozone gas and beneficial microorganisms as a pre plant soil treatment for tomatoes and strawberries. Pest Management Grants Final Report (California Deptt Pesticide Regulation, California, USA), p 18
Rai N, Yadav DS (2005) Advances in vegetable production. Researcher Book Center, New Delhi
Sharma P, Jha AB, Dubey RS, Pessarakli M (2012) Reactive oxygen species, oxidative damage, and antioxidative defense mechanism in plants under stressful conditions. J Botany 1–26
Singh HP (2010) Ongoing research in abiotic stress due to climate change in horticulture. In: Curtain raiser meet on research needs arising due to global climate scenario, Baramati, Maharashtra, 29–30 Oct 2010, pp 1–23
Stewart CA (1998) Impact of ozone on the reproductive biology of Brassica campestris L. and Plantago major L., Ph D thesis, University of Loughborough
Suslow TV (2004) Ozone applications for postharvest disinfection of edible horticultural crops. Publication 8133. University of California, Division of Agriculture and Natural Resources, Oakland
Ueda Y, Uehara N, Sasaki H, Kobayashi K, Yamakawa T (2013) Impacts of acute ozone stress on superoxide dismutase (SOD) expression and reactive oxygen species (ROS) formation in rice leaves. Plant Physiol Biochem 70:396–402
Vingarzan R (2004) A review of surface ozone background levels and trends. Atmos Environ 38:3431–3442
Wagg S, Mills G, Hayes F, Wilkinson S, Cooper D, Davies WJ (2012) Reduced soil water availability did not protect two competing grassland species from the negative effects of increasing background ozone. Environ Pollut 165:91–99
Yadav RK, Kalia P, Singh SD, Varshney R (2012) Selection of genotypes of vegetables for climate change adaptation. In: Climate change impact, adaptation and mitigation in agriculture: methodology for assessment and application. Division of Environmental Sciences Indian Agricultural Research Institute, New Delhi, vol 110, p 012
Yendrek CR, Ainsworth EA (2015) A comparative analysis of transcriptomic, biochemical, and physiological responses to elevated ozone identifies species-specific mechanisms of resilience in legume

# Chapter 2
# The Combined Effect of Tropospheric Ozone and Elevated Concentration on Agricultural Crops

**Ashish K. Chaturvedi, Amit Kumar, Divya Singh, Athul Mohanan, Chongtham Allaylay Devi, and Ashutosh K. Pandey**

**Abstract** Significant contributors to both regional and global climate change include carbon dioxide ($CO_2$) and tropospheric ozone ($O_3$), which have a significant effect on crop plant development and productivity and are anticipated to have an impact on food production. Plants absorb $O_3$ through their leaves, which causes an excess of reactive oxygen species (ROS) to build up in the mesophyll and protect their cell walls. ROS can impede photosynthetic electron transport and harm the ultrastructure of chloroplasts. $O_3$ has the ability to cause stomatal closure and modify stomatal conductance, which can impede the fixation of $CO_2$. Thus, there is a detrimental impact on the growth and reproductive capabilities of plants. Therefore, the biggest effects of $O_3$ stress on plants are a decrease in crop output and a degradation in crop quality. Increased atmospheric $CO_2$ concentrations will probably lessen the harmful effects of $O_3$ on plants, even though elevated ozone may have a direct impact on the many development and metabolic processes in agricultural crops such as pollination and seed set increased in response to elevated $CO_2$, conversely, $O_3$ adversely impact on plant reproductive features, outweighing the benefits of $CO_2$. Present study highlights the significant risks to agricultural crop production, including the possibility

A. K. Chaturvedi (✉) · A. Mohanan
Centre for Water Resources Development and Management, Kozhikode, Kerala, India
e-mail: ashispc@gmail.com; ashispc@cwrdm.org

A. Kumar
ICAR-National Bureau of Plant Genetic Resources, Regional Station, Jodhpur, Rajasthan, India

D. Singh
CSB- Regional Sericultural Research Station, Central Tasar Research and Training Institute, Bhimtal, Nanital, India

C. A. Devi
Amity Institute of Horticultural Studies and Research, Amity University, Noida, Uttar Pradesh, India

A. K. Pandey (✉)
Department of Life Sciences, Sharda School of Bio-Science & Technology, Sharda University, Greater Noida, Uttar Pradesh, India
e-mail: Aashu.p20@gmail.com; Ashutosh.pandey1@sharda.ac.in



A. K. Pandey and E. Oksanen (eds.), *Tropospheric Ozone and Food Security*,
https://doi.org/10.1007/978-3-032-03608-7_2

that future yield improvements brought about by higher $CO_2$ concentrations could be offset by a broad increase in $O_3$ concentrations. It also demonstrated how elevated $O_3$ and elevated $CO_2$ affect plant physiology, including disease, growth, yield and the biochemical and nutritional quality of agricultural crops.

**Keywords** Biotic stress · Carbon dioxide · Ozone · Climate and agricultural crops

## 2.1 Introduction

Carbon dioxide ($CO_2$) and tropospheric ozone ($O_3$) are critical for atmospheric health, both of which have significant effects on agricultural plant development and productivity and are predicted to have an impact on food supply, especially, in the South Asian subcontinent (IPCC 2013; Ghude et al. 2014). Climate trends have revealed a fast rise in the concentrations of $CO_2$ and $O_3$ in several agricultural regions across the world over the last few decades (Lobell and Gourdji 2012). Plant productivity is often stimulated by $CO_2$ (Ainsworth and Long 2005), however $O_3$ is phytotoxic to various plant species (Agathokleous et al. 2018). The ideal conditions for $O_3$ generation are higher irradiance, urbanization, raised temperatures, and growing quantities of precursors emitted from metropolitan areas. According to the IPCC Fifth Assessment Report, there is a projected elevation of approximately 8 ppb (25% of present concentration) in background tropospheric $O_3$ and an estimated increase in $CO_2$ concentration to 700 ppm by the year 2100 (IPCC 2013). Surface $O_3$ will have experienced its most significant rise in the South Asian region by the year 2030 (an average annual increase of 7.2 ppb) (Lenka and Lenka 2012). Plants growing in the field, on the other hand, are frequently counter multiple abiotic stress at a given time (i.e. stress combination), putting new-fangled exclusive metabolic adjustments plants (Zandalinas et al. 2021; Rivero et al. 2022). The plant metabolome significantly influences the plant's characteristics, particularly during stress, shaping its phenotype. Thus, a better understanding of the metabolic retort of plants to varied stress provocations in the field is critical for enhancing crop stress tolerance (Fang et al. 2019; Carrera et al. 2021; Sharma et al. 2021a, b). The greatest flux involved with the global carbon cycle is photosynthesis (Friedlingstein et al. 2019).

Anthropogenic activities are generating global changes that have far-reaching consequences for many areas of ecosystem functioning, including photosynthetic rates (Fernåndez Martnez et al. 2019). In agricultural environment, interactions between $O_3$ and $CO_2$ might occur. elevated $CO_2$ typically alleviates or adjourn the adverse impacts of $O_3$ on crop productivity (Kumari et al. 2013; Mishra et al. 2013).

The presence of both biotic and abiotic stress factors can alter the signalling pathways responsible for plant defence mechanisms, making plants more susceptible to endemic infections as well as potentially novel pathogens. It is crucial to comprehend the impact of climate change on the occurrence and intensity of diverse biotic stress factors to develop resilience in plants against changing climatic conditions (Leisner and colleagues 2022). Plants respond to individual biotic and/or abiotic challenges, as well as concurrent biotic as well as abiotic pressures, in intricate and exclusive ways across cellular, transcriptional, and physiological mode (Zhang et al. 2022). Plants may adopt varies strategies in order to cope with abiotic stresses such as, stomatal closure, reduced photosynthesis, upregulation of antioxidative defence system, and increased root length. Plants respond differently to different diseases (bacterial, fungal, and viral) based on pathogens nature (biotrophic or necrotrophic) leading to stomatal closure, decreased photosynthesis and water use efficiency, imbalance in the ROS detoxification, phytoalexin synthesis, and localized cell death. Both biotic and abiotic stress may collectively decrease productivity and crop quality (Bilgin et al. 2010; Grimmer et al. 2012; Kissoudis et al. 2014).

## 2.2 Higher $O_3$ in Atmosphere: Threat to Plants Survival

The one of the sources of tropospheric ozone is the downward movement of the stratospheric $O_3$ in the environment. It contributes about 10% of total tropospheric $O_3$ chemical synthesis (Stohl et al. 2003). Plants absorb $O_3$ along with $CO_2$ during photosynthesis. It disrupts plant physiological systems by elevating the concentration of free radicals in cell fluids. Other effects of tropospheric $O_3$ include decreased plant growth, seed formation, and leaf area (Mills et al. 2011). Prolonged exposure to high concentration of ozone can interfere with crucial vital processes such as stomatal structure, functionality, photosynthesis, respiration, and volatile organic carbon (VOC) emissions (Monks et al. 2015). Atmosphere in Indian sub-contents is burdened with higher tropospheric concentration of $NO_2$ and $O_3$ (Lu et al. 2018). According to Lal et al. (2012), industrial and anthropogenic emissions caused a positive trend in tropospheric ozone in north-east India from 1979 to 1992. In 2003, model simulations and measurements revealed an increase in ozone in the Indo-Gangetic Plain (IGP) (Roy et al. 2008). In India, there has been a steady increase in TPO. During this period, IGP has the highest TPO concentrations (35–50 DU). This is a result of various human-induced activities such as traffic-related pollution, industrial emissions, a dense population, and natural sources including biogenic and pyrogenic factors (Rathore et al. 2023).

The surge in the $O_3$ concentrations has been reached up to 40 ppb from 10 ppb since eighteenth century primarily because of escalated anthropogenic emissions. Meanwhile, atmospheric $CO_2$ concentration have surged from 320 ppm in 1960 to 412 ppm by 2019 (NOAA 2020). Tropospheric ozone ($O_3$) is produced as a byproduct of $NO_x$ photochemical oxidation in the presence of CO, $CH_4$, and non-methane hydrocarbons (Myhre et al. 2013). Tropospheric ozone ($O_3$) is a Short-Lived Climate Forcers that contributes to global and Arctic warming (AMAP 2015; Sand et al. 2016; von Salzen et al. 2022). Since the mid-1990s, a mix of relatively weak positive and negative trends (+1 to 1 ppbv per decade) in the NH at the surface as well as in free troposphere have been reported, with the greatest increases over south and eastern Asiatic region, due to enhanced anthropogenic emissions (Cooper et al. 2020; Wang et al. 2021).

Most of the chemistry-climate models (CCMs) indicate that the $O_3$ layer in the tropical stratosphere (lower boundary) will continue to decrease throughout the twenty-first century (Zubov et al. 2013; Banerjee et al. 2016; Dhomse et al. 2018; Keeble et al. 2021). This reduction in the thickness is determined by the GHG climate scenario (Morgenstern et al. 2018; Dhomse et al. 2018; Keeble et al. 2021; Shang et al. 2021). The trends of future evolution of atmospheric $O_3$ on a global/regional scale using CCMs on shared socioeconomic pathways scenarios, which is developed by a collaborative group of worldwide climate scientists, economists and energy systems in order to give the climate change research community a toolkit for conducting comprehensive analysis across multiple disciplines (Keeble et al. 2021; Shang et al. 2021; Revell et al. 2022), remain uncertain as these assessments do not bear the statistical support such as multivariate regression analysis.

Tropospheric $O_3$ is a noteworthy secondary greenhouse gas with the ability to generate positive radiative forcing in the upper atmosphere. Its secondary air pollutants that can impairs human health by producing respiratory disease, degrades vegetation, and hinders the biosphere from absorbing carbon (Sitch et al. 2007). According to IPCC, AR6 reported that Southeast Asia had the largest increase in $O_3$ levels by 4–6 ppb per decade since 1994 than East Asia or South Asia (Szopa et al. 2021). This estimation takes into account the surface observations (Cooper et al. 2020) and aircraft measurements (Gaudel et al. 2020). The increase in $NO_x$ emissions has resulted in an unanticipated increase in $O_3$ concentrations in Southeast Asia (Zhang et al. 2021). Tropospheric $O_3$ concentration persistently increase in many crop growing regions throughout the world as a result of inadequate global control of precursor pollutants and short and long-distance pollution transport (Lin et al. 2017). Secondary air pollutants, such as tropospheric $O_3$, are also expected to rise this century (IPCC 2021). While the expected negative effects of global warming on agricultural output and the accompanying socioeconomic consequences on food systems have been thoroughly studied (e.g. Cushman et al. 2022; Slattery and Ort 2019). These studies indicate that $O_3$ currently poses a significant threat to crop productivity and will persist as a serious concern in the future. Ozone leads to diverse damages in agricultural crops such as visible injury, decreased photosynthesis rate, changes in carbon allocation and decline in both quantity and quality of yield (Emberson et al. 2018). $O_3$-induced decreases in Ps rate have been linked to decreases in stomatal conductance (Vandermeiren et al. 2002), photosynthetic pigment (Nie et al. 1995), Rubisco activity and/or content (Long and Naidu 2002), photochemical efficiency, and enzymatic activities (Vandermeiren et al. 2005). A decline in chlorophyll concentration is considered to be a sign of metabolic damage induced by $O_3$ stress (Rai et al. 2011). The chlorophyll content decreased over time, notably accelerated in the presence of $ACO_2 + EO_3$, attributed to increased $O_3$ penetration into leaves and enhanced leaf senescence. Interestingly, Bindi et al. (2002) results show that total chlorophyll content increased under $eCO_2 + AO_3$ and $eCO_2 + eO_3$ compared to $ACO_2 + AO_3$. Reduced nitrogen uptake under $eCO_2$ due to lower transpiration rates leads to decreased protein concentration (Polley et al. 1999). $O_3$-induced ROS alters the structures of cellular proteins (Pell and Dann 1991). These physiological responses result in significant reduction in crop yield with rising $O_3$

(Ainsworth 2017a, b). Additionally, under $O_3$ stress, stomatal conductance and photosynthesis in mature leaves can become uncoupled, often causing greater reductions in measured photosynthetic rate than in stomatal conductance (Lombardozzi et al. 2012).

The experiments indicate that elevated $O_3$ can detrimentally affect both the photosynthetic properties and biomass production of maize (*Zea mays* L.), a model $C_4$ plant (Bagaed et al. 2015; Yendrek et al. 2017). The distinctive ratio of photosynthetic $CO_2$ assimilation to stomatal conductance, and confinement of Rubisco carboxylation in the bundle sheath cells, suggests that $C_4$ photosynthesis might react differently to elevated $O_3$ compared to $C_3$ photosynthesis. Notably, increased $O_3$ accelerates the decline in photosynthetic capacity (Yendrek et al. 2017; Cook et al. 2014). Simultaneously, atmospheric concentration of $CO_2$ is also rising. Although this is not anticipated to directly stimulate photosynthesis in $C_4$ crops like maize (Leaky et al. 2006; Leakey 2009), the elevation in $CO_2$ could decrease stomatal conductance, potentially reducing the flux of $O_3$. Consequently, further experimental investigations are required to comprehend the interplay between increasing $O_3$ concentration and the escalating challenges of drought, heat stress and rising atmospheric $CO_2$ (Yendrek et al. 2017). The rise in MDA content is directly correlated with the $O_3$ exposure level (Sarkar et al. 2010). Moreover, higher solute leakage observed under these conditions indicate additional membrane- level damage (Iglesias et al. 2006). Elevated concentration of free radicals and ROS triggers a heightened activity of antioxidant enzymes, serving as a defence mechanism against oxidative stress induced by $O_3$ (Severino et al. 2007; Mishra et al. 2013). Furthermore, photosynthetic pigment loss especially chlorophyll was also lower under $eCO_2 + eO_3$ than $ACO_2 + AO_3$. Based on this output, it can be envisaged that elevated $CO_2$ will augment the antioxidative system to avert the ROS effects produced by ambient and elevated $O_3$ (Kumari et al. 2015a, b).

Primarily $O_3$ enters in the plant through stomata, however plant absorbs $O_3$ by both stomatal and non-stomatal pathways (Ainsworth 2017a, b). In $C_3$ plants, elevated level of $O_3$ reduces photosynthesis and stomatal conductance whereas increases rate of respiration (Ainsworth 2008; Feng et al. 2008; Morgan et al. 2003). This decrease in primary metabolism can dimmish plant growth rates, leaf area and accumulation of biomass (both above and below ground) (Ainsworth 2008). $O_3$ has a direct impact of plant reproductive development (Black et al. 2000; Leisner and Ainsworth 2012). During reproductive stages, $O_3$ exposure can lead to a more significant decline in photosynthesis compared to vegetative growth in certain plants (Ainsworth 2008; Feng et al. 2008; Morgan et al. 2003). Under high level of $O_3$, plants become more susceptible to attacks by necrotrophic pathogens, root-rot fungi and insect like bark beetles. Interestingly, prior exposure to elevated $O_3$ might reduce obligate biotrophic infections (Karnosky et al. 2002; Sandermann 2000). Additionally, $O_3$ exposure boosts the activity of various enzymes in phenylpropanoid, flavonoid and lignin pathways that are crucial for plant defence mechanisms (Kangasjarvi et al. 1994).

## 2.3 Rising $CO_2$: Affecting Plant Productivity

Atmospheric $CO_2$ concentrations are projected to increase at an alarming rate, potentially doubling by the end of this century (Högy et al. 2010; Leaky et al. 2009; IPCC 2013). This persistence increases in atmospheric $CO_2$, attributed to human activities and industrialization, has escalated from 316 ppm in 1959 (Keeling et al. 2009) to 400.14 ppm in 2015 (ESRL 2015) and could reach potentially reach 570 ppm by 2050 (IPCC 2013). In 2022, the global average $CO_2$ concentration reached a new high record of 417.06 ppm. Between 2021 and 2022, there was a 2.13 ppm increase in atmospheric $CO_2$ concentration. Presently, atmospheric $CO_2$ stands at 50% higher than pre-industrial revolution concentration. Over the past six decades, the annual rate of $CO_2$ rise has surged approximately 100 times faster than natural increases in previous eras, similar to that observed at the end of the last ice age spanning 11,000–17,000 years ago (Annual report from NOAA's Global Monitoring Lab). Since the mid-twentieth century, fossil fuel burning emissions have steadily increased each decade, escalating from nearly 11 billion tonnes of $CO_2$ annually in the 1960s to an estimated 36.6 billion tonnes in 2022 as per Global Carbon Budget 2022. According to the observations of NOAA Global Monitoring Lab in the year 2021, $CO_2$ alone responsible for 2/3rds of the overall warming impact caused by all anthropogenically produced greenhouse gases.

Photosynthesis serves as the gateway for carbon to enter vegetation, playing a pivotal role in determining food production. Anthropogenic activities are profoundly affecting ecosystem functions globally, including the rate of photosynthesis (Fernández Martínez et al. 2019). Notably, the rapid increase in global temperatures and atmospheric $CO_2$ concentration exceeds earlier predictions (Schwalm et al. 2020). Higher concentrations of $CO_2$ in the atmosphere stimulate the growth and yield of $C_3$ crops due to higher photosynthesis rate and reduced photorespiration (Cure and Acock 1986; Poorter 1993; Rogers and Dahlman 1993), leading to increased grain production (Ainsworth et al. 2008). However, this increase comes with a trade-off major crop plants exhibit decreased nutritional value under these conditions (Myers et al. 2014; Soares et al. 2019). Moreover, studies indicate that the detrimental effects of elevated $CO_2$ on nutritional quality can be moderated under higher temperatures (Köhler et al. 2019). Additionally, photosynthetic acclimation involves a shift in nitrogen allocation away from ribulose-bisphosphate carboxylase/oxygenase towards enzymes related to the thylakoid electron transport chain and the regeneration of inorganic phosphate (Pi) (Sage et al. 1989; Webber et al. 1994).

Elevated $CO_2$ being a photosynthetic substrate, enhance the $C_3$ photosynthetic output by 20–45% by saturating the Rubisco and decreasing photorespiration activities (Leakey et al. 2009a, b; Walker et al. 2016), which is not a case for $C_4$ plants (Leakey et al. 2009a, b). As a prime effect, $eCO_2$ decreases stomatal density and/or stomatal aperture, thus disease incidences and infection process through stomata need to understand further under these circumstances (Ainsworth and Rogers 2007; Eastburn et al. 2011; Li et al. 2015). Additionally, elevated $CO_2$ may decrease the leaf and seed nutrient content due to reduced transpiration and reduced bulk flow (Ebi and Loladze 2021; Ebi et al. 2021; Loladze et al. 2019).

## 2.4 How Does Interaction May Affect

The interactive effects of the elevated $CO_2$ and elevated $O_3$ are highly complicated and subjective to selective as well as combined factor response with dominant physiological process such as the crop types and response to the different stress, proximal environmental condition and threshold of the $CO_2$ and $O_3$. Elevated $CO_2$ as alone is responsible to increase C fixation rate in $C_3$ plants and reduces transpiration losses. Elevated $O_3$ as alone is responsible to reduces the photosynthesis by damaging the plant tissues. Thus, elevated $CO_2$ bears the positive effects and $eO_3$ bears the negative effects on crop yield. Fiscus et al. (1997) observed that under doubled ambient $CO_2$ concentration helps to reduce the $O_3$ flux (up to 35%) at soyabean canopy. Under elevated $O_3$ the photosynthetic rate, photosynthetic pigments, Fv/Fm ratio, protein functioning, mesophyll activities reduce and increases in oxidative stress, total phenolics has been observed through different investigations.

There are evidences that elevated $CO_2$ is able to compensate the negative effects of the $eO_3$ in agricultural crops, these, compensation is subjective to the relative elevation in the concentration of the $CO_2$ and $O_3$ and specific to crop growth stage and exposure time-period (Kumari et al. 2015a, b; Mina et. al. 2017; Bhatia et al. 2021a, b). However, the pronounced effects on the other factors such as nutrients dynamics, growth regulator dynamics, response to the particular physiological and biochemical pathways and duration of oxidative stress, etc. need to be understand fully.

## 2.5 Effects of Carbon Di Oxide on Crop Morphology and Anatomy

Elevated $CO_2$ influences the growth and development by inducing the morphological and anatomical changes, which directly yields the qualitative and quantitative output of the agroecosystem. Zhang et al. (2021) found the higher biomass production in tobacco under elevated $CO_2$ condition. Significant research reports assessing role of elevated $CO_2$ on crop growth and yield has been mention in Table 2.1.

**Table 2.1** Significant research reports assessing role of elevated $CO_2$ on crop growth and yield

| Location | Crop | $CO_2$ concentration (ppm) | Experimental condition | Effects | Mechanism/mode of actions | Reference |
|---|---|---|---|---|---|---|
| University of Copenhagen, Denmark<br>55°67′ N, 12°30′ E | Tobacco ($C_3$ plant) | $eCO_2$: 800 | Climate chamber | • Increased biomass (leaf and stem),<br>• $P_n$ and water use efficiency (WUE);<br>• Reduced $g_s$, water loss per unit $CO_2$ assimilation and $T_r$ | • Decrease in stomatal density or stomatal pore size is attributed to decline in $g_s$ that partly offset the stimulation of Rubisco carboxylation, decreases $CO_2$ diffusion into leaf and hence lowers $C_i$ leading to increased WUE<br>• Enhanced $P_n$ increases biomass and may also induce accumulation of solutes leading to slight osmotic stress condition that raises WUE of plant | Zhang et al. (2021) |

(continued)

**Table 2.1** (continued)

| Location | Crop | $CO_2$ concentration (ppm) | Experimental condition | Effects | Mechanism/mode of actions | Reference |
|---|---|---|---|---|---|---|
| Hebei University of Engineering, Handan, China | Soybean ($C_3$ plant) | $eCO_2$: 600, 800, 1000, 1200, 1400 and 1600 | Walk-in environmental chamber | • Down-regulation of $P_n$, maximum carboxylation rate ($V_{cmax}$) and maximum electron transport rate ($J_{max}$) at high $CO_2$ concentration (> 600 ppm);<br>• Reduced stomatal density, stomatal area, stomatal conductance and mesophyll size | • The photosynthesis may down-regulated due to insufficient nitrogen availability, lower Rubisco enzyme activity, reduced stomatal conductance and accumulation of excessive carbohydrates<br>• The decreasing $V_{cmax}$ and $J_{max}$ indicate that high $CO_2$ concentration may influence both light and dark reactions of photosynthesis<br>• A decline in mesophyll cell area has restricted space to accommodate chloroplast<br>• A positive correlation between mesophyll cell area and photosynthesis indicates that reduction in spongy and palisade cell area in mesophyll affects $P_n$ | Zheng et al. (2019) |

(continued)

**Table 2.1** (continued)

| Location | Crop | $CO_2$ concentration (ppm) | Experimental condition | Effects | Mechanism/mode of actions | Reference |
|---|---|---|---|---|---|---|
| Tsukuba FACE experimental facility, Tsukubamira, Ibaraki, Japan 3°8′ N, 139°60′ E | Rice varieties: Koshihikari and Takanari ($C_3$ plant) | $eCO_2$: 576–578 | FACE | • Decreased TCA cycle enzymes and $CO_2$ efflux rate on the basis of leaf dry weight at all growth stages in both varieties<br>• As compared to Koshihikari, Takanari leaves having high $CO_2$ efflux rate, major amino acids (glutamate, glutamine, aspartate, glycine, alanine, valine) and total amino acids, nitrogen content and low organic acid content | • The high major amino acid contents and low organic acid contents in Takanari leaves indicate utilization of organic acids in nitrogen assimilation that ultimately encourages high photosynthesis rate<br>• A decline in the TCA cycle enzyme activity in Koshihikari leaves may lead to reduce $CO_2$ efflux rates under $eCO_2$ condition, however, decreased $CO_2$ efflux rates in Takanari leaves could not be explained as the measured enzyme activities was maximum | Noguchi et al. (2018) |

(continued)

**Table 2.1** (continued)

| Location | Crop | $CO_2$ concentration (ppm) | Experimental condition | Effects | Mechanism/mode of actions | Reference |
|---|---|---|---|---|---|---|
| Shanxi Agricultural University, Taigu, Shanxi, China 37.42° N, 112.55° E | Foxtail millet ($C_4$ plant) | $eCO_2$: 590 | OTC | • High biomass (above ground), WUE, $WUE_i$ and $P_n$<br>• Increased $\phi_{PSII}$, qP, ETR, $g_s$ and $T_r$ in first year while decreased in 2nd year | • Increased leaf weight per unit area and production of panicle contribute to higher above ground biomass under elevated $CO_2$<br>• Higher WUE and $WUE_i$ contributes to positive growth. The high photosynthesis is correlated with increased ETR and $\phi_{PSII}$<br>• Enhanced PSII activity generates ATP and NADPH that are utilized in the fixation of additional carbon by photosynthesis | Li et al. (2019a, b, c) |
| National Institute of Crop Science, Wanju, Jeollabukdo, Korea 35°84′34″ N, 127°04′44″ E | Potato ($C_3$ plant) | $eCO_2$: 800 | KSPAR | • Lower biomass and leaf area at early growth stage whereas relatively higher biomass and canopy $P_n$ at late growth phase, slight increase in tuber yield<br>• Decline in $g_s$ and chlorophyll content at late growth phase | • The non-significant increase in tuber yield may be the effect photosynthesis acclimation due to long-term exposure of $eCO_2$ or relatively high density of plants followed by decreased chlorophyll content, reduced $g_s$ and Rubisco activation state | Lee et al. (2020) |

(continued)

**Table 2.1** (continued)

| Location | Crop | $CO_2$ concentration (ppm) | Experimental condition | Effects | Mechanism/mode of actions | Reference |
|---|---|---|---|---|---|---|
| State Key Laboratory Genetics and Germplasm Enhancement, Nanjing Agricultural University, China 32°02′ N, 118°50′ E | Carrot ($C_3$ plant) | e$CO_2$: 3000 | Chamber | • Reduced root/shoot ratio and xylem area | • e$CO_2$-induced stomatal closure reduces $CO_2$ fixation. The alteration in $CO_2/O_2$ ratio causes morphological changes. The newly synthesized xylem in root is different from the constitutive xylem. The reduction in xylem region shows growth inhibition by e$CO_2$ at 3000 ppm | Wu et al. (2017) |

(continued)

**Table 2.1** (continued)

| Location | Crop | $CO_2$ concentration (ppm) | Experimental condition | Effects | Mechanism/mode of actions | Reference |
|---|---|---|---|---|---|---|
| Faculty of Sciences, University of Copenhagen, Taastrup, Denmark | Amaranthus ($C_4$ plant) | $eCO_2$: 800 | Greenhouse cell | • Enhanced plant water consumption, $P_n$ and WUE, no difference in leaf area and plant biomass between $aCO_2$ and $eCO_2$<br>• Declined $g_s$, stomatal density | • The ability of plant to concentrate $CO_2$ and reducing its loss by leakage in bundle sheath cells increases $P_n$<br>• Decreasing $g_s$ in response to short-term exposure to $eCO_2$ may be due to stomatal closure induced by interaction of carbon anhydrases with ABA<br>• However, long-term exposure decreases stomatal density and shows ABA-induced stomatal response. Low $g_s$ conserves soil water that prolonged the duration of plant water use | Wei et al. (2022) |

(continued)

**Table 2.1** (continued)

| Location | Crop | $CO_2$ concentration (ppm) | Experimental condition | Effects | Mechanism/mode of actions | Reference |
|---|---|---|---|---|---|---|
| Institute of Environment and Sustainable Development in Agriculture, Chinese Academy of Agricultural Sciences, Changping, Beijing, China 40.13° N, 116.14° E | Mung bean ($C_3$ plant) | $eCO_2$: 550 | Mini FACE | • High biomass and yield, $P_n$, WUE, NPQ, no significant difference in the maximum velocity of carboxylation and maximum rate of electron transport<br>• Low $g_s$, $T_r$, $F_v'/F_m'$ and $\phi_{PSII}$ | • Despite of high $P_n$, no significant changes in maximum velocity of carboxylation and maximum rate of electron transport<br>• High reduction in $F_v'/F_m'$ and $\phi_{PSII}$ at pod maturity state indicates that photosynthetic activity of leaves becomes down-regulated because of restriction to new sink development at pod maturity stage<br>• Source-sink relationship affects photosynthetic activity of leaves | Ji et al. (2015) |

(continued)

**Table 2.1** (continued)

| Location | Crop | $CO_2$ concentration (ppm) | Experimental condition | Effects | Mechanism/mode of actions | Reference |
|---|---|---|---|---|---|---|
| Division of Plant Physiology, Indian Agricultural Research Institute, **New Delhi, India** | chickpea desi (Pusa 1103) and kabuli (Pusa 1105) | $eCO_2$ 570 ± 45 | OTCs | • Plant growth, photosynthesis and carbohydrate assimilation increased<br>• Increased Starch and sugar content and rubisco activity<br>• Expression of rbcS and rbcL genes was higher under e $CO_2$ during flowering stage<br>• Shoot biomass and seed yield increased | • Glycolate oxidase activity decreased under $eCO_2$ and the reduction was greater in desi compared to kabuli genotype, suggesting suppressed photorespiration in both the chickpea<br>• Chickpea crop can perform better under $eCO_2$ due to increased rate of photosynthesis and suppressed photorespiration<br>• Between two types of chickpeas, desi may respond better to $eCO_2$ owing to its higher sink potential than kabuli types | Rai et al. (2016) |

Table covers the targeted crops, effects on the plant growth and possible mechanism/mode of action along the study location, experimental setup with exposed $CO_2$ concentrations in individual research works

$P_n$: net photosynthetic rate; $g_s$: stomatal conductance; $C_i$: intercellular $CO_2$ concentration; $C_a$: external $CO_2$ concentration; $T_r$: transpiration rate; $WUE_i$: intrinsic water use efficiency; $WUE_n$: instantaneous water use efficiency; $\phi_{PSII}$: effective quantum yield of PSII photochemistry; qP: photochemical quenching coefficient; ETR: highest photosynthetic electron transport; $F_v/F_m$: maximum quantum efficiency of PSII; $F_v'/F_m'$: intrinsic efficiency of PSII

### 2.5.1 Leaf Growth and Expansion

Elevated $CO_2$ concentration can stimulate leaf growth and expansion, resulting in increased leaf area. This has the potential to increase crop productivity by improving photosynthesis and carbon assimilation. Li et al. (2021a, b) suggested an increase of 17% in leaf area index in $C_3$ crops through a metanalysis investigation of 110 studies. Liu et al. (2021) suggested that elevated $CO_2$ increased leaf size, thickness, and nitrogen content, as well as photosynthetic rate and biomass production. Liu et al. (2021) advocated that under elevated $CO_2$ resulted in significant increase in the biomass of organs serving for carbon sink but also increased both weight and leaf area of flag leaf in plants. De Souza et al. (2020) observed delayed leaf-senescence, enhanced leaf size in soyabean under elevated $CO_2$ conditions. Wang et al. (2021) observed the enhancement in the leaf area (length and width both) and number of cells in leaf in maize under elevated $CO_2$ conditions. Singh et al. (2020) observed enhanced leaf area and biomass and grain yield in wheat under elevated $CO_2$ conditions.

### 2.5.2 Reduced Leaf Thickness

Elevated $CO_2$ concentration can reduce leaf thickness in some crops by inhibiting the production of structural carbohydrates. This has the potential to change the balance between carbon assimilation and allocation to structural tissues, potentially affecting crop growth and development. Effect of elevated $CO_2$ concentration on leaf thickness in soybean plants showed that exposure to elevated $CO_2$ concentration led to a decrease in leaf thickness, which could contribute to improved photosynthetic efficiency and reduced water loss (Porter and Nagel 2000). Miglietta et al. (2005) suggested that elevated $CO_2$ decreased leaf thickness in grapevine which could contribute to improved grape and wine quality. Martínez-Lüscher et al. (2019) suggested that exposure to elevated $CO_2$ concentration led to a decrease in leaf thickness, which could contribute to improved water use efficiency and growth through grapevine example.

### 2.5.3 Increased Root-to-Shoot Ratio

Elevated $CO_2$ conditions can promote root growth and development, leading to a larger root-to-shoot ratio. This has the potential to increase agricultural output by improving fertilizer and water uptake (Poorter and Nagel 2000; Kimbell et al. 2002). Cotrufo et al. (1998) demonstrated that elevated $CO_2$ concentration decreases the nitrogen concentration of plant tissues while increasing the root–shoot ratio, which could lead to improved nutrient use efficiency in wheat. Pal et al. (2019) also suggested the positive impact of elevated $CO_2$ in the root–shoot ratio, architecture, and morphology of sorghum plants under water stress and found helpful in addressing water use efficiency in water stress environments.

### 2.5.4 *Stomatal Density*

Higher stomatal density is generally considered for higher water loss through transpiration. The decrease in stomatal density and or their opening time is directly affecting the transpiration losses of any crops. This transpiration loss is highly significant under dry environments or water stress conditions. Hetherington and Woodward (2003a, b) suggested that elevated $CO_2$ helps plants for their internal water balance and managements through lowering the stomatal density. Lake and quick, (2006) also suggested that stomatal conductance have significant role in plant water use efficiency and growth which chiefly governed by environmental conditions. (Franks and Beerlings 2009) suggested that change in the stomatal density is a time-driven process which depends on the prevailing environmental condition such as elevated $CO_2$. Li et al. (2019a, b, c) also provides the similar evidences in immature apple leaves under $eCO_2$ conditions. Liu et al. (2015) and Li et al. (2019) concluded the same for saline environments and drought stress through sorghum and cotton examples respectively. Bunce, (2020) and Zheng et al. (2019) also advocates that elevated $CO_2$ concentration is responsible to the reduces stomatal density, stomatal area, stomatal conductance and mesophyll size in potato and soyabeans respectively.

### 2.5.5 *Effects of Carbon Di Oxide on Crop Physiology*

There are plenty of the investigation has been carried out related to the positive influences of the elevated $CO_2$ on the crop physiology. These positive results include the increased physiological growth and reproductive development such as increased in biomass, seed mass, leaf numbers, leaf area, number of tillers, number of flowers, seeds, nutrient absorption etc. (Jablonski et al. 2002; Bhatia et al. 2013, 2021a, b; Tomer et al. 2015; Mina et al. 2010). As a result of elevated $CO_2$ reproduction is often increased, due to increased allocation of resources and a higher number of nodes (and branches) that enhances the available sites for flower and fruit set. Ward and Strain (1997), suggested that the increased reproductive capacity under elevated $CO_2$ can also be a function of increased plant size rather than increased biomass allocation to reproductive structures. Increased number of fully filled pollen grains per flower, reduced viable pollen grains. Production of large number of pollens with low viability indicates that quality of pollens under elevated $CO_2$ has been compromised (Marshall et al. 2010). Earlier studies suggest that under elevated $CO_2$ increases reproductive output (flower number, fruit number, and seed production) of herbaceous species (Garbutt and Bazzaz 1984, Curtis et al. 1994, Curtis et al. 1996, Ward and Strain 1997). However, Billings and Billings (1983) observed opposite results in alpine and arctic annual, e.g. *Saxifraga flagellaris* Willd., shows reduction in reproductive response under elevated $CO_2$ conditions. The following are some of the effects of increased $CO_2$ on crop physiology:

#### 2.5.5.1 Increased Photosynthesis

$CO_2$ is a major substrate for photosynthesis, and higher $CO_2$ concentration in the atmosphere can boost photosynthesis rates in many crop species. This has the potential to increase biomass production and yield. Increased shoot fresh weight, leaf area, chlorophyll content, fruit yield, net assimilation rate, $P_n$ and decreased $g_s$ High $P_n$, net assimilation rate and fruit yield of the plants grown under elevated $CO_2$ indicates greater sink capacity of fruit in comparison to shoot (Mavrogianopoulos et al. 1998). Ainsworth and Long (2005) compared the 32 FACE (Free-Air $CO_2$ Enrichment) experimental data related to soybean, wheat, rice, maize, cotton, and sorghum and to determine the effects of elevated $CO_2$ on photosynthesis, canopy properties, and plant production. Ainsworth and Long, (2005) suggested that elevated $CO_2$ led to increased photosynthesis (31%) in all six crops, and concluded that elevated $CO_2$ increased the plant production (up 23%) in most of the crops. The legumes are having the potential to use the elevated $CO_2$ concentration through stimulated photosynthesis process along with increased $N_2$ fixation (Rogers et al. 2009). Legume species have benefit to acquire more N and produce more seeds at elevated $CO_2$ than non-nitrogen fixing species (Miyagi et al. 2007) however, Hasegawa et al. (2013) provide different observation and suggested that elevated $CO_2$ does not necessarily enhance plant growth or crop yield and also showed large variation even within same species. Enhanced $P_n$ increases biomass and may also induce accumulation of solutes leading to slight osmotic stress condition that raises WUE of plant (Zhang et al. 2021). Enhanced PSII activity generates ATP and NADPH that are utilized in the fixation of additional carbon by photosynthesis (Li et al. 2019a, b, c). Kumar et al. (2012) suggested that under elevated $CO_2$ conditions, due to increased photosynthetic activity, and increased availability of assimilates and N can prolong, stay green character, delayed leaf senescence and crop maturity.

### *2.5.6 Changes in Plant Growth and Development*

Elevated $CO_2$ concentration can affect plant growth and development, such as, increased crop height and biomass and altered flowering and fruiting timing. Vanaja et al. (2007) reported that in black gram (*Vigna mungo* L.) at 700 ppm $CO_2$ concentration, showed an improved, length and weight of root and shoot, root-to-shoot ratio, leaf area and plant weight. The root length/root weight ratio was found reduced and root weight/root length ratio was however, significantly increased under elevated condition in mung bean. The $CO_2$ enrichment ($580 \pm 20$ μmol) caused greater expansion of leaves resulting in increased in area per unit leaf at all growth stages as well as in increase in vegetative and reproductive dry matter in chickpea (Saha et al.,2015). A positive effect of elevated $CO_2$ on photosynthesis and growth are also reported in some pulse crops such as cowpea (Mbikayi et al. 1983), green gram (Srivastava et al. 2001), black gram (Vanaja et al. 2007), lentil (Nasser et al. 2008) and soybean (Ainsworth et al. 2002). If chickpea was exposed at elevated $CO_2$ of 550 $\pm$ 50 ppm increased shoot length (Prasad et al. 2006), total number of branches, leaf area plant$^{-1}$, seed yield, increased leaf and shoot dry weight (Pal et al. 2004). Imai et al. (1985) reported increased harvest index (HI) under elevated $CO_2$, whereas Baker et al. (1992), Allen et al. (1996), Ziska et al. (1997) and Kim et al. (2001) reported an opposite trend in harvest index. In pigeon pea, the total biomass increases (91.3%) at elevated (550 $\pm$ 50 ppm) $CO_2$. The biomass response of initial growth stages of mung bean at elevated $CO_2$ was higher (55%) because of higher relative growth rate (Das et al. 2002). The plant biomass decreased towards the end of the growing period (Fangmeier et al. 1996; Pinter et al. 1996; Mulholland et al. 1997). Elevated $CO_2$ increases carbon fixation enables plants to utilize more photosynthates for its growth and leaf area in chickpea. Increase in leaf area also an additional factor for carbon fixation. Moreover, there is possibility of increased nitrogen fixation under elevated $CO_2$ in chickpea (Singh and Diwakar 1995), eventually, utilized by the plants to meet growth demands (Gamper et al. 2005). Increased plant height, leaf area, root, shoot, total biomass is observed in Wheat (Pal et al. 2005). High $P_n$ under elevated $CO_2$ influences growth and productivity, elevated $CO_2$ may reduce $T_s$ due to decline in $g_s$. Enhanced Pn and reduced $T_s$ increases WUE (Mamatha et al. 2014). Higher WUE and $WUE_i$ contributes to positive growth. The high photosynthesis is correlated with increased ETR and $\phi_{PSII}$ (Li et al. 2019a, b, c). Plant growth, photosynthesis and carbohydrate assimilation increased in chickpea (Rai et al. 2016).

### *2.5.7 Yield and Yield Contributing Parameters*

Higher leaf area with increased photosynthesis rate led to increment in plant biomass and yield (Ruhil et al. 2014). The elevated $CO_2$ concentration significantly enhance the seed yield of mung bean ranged from 19 to 23% at 575–630 ppm elevated $CO_2$ level but no effect on harvest index. The grain yield showing 1.5 times increment with elevated $CO_2$, improves its yield potential (Uprety et al. 1995). The increased seed yield, pod weight plant$^{-1}$ and total seed weight were reported in chickpea, mung bean, lentil (*Lens culinaris*) and soybean (Uprety et al. 1995; Ainsworth et al. 2002; Ziska and Blowsky 2007; Naseer et al. 2008; Twine et al. 2013; Saha et al. 2012). The 7.2% increment in number of filled seed rather than increment in weight of filled seed showing it to be an important yield contributing component with elevated $CO_2$. Developing seeds can represent a large carbon sink for the plant and elevated $CO_2$ can also lead to heavier seeds reported by Ziska and Bunce (2007) in soybean and Vanaja et al. (2007) in black gram but no such effect was observed in chickpea (Saha et al. 2015). Increase in yield attributes viz. total number of pods, pod dry weight per plant also reported in different crops in: broad bean (*Vicia faba* L.) (Wu and Wang

2000), rice (Kim et al. 2003), wheat (Wu et al. 2004; Franzaring et al. 2010), rapeseed (*Brassica napus* L. cv Mozart) (Franzaring et al. 2011) and barley (Manderscheid et al. 2009).

In pigeon pea, under elevated $CO_2$ condition pods plant$^{-1}$ decreased, test weight and harvest index are decreased by 5.1% and 19–38% respectively (Saha et al. 2012). Increased carbon concentration showed no effect seed nutrient content, unlike previous reports in case of legumes (Miyagi et al. 2007). Mayer et al. (2014) have investigated that zinc and iron concentration were reduced in the grains of wheat, rice and soybean (Phaseolus vulgaris) that were grown in atmospheres containing the elevated concentration of $CO_2$. The nutritional quality of mung bean leaves declines substantially due to a dilution of nitrogen under elevated $CO_2$ in atmosphere. Prasad et al. (2002) suggested an increase of 24% in yield of kidney bean exposure to 700 ppm $CO_2$. In addition, yield of pulses is also enhanced by nitrogen fixation. An increased amount of $N_2$ fixation (20–86%) in chickpea and field pea (*Pisum sativum*) (44–51%) reported by Lam et al. in 2012. elevated $CO_2$ increase carbon supply from leaves to nodule that increases nodule size and nodule number per plant. Increased carbon supply to nodule also stimulated nitrogenase activity and there is direct co-relation between carbon supply and nitrogenase activity (Tissue et al. 1997; Cen and Layzell 2004).

### 2.5.8 Reduced Stomatal Conductance

Elevated $CO_2$ concentration can cause crop stomata to partially close, reducing water loss through transpiration and may improve the crop water use efficiency. Higher atmospheric $CO_2$ concentration influence positively through a decrease in stomatal conductance. Reduction in transpiration due to elevated $CO_2$ was observed in mung bean. elevated $CO_2$ increase stomatal conductance in mung bean. Morison, (1987) suggested that stomatal aperture is reduced in all species regardless of photosynthetic rate increased and there was no evidence that stomata acclimate in response to elevated $CO_2$. An investigation on tobacco revealed that the decreased stomatal density or stomatal pore size resulted in the decline in stomatal conductance, which partly offset stimulation of rubisco carboxylation (Zhang et al. 2021). Increased WUE can be attributed to reduced $g_s$ due to low stomatal density and stomata closure under elevated $CO_2$ condition (Maroco et al. 1999). The decrease in $g_s$ and $T_r$ may be the result of $CO_2$-induced partial closure of stomata (Prior et al. 2010). Stomatal density is declined and which results in the reduction in stomatal conductance (Wei et al. 2022).

### 2.5.9 Transpiration and WUE

It has been envisaged that the evapotranspiration and water use efficiency of crop will be affected under present climate change scenario. Transpiration remains to be the vital factor to understand soil–water–plant relationship and their interactions (Mo et al. 2007). Mung bean varieties significantly increase in rate of photosynthesis, stomatal resistance, leaf water potential, root growth and leaf dry matter production and decrease in rate of $CO_2$ efflux and rate of transpiration (Uprety et al. 1995). Pigeon pea plants when grown under elevated $CO_2$ showed higher transpiration rates since plants extracted more moisture, and even with higher amount of water use, plants had 22.1% higher water use efficiency as compared to than ambient grown plants (Saha et al. 2012). The enhanced $CO_2$ significantly depressed the rate of transpiration but significant increase in leaves water potential in mung bean. Increasing $CO_2$ causes partial closure of stomata so as to fall in stomatal conductance. WUE and transpiration of leaves drop off due to decrease in stomatal conductance, but transpiration decrease is not in direct proportion to the decrease of stomatal conductance because leaf temperature increases up to 1–2 °C which in turn increase vapour pressure which is the driving force for transpiration. This slight increase in temperature, crop transpiration might decrease slightly in elevated $CO_2$, but WUE will increase if temperatures rise (Allen et al. 2003).

### 2.5.10 ROS and Antioxidant Enzymes

Elevated $CO_2$ concentration not only effect the primary metabolism, but also triggers increased $pCO_2/pO_2$ ratio, with consequent reduction of ROS (Halliwell and Gutteridge 1989), formation in the plant cell, and effective elimination of photorespiration in $C_3$ plants (Bowes 1991). As a consequence, to ROS imbalance in the cell, the antioxidant defence system which consists of, metabolites, as most abundant glutathione (GSH), enzymes responsible for the scavenging of oxidants, superoxide dismutase (SOD), nonspecific peroxidases (i.e. guaiacol peroxidase, GPX,) and catalase (CAT); and antioxidant-regenerating enzymes of the ascorbate–glutathione cycle, such as the NADPH-dependent glutathione reductase (GR) decline (Badiani et al. 1993; Foyer et al. 1994). ROS imbalance can lead peroxidation of membrane lipid, and thiobarbituric acid reactive substances (TBARS) are produced (Esterbauer 1982).

### 2.5.11 Changes in Nutrient Content

Elevated $CO_2$ concentration can alter crop nutrient content, often lowering the concentration of essential nutrients like nitrogen and phosphorus especially in non-leguminous crops, which can have substantial implications for human and animal nutrition. The photosynthesis may down-regulated due to insufficient nitrogen availability, lower Rubisco enzyme activity, reduced stomatal conductance and accumulation of excessive carbohydrates (Zheng et al. 2019). Higher rate of photosynthesis causes increment in leaf thickness. Elevated $CO_2$ enhances increased photosynthesis ability per unit of N uptake even if N is limiting (Pal et al. 2005). The alterations in macro and micro-nutrient content in leaves affect the nutritional quality of plant (Jain et al. 2007). Increased C/N ratio and which speed up the senescence process and may raise nitrogen availability through release of rubisco and nitrogen from old leaves (de la Mata et al. 2012). Seed protein, macro and micronutrients decreased drastically under elevated $CO_2$. Elevated $CO_2$ may dilute the accumulation of higher carbohydrate and biomass of seed proteins and macro and micronutrients (Pal et al. 2014).

### 2.5.12 Increased Susceptibility to Pests and Diseases

Some research suggests that elevated $CO_2$ concentration can make crops more susceptible to pests and diseases, though the mechanisms behind this are unknown. Even with the introduction of various pest and disease management measures, estimates indicate that 10–16% of crop production losses occur due to plant pathogen globally. Elevated $CO_2$ effects pests indirectly, by changing the architecture of their host plant and weakening its defences system against stress conditions. Elevated $CO_2$ and high temperatures could also provide conducive environment for pathogens like fungi. Predicting the impacts of climate change on plant disease incidence is complex and challenging, due to interactions among multiple factors such as, plants, pathogens and the environment (Brajesh et al. 2023). Entomologists suggested that elevated $CO_2$ may impact the distribution, abundance and performance of insects that feed on plants (Lincoln et al. 1984, 1986; Fajer et al. 1989). Under the present climate change scenario even the predators, parasites and pathogens that maintain certain level of control over insect populations (Stiling et al. 1999). Plants under elevated $CO_2$ showed, lower concentration of nitrogen and higher C: N ratios which have been associated with compensatory feeding and subsequent increased damage or defoliations (Lincoln et al. 1986, 1993; Fajer et al. 1989; Lindroth et al. 1993, 1995; Salt et al. 1995; Docherty et al. 1997; Kinney et al. 1997; Williams et al. 1997).

## 2.6 Effects of Tropospheric Ozone on Crop Morphology

Tropospheric $O_3$ is a secondary air pollutant that can have serious consequences for crop morphology, resulting in decreased crop productivity and quality. Some of the effects of tropospheric $O_3$ on crop morphology includes regulation of enzyme activity, determination of protein structure and control of transcription factor activity depends upon the maintenance of cellular thiol redox. Thiol antioxidants may have several mechanisms of action especially function as a component of thiol/disulfide redox buffer. In cell membrane, $O_3$ generates ozonide, $H_2O_2$ and reactive aldehydes by reacting with polyunsaturated fatty acids. Aldehydes may damage proteins by reacting with some amino acids leading to increase in carbonyl groups content. The reduced production of phenolic compounds may be due to limited carbon fixation and energy generation that is used to fulfil basic needs for survival or irreversible damage to enzymes involved in the synthesis of phenolic compounds (Bortolin et al. 2014). Significant research reports assessing role of elevated $O_3$ on crop growth and productivity has been mentioned in Table 2.2.

### *2.6.1 Decreased Leaf Area and Reduced Photosynthesis*

Tropospheric $O_3$ can cause visible injury to crop leaves, resulting in a decrease in leaf area. This has the potential to reduce crop productivity by limiting photosynthesis and carbon assimilation. Reduced leaf area index results in the reduced radiation use efficiency and above ground biomass. Harvest index is also reduced due to reduced yield (Singh et al. 2021; Chaudhary and Rathore 2022). The leaf area and leaf area index were decreased with total chlorophyll, carotenoid, carbohydrates, and carotenes content. Reducing carotenoid may protect PSII from photoinhibition (Chaudhary and Rathore 2022). Ozone influences chlorophyll indirectly by accelerating leaf senescence that enhances deterioration of chlorophyll. Decline in carotenoid and protein content may be attributed to free radicals-induced oxidative destruction (Kumari et al. 2013).

### *2.6.2 Reduced Plant Height*

Tropospheric $O_3$ can inhibit stem elongation, causing plants to grow shorter. This can have an impact on crop architecture, potentially resulting in lower crop yield. Plant height and weight is reduced at high concentration of $O_3$ (Dhevagi et al. 2022). Plant height is decreased and biomass also reduced accordingly. This may due to the down regulation of photosynthesis and stomatal conductance, translating to reduce in crop productivity (Al-Muwayhi et al. 2014).

**Table 2.2** Significant research reports assessing role of elevated $O_3$ on crop growth and yield

| Location | Crop/varieties | $O_3$ concentration (ppb) | Experimental condition | Effects | Mechanism/mode of action | References |
|---|---|---|---|---|---|---|
| Wetland farm, Tamil Nadu Agricultural University, Coimbatore, India 11.00° N, 76.92° E | Blackgram variety: CO 6, VBN 1, VBN 2, VBN 3, VBN 4, VBN 5, VBN 6, VBN 7, VBN 8 | 50 and 100 | OTC | Early necrotic spots on younger leaves<br>• Increased leaf injury index at high concentration of ozone,<br>• Reduced plant height, weight,<br>• Reduced chlorophyll content, $P_n$ and $g_s$ | • Increases ROS production causing leaf injury<br>• Chloroplast damage inhibiting chlorophyll synthesis<br>• Restriction of the entry $O_3$ though stomatal closure decreasing $g_s$<br>• Stomatal closure declines entry of $CO_2$ reducing Photosynthetic C assimilation thereby reduced growth and biomass | Dhevagi et al. (2022) |
| ICAR-Indian Agricultural Research Institute, New Delhi, India 28° 35′ N, 77° 12′ E | Chickpea | 60 | FAOCE | • Early maturity<br>• Reduced leaf area index, radiation use efficiency, above ground biomass, yield, harvest index | • Reduced crop duration by escalating all the phenological stages<br>• Higher Ozone increases temperature resulting in early maturity of crop<br>• Decline in transpiration per unit leaf due to stomatal closure<br>• Decrease in leaf area index contribute towards low radiation use efficiency and crop yield | Singh et al. (2021) |

(continued)

**Table 2.2** (continued)

| Location | Crop/varieties | $O_3$ concentration (ppb) | Experimental condition | Effects | Mechanism/mode of action | References |
|---|---|---|---|---|---|---|
| Central University of Gujarat, Gandhinagar, Gujarat, India | Cotton | 75 | OTC | • Visible injury, chlorosis, necrotic spots<br>• Early leaf senescence<br>• Reduced leaf area, biomass, absolute growth rate, relative water content, yield and harvest index | • Higher injury in vegetative stage of plant suggesting attention for the reproductive success rather defence mechanism<br>• Reduction in leaf area and leaf weight may increase leaf thickness indicating protection against ozone<br>• Reduction in relative water content enhances membrane damage by lipid peroxidation | Chaudhary and Rathore (2022) |

(continued)

**Table 2.2** (continued)

| Location | Crop/varieties | $O_3$ concentration (ppb) | Experimental condition | Effects | Mechanism/mode of action | References |
|---|---|---|---|---|---|---|
| Central University of Gujarat, Gandhinagar, Gujarat, India 23.2156° N, 72.6369° E | Groundnut cultivars: TG-37A, TPG-41, TAG-24, GG-20, Dh-86 | 75 | OTC | • Early leaf senescence, reduction in leaf area, leaf area index, dry matter, total chlorophyll, carotenoid, carbohydrates and protein content and yield<br>• Increased root/shoot ratio | • Increased stem length and decreased stem mass causing weaker stem under elevated ozone<br>• Enhanced root/shoot ratio indicates distribution of photosynthate towards underground part of plant<br>• Reduced carbohydrates, chlorophyll pigment synthesis and disruption in membrane integrity reduces photosynthesis and influences biomass, carbohydrates and protein content<br>• Reducing carotenoids may protect PSII from photo-inhibition<br>• Cumulative physiological dysfunctioning causes decline in yield depending on the genetic makeup | Chaudhary and Rathore (2021a) |

(continued)

**Table 2.2** (continued)

| Location | Crop/varieties | $O_3$ concentration (ppb) | Experimental condition | Effects | Mechanism/mode of action | References |
|---|---|---|---|---|---|---|
| Central University of Gujarat, Gandhinagar, Gujarat, India 23.2156° N, 72.6369° E | Groundnut cultivars TG-37A, TPG-41, TAG-24, GG-20, Dh-86 | 75.91 | OTC | • Inter-venial chlorosis and necrosis on adaxial surface of leaves<br>• Increased number of stomata and stomatal index, number of epidermal cells, stomatal closure<br>• Reduced size of xylem vessel | • Mature leaves show higher injury compared to younger leaves suggesting long duration increases leaf injury<br>• Enhanced stomata number and stomatal density suggests reduction in size of epidermal cells<br>• Stomatal closure as a protective mechanism against $O_3$ stress<br>• Stomata closure under elevated ozone may reduce evapotranspiration and WUE<br>• Decline in vessel size indicates reduction in water transport | Chaudhary and Rathore (2021b) |

(continued)

**Table 2.2** (continued)

| Location | Crop/varieties | $O_3$ concentration (ppb) | Experimental condition | Effects | Mechanism/mode of action | References |
|---|---|---|---|---|---|---|
| Tangjiapu, Northwestern of Beijing, China 40° 47′ N, 116° 34′ E | Maize cultivar ZD958 | 40, 60, 80 | OTC | • Down-regulation of light-saturated net photosynthesis in both heading and grain-filling stage,<br>• Reduced PS II efficiency<br>• Increased leaf senescence and $C_i$<br>• No significant effect on leaf mass per unit area (LMA) and root/shoot ratio | • Decline in photosynthesis was attributed to restricted activity of PEPc and Rubisco enzyme<br>• Higher $C_i$ suggests the impairment of biochemical processes to photosynthesis rather stomata closure<br>• Down-regulation of $F_v'/F_m$, qP $\phi_{PSII}$ and chlorophyll indicates chloroplast damage<br>• Increased production of ROS damages membrane integrity and chloroplast structure leading to down-regulation of chlorophyll synthesis and photosynthetic rate<br>• Non-significant changes on LMA and root/shoot ratio suggests no significant alteration of biomass between aboveground and belowground parts of the cultivars | Peng et al. (2020) |

(continued)

**Table 2.2** (continued)

| Location | Crop/varieties | $O_3$ concentration (ppb) | Experimental condition | Effects | Mechanism/mode of action | References |
|---|---|---|---|---|---|---|
| Banaras Hindu University, Varanasi, India 25° 81′ N, 83° 1′ E | Mungbean cultivars: HUM-2, HUM-4, HUM-6, HUM-12 and HUM-16 | Ambient + 10 ppb | OTC | • Inter-venial chlorosis, High malondialdehyde (MDA),<br>• Decreased chlorophyll and carotenoid content, $P_n$, $C_i$, $g_s$, WUE | • ROS production in sub-stomatal cavity<br>• Increased ROS react with unsaturated fatty acids leading to lipid peroxidation and membrane damage<br>• Chloroplast damage causing reduced chlorophyll and carotenoid content due to pigment biosynthesis<br>• Stomatal closure under high ozone concentration leads to decreasing $g_s$, decline in entry of $CO_2$ ultimately reduces $P_n$<br>• Elevated $O_3$ leads in reduction of $CO_2$ fixation which may decrease ATP and NADPH demand leading to over- decline of PSII in HUM-2 cultivar | Mishra and Agrawal (2015) |

(continued)

**Table 2.2** (continued)

| Location | Crop/varieties | $O_3$ concentration (ppb) | Experimental condition | Effects | Mechanism/mode of action | References |
|---|---|---|---|---|---|---|
| Banaras Hindu University, Varanasi, India 25° 81′ N, 83° 1′ E | Mungbean cultivars: HUM-1, HUM-2, HUM-6, HUM-23, HUM-24 and HUM-26 | Ambient + 10 ppb | OTC | • Foliar injury, accumulation of ROS, increased membrane damage and solute leakage<br>• Negative effects on seed quality and yield | • $eO_3$ reduces photosynthesis and induced foliar injury. Decrease in photosynthesis limited carbohydrates availability and alters metabolic activity that further declines plant growth<br>• Decline in leaf area attributed to reduction in cell number due to alteration in cell division, expansion or change in cell structure<br>• Increased lipid peroxidation under high ozone alters membrane permeability influencing cellular events linked with foliar injury and accelerates senescence | Chaudhary and Agrawal (2015) |

(continued)

**Table 2.2** (continued)

| Location | Crop/varieties | $O_3$ concentration (ppb) | Experimental condition | Effects | Mechanism/mode of action | References |
|---|---|---|---|---|---|---|
| Air Pollution Laboratory, Centre of Excellence in Environmental Studies, King Abdulaziz University, Saudi Arabia | Radish cultivars Egyptian B and western CB, PR and SG | 80 | Closed fumigation chambers housed in glasshouse | • Foliar injury (chlorotic and necrotic spots),<br>• Increased $g_s$ (except in PR and SG cultivars), MDA content, hydrogen peroxide content, antioxidant enzymes (CAT, SOD and GR)<br>• Decreased $P_n$, $g_s$ (except in B and CB cultivars), $\phi_{PSII}$ (except SG cultivar),<br>• Declined $F_v/F_m$ in B and CB cultivars, no significant effect on other cultivars | • Chlorosis and necrosis spots indicate damage of cell membranes, epidermal cells and consequently leakage of ions<br>• High $g_s$ in B and CB cultivars allow high ozone influx that reduces $P_n$ leading to decrease in photochemistry of PSII and carboxylation efficiency<br>• Exposure to elevated ozone may induce partial stomatal closure in resistant cultivars while increase $g_s$ in sensitive cultivars | Hassan et al. (2018) |

(continued)

Table 2.2 (continued)

| Location | Crop/varieties | $O_3$ concentration (ppb) | Experimental condition | Effects | Mechanism/mode of action | References |
|---|---|---|---|---|---|---|
| Wetland farm, Tamil Nadu Agricultural University, Coimbatore, India 11.00° N, 76.92° E | Rice cultivars Rice CO 51, CO 47, ADT 36, ADT 37, ADT 43, ADT (R) 45, ADT (R) 48, ASD 16, ASD 18, MDU 5, Rice MDU 6, Rice TPS 5, TRY (R) 2, Anna (R) 4, PMK (R) 3 | 50 | OTC | • Foliar injury and reduction in biomass and yield<br>• Decreased $g_S$, $P_n$, chlorophyll content, ascorbic acid content<br>• Increased MDA and proline content | • Foliar injury induced by ROS through the breakdown of ozone in the apoplast during gas exchange process<br>• These ROS controls the activities of ion channels in guard cells and protein kinase in stomata leading to reduced $g_s$ that limits photosynthetic fixation of $CO_2$<br>• Degeneration of chloroplasts reduces chlorophyll content affects carbon assimilation<br>• Increment of MDA content indicates ROS-induced membrane damage. Proline provides protection against hydroxyl radicals and singlet oxygen species as a defence response of the plant<br>• Ascorbic acid content decreases due to ROS-induced depletion of ascorbic acid pool in plants indicating susceptibility towards ozone<br>• Decline in availability of carbohydrates may affect the grain-filling process reducing grain yield | Ramya et al. (2021) |

(continued)

**Table 2.2** (continued)

| Location | Crop/varieties | $O_3$ concentration (ppb) | Experimental condition | Effects | Mechanism/mode of action | References |
|---|---|---|---|---|---|---|
| Indian Space Research Organization's Climate Change Observatory, Woodhouse farm, Horticultural Research Station (TNAU), Ooty 11.4° N, 76.7° E | Turnip | 150 and 200 | OTC | • Necrotic spots on adaxial surface of leaf, decreased plant height, tuber weight, tuber size, chlorophyll content, N, K, Fe, Mn, Zn, Cu<br>• Increased activities of antioxidant enzymes | • The ozone sensitivity is linked with several physiological and biochemical alterations in field crops<br>• Elevated ozone influences metabolic processes that reduce carbon assimilation, transpiration rate, leaf area, growth and dry matter production<br>• Foliar injury is caused by the collapse of palisade cells leading to the formation of lesions on adaxial surface between the veins that consequently become chlorotic and gradually necrotic<br>• It further reduces leaf area for photosynthetic carbon assimilation leading to limitation of growth and tuber maturation<br>• Alterations of chlorophyll content by elevated ozone may decline N and K nutrient level. Lower carbohydrate content in roots can reduce uptake of nutrients (Cu, Fe, Mn and Zn) | Kovilpillai et al. (2023) |

(continued)

**Table 2.2** (continued)

| Location | Crop/varieties | $O_3$ concentration (ppb) | Experimental condition | Effects | Mechanism/mode of action | References |
|---|---|---|---|---|---|---|
| Department of Botany, Banaras Hindu University, Varanasi, India | Wheat cultivars (old) HUW234 and HD3118 (modern) | Ambient + 20 ppb | OTC | • Lower $g_s$, chlorophyll, carotenoid, starch and protein content, yield and grain quality, photosynthetic rate, $C_i$, WUE, $V_{cmax}$, electron transport rate, $F_v/F_m$<br>• Higher total soluble sugar and reducing sugar content | • High reduction in photosynthesis rate, chlorophyll fluorescence, $V_{cmax}$ and electron transport rate in old cultivar as compared to modern cultivar<br>• High stomatal conductance and density in modern cultivar may increase ozone uptake triggering damage to adjacent stomatal cells, carotenoid and chlorophyll pigments coupled with rate of photosynthesis and photosynthetic nitrogen use efficiency<br>• The reduction in photosynthetic rate may be due to limitations of Rubisco and RuBP activity with stomatal closure and damage of carotenoid and chlorophyll pigments | Yadav et al. (2020) |

(continued)

**Table 2.2** (continued)

| Location | Crop/varieties | $O_3$ concentration (ppb) | Experimental condition | Effects | Mechanism/mode of action | References |
|---|---|---|---|---|---|---|
| FACE facility, Champaign. IL 40° 02′ N, 88° 14′ W | Sorghum genotypes ($C_4$ plant) | 100 | FACE | • No visual foliar injury<br>• No effect on net $CO_2$ assimilation and stomatal conductance, chlorophyll fluorescence, stomatal sensitivity to environment and plant biomass<br>• Reduction in maximum carboxylation capacity of phosphoenolpyruvate | • Continuous development of vegetative tissue even after 10 weeks may maintain plant biomass under elevated ozone<br>• The detoxification capability of plant is associated with the efficiency of photosynthetic $CO_2$ assimilation providing energy for detoxification and repair processes<br>• A balance between $P_n$ and $g_s$ determines the extent of response to ozone stress | Li et al. (2020) |

(continued)

**Table 2.2** (continued)

| Location | Crop/varieties | $O_3$ concentration (ppb) | Experimental condition | Effects | Mechanism/mode of action | References |
|---|---|---|---|---|---|---|
| Botanical Garden, Banaras Hindu University, India 24° 14′ N, 82° 03′ E | Indian bean varieties: Kashi Sheetan and Kashi Haritima | Ambient + 30 ppb | OTC | • Higher endogenously generated $H_2O_2$, lipid peroxidation, activities of antioxidant enzymes (APX and GR)<br>• Lowe $P_n$, $g_s$, $C_i$ and yield | • Excessive generation of ROS due to oxidative stress under elevated ozone may increase localization of $H_2O_2$ in leaves and lipid peroxidation. Increment in the activities of APX and GR indicates defence response to scavenge $H_2O_2$ in order to restrict ozone stress<br>• Reduction in $P_n$ may be the effect of chloroplast deterioration preventing the production of chlorophyll and subsequently decreased carbon assimilation leading to curtail in yield | Sahoo et al. (2023) |

(continued)

**Table 2.2** (continued)

| Location | Crop/varieties | $O_3$ concentration (ppb) | Experimental condition | Effects | Mechanism/mode of action | References |
|---|---|---|---|---|---|---|
| Qianyanzhou ecological station, Chinese Academy of Sciences 115° 03′ 29.2″ E, 26° 44′ 29.1″ N | *Phoebe bournei* and *Phoebe zhennan* | ~ 100 and 150 ppb | OTC | • Decreased $P_n$, root, leaf and total biomass<br>• No significant effect on root/shoot ratio<br>• Increased water-soluble carbohydrates (WSC) in leaves of both species<br>• Enhanced total non-structural carbohydrates (TNC) in leaves of both species in response to ~ 100 ppb $O_3$ treatment<br>• Increased leaf starch content in *P. bournei* under ~ 100 ppb $O_3$ treatment<br>• Enhanced TNC and starch in roots of both species under ~ 100 ppb $O_3$ treatment | • Decreased $P_n$ results in lower biomass. Enhancement in TNC under ~ 100 ppb $O_3$ treatment suggests that although photosynthesis is inhibited, the energy is stimulated to support plant growth, thus, carbon reserves can accumulate under less severe stress<br>• Increased WSC and TNC under ~ 100 ppb $O_3$ treatment indicate that plant may reserve energy for growth and development under moderate stress<br>• Increased starch content in *P. bournei* suggests reduction or inhibition of photoassimilates transport from leaves that may act as carbon source for respiration under $O_3$ stress<br>• The accumulation of starch in roots may be associated with reduced mycorrhizal infection that may inhibit starch hydrolysis to soluble sugars | Chen et al. (2018) |

Table covers the targeted crops, effects on the plant growth and possible mechanism/mode of action under $O_3$ injury along the study location, experimental setup with exposed $O_3$ concentrations in individual research works

$P_n$ net photosynthetic rate, $g_s$ stomatal conductance, $C_i$ intercellular $CO_2$ concentration, $C_a$ external $CO_2$ concentration, $T_r$ transpiration rate, $F_v/F_m$ maximum quantum efficiency of PSII, $F_v'/F_m'$ actual photochemical efficiency of PSII in the saturated light, *qP* quenching of photochemical efficiency of PSII, $\phi_{PSII}$ quantum yield of non-cyclic electron transport, *LMA* leaf mass per unit area, *FW* fresh weight, *DW* dry weight, *CAT* catalase, *SOD* superoxide dismutase, *GR* glutathione reductase, *APX* Ascorbate peroxidase, *ROS* reactive oxygen species, $V_{cmax}$ maximum rate of Rubisco carboxylation, $J_{max}$ maximum rate of electron transport, $A_{sat}$ light-saturated $CO_2$ assimilation rate at ambient $CO_2$

### 2.6.3 Leaf Cuticle Thickening

Tropospheric $O_3$ can stimulate cuticle production on crop leaves, resulting in thicker leaves. This can limit photosynthesis and reduce gas exchange, potentially resulting in lower crop productivity. The alterations in stomata number, thickness of leaf and palisade layer shows sensitivity towards $O_3$ (Al-Muwayhi et al. 2014). Leaf is observed thickened for the protection against $eO_3$ (Chaudhary and Rathore 2022).

### 2.6.4 Reduced Root Growth

Tropospheric $O_3$ has the potential to limit nutrient and water uptake by inhibiting root growth and development, which effects the crop productivity and yield. The root–shoot ratio found increased and may be due to distribution of photosynthate towards underground part of plant (Chaudhary and Rathore 2022). Higher reduction in root biomass in comparison to shoot shows major allocation of photosynthate to aboveground parts for conquering $O_3$ damage (Mina et al. 2010). Root development is dependent on current photosynthate, reduction in photosynthesis and allocation of carbon may affect the development process.

### 2.6.5 Affecting Reproductive Development

Tropospheric $O_3$ can affect reproductive development in some crops by altering flower and fruit formation, as well as seed size and quality. Foliar injury and accumulation of ROS leads to the solute leakage due to membrane damage. Which reduce photosynthesis and results in the limited carbohydrate availability and metabolic activity. This activity was promoted by increased lipid peroxidation and reduced leaf area (Chaudhary and Agrawal 2015). Foliar injury leads to the reduction in biomass and which leads to the reduction in yield (Ramya et al. 2021). Overall reduction in the yield may be the effect of $O_3$-induced pollen injury or decline in availability of photosynthate for development of fruit (Mina et al. 2010). Decreased fruit weight, fruit number was observed (Bortolin et al. 2014).

### 2.6.6 Presence of Necrotic Spots and Chlorosis

The leaf injury index was increased at $eO_3$. Early necrotic spots on younger leaves happens (Dhevagi et al. 2022). Formation of dead cell lesions in leaves are also observed (Al-Muwayhi et al. 2014). Early leaf senescence was observed with chlorosis and necrosis (Chaudhary and Rathore 2022; Kovilpillai et al. 2023). Interveinal chlorosis and necrosis on adaxial surface of leaves is found especially at $eO_3$ level and mature leaves are more affected than young leaves (Chaudhary and Rathore 2021b; Mishra and Agrawal 2015; Borowiak et al. 2010). Chlorosis and necrosis spots indicate the damage of cell membranes, epidermal cells and therefore ionic leakages (Hassan et al. 2018).

### *2.6.7 Effects of Tropospheric Ozone on Crop Physiology*

Tropospheric $O_3$ can have a number of deleterious effects on the physiology of crops, including the following.

#### 2.6.7.1 A Slower Rate of Photosynthesis

Tropospheric $O_3$ has the potential to cause damage to the chloroplasts, which slow down the photosynthetic process and reduces the growth rate and finally crop yield. Limited availability of carbohydrates due to decreased rate of photosynthesis may reduce seed yield (Sun et al. 2014a, b). The reduction in photosynthetic rate may be due to limitations of Rubisco and RuBP activity with stomatal closure and damage of carotenoid and chlorophyll pigments (Yadav et al. 2020). (Dhevagi et al. 2022). Elevated $O_3$ leads in reduction of $CO_2$ fixation which may decrease ATP and NADPH demand leading to over- decline of PSII. (Mishra and Agrawal 2015). Reduced PSII efficiency leads to down-regulation of light-saturated net photosynthesis in both heading and grain-filling stage. Decline in photosynthesis was attributed to restricted activity of PEPc and Rubisco enzyme also chloroplast is damaged may due to increased production of ROS. Peng et al. (2020) suggested that higher $C_i$ suggests the impairment of biochemical processes to photosynthesis rather stomata closure. Chloroplast damage causing reduced chlorophyll and carotenoid content due to pigment biosynthesis.

#### 2.6.7.2 Decreased Stomatal Conductance

$O_3$ can cause damage to the stomata, and affects gaseous exchange. The reduction of stomatal conductance was very common in the effects related to elevated $O_3$. Some studies were mentioned here (Dhevagi et al. 2022; Mishra and Agrawal (2015);; Hassan et al. 2018; Ramya et al. 2021; Sun et al. 2014a, b; Yadav et al. 2020; Sahoo et al. 2023). ROS controls the activities of ion channels in guard cells and protein kinase in stomata leading to reduced $g_s$ that limits photosynthetic fixation of $CO_2$ (Ramya et al. 2021). Enhanced stomatal limitation and citrate content reduced photosynthesis and stomatal conductance leads to reduced growth (Sun et al. 2014a, b).

### 2.6.8 *An Increase in the Level of Oxidative Stress*

$O_3$ can cause plants to experience oxidative stress because it generates ROS, which may damage to DNA, lipids, and proteins and decrease the growth rate, yield, and overall quality of the crops. Increases ROS production causes leaf injury (Dhevagi et al. 2022). Increase in proline content indicates the activation of non-enzymatic antioxidant mechanism of defence (Al-Muwayhi et al. 2014). Foliar injury induced by ROS through the breakdown of $O_3$ in the apoplast during gas exchange process. The ROS plays an important role in regulating ions present in the guard cells and protein kinase in stomata leading to reduced $g_s$ that limits photosynthetic fixation of $CO_2$. Increment of MDA content indicates ROS-induced membrane damage. Proline provides protection against hydroxyl radicals and singlet oxygen species as a defence response of the plant. Ascorbic acid content decreases due to ROS-induced depletion of ascorbic acid pool in plants indicating susceptibility towards $O_3$ (Ramya et al. 2021). Higher endogenously generated $H_2O_2$, lipid peroxidation, activities of antioxidative enzymes (APX and GR). Excessive generation of ROS due to oxidative stress under elevated $O_3$ may increase localization of $H_2O_2$ in leaves and lipid peroxidation. Increment in the activities of APX and GR indicates defence response to scavenge $H_2O_2$ in order to restrict ozone stress (Sahoo et al. 2023).

### 2.6.9 *Changes in the Timing and Pattern of Plant Development*

$O_3$ has the ability to change plant phenology and its development, and finally plant morphology, for examples reduced root growth, reduced stem elongation, and altered leaf size and shape. All the phonological stages were escalated due the $eO_3$ level and later affected the crop duration negatively. Ozone exposure led to early flowering of plants. $O_3$ may also affects the reproductive process and capacity in plants, however, the directional and proportional impact is considerable varied based on the plant age and respective concentration levels (Laura Duque et al. 2021). Number of fruits/seeds/total seed weight may differ accordingly based on the exposure time to the life cycle of the plants, and its impacts may be differed accordingly (Laura Duque et al. 2021).

### 2.6.10 *Decreased Nutrient Uptake*

$O_3$ can cause damage to the root system, which in turn decreases the plant's ability to efficiently absorb nutrients. This can result in a decreased uptake of essential nutrients such as nitrogen, phosphorus, and potassium. This can result in a deficiency of nutrients as well as a decreased crop yield.

### *2.6.11 An Increased Propensity to Be Attacked by Pests and Diseases*

$O_3$ has the potential to weaken the plant's defence mechanism and make plants susceptible to pest and disease. Alterations of chlorophyll content by elevated $O_3$ may decline N and K nutrient level. Lower carbohydrate content in roots can reduce uptake of nutrients (Cu, Fe, Mn and Zn) (Kovilpillai et al. 2023). Long-term exposure of $eO_3$ disrupts important vial process in crops/plants i.e.., respiration, water and nutrient uptake, and translocation of assimilates as well as metabolites. These changes can significantly reduce the plant vigour and negatively affects the growth and reproductive capabilities of plants (Nowroz et al. 2024). Elevated $O_3$ may significantly alter the leaf C: P ratio. Hou et al. (2022) elaborate the varied impact of $O_3$ on N, C, and P stoichiometry in fine roots and leaves among different tree species. The elevated $O_3$ may augmented the N content while decline the C: N ratio in fine roots (Hou et al. 2022).

## 2.7 Synergistic of Carbon Dioxide and Tropospheric Ozone on Crop Morphology and Physiology

The combination of elevated $O_3$ and $CO_2$ affect plant in different ways. It is visible that the elevated $CO_2$ mainly affect positively and elevated $O_3$ affects negatively. Root and shoot length were decreased with the $eO_3$ (Kumari et al. 2013), and which results in the decreased shoot–root ratio also (Mishra et al. 2013). Leaf area and leaf area index was also affected directly with elevated $O_3$ and $CO_2$. The specific leaf weight is reduced according to Heagle et al. (1998). Decreased stomatal conductance is happened (Bhatia et al. 2021a, b; Kumari et al. 2013). The effect may be positively such as increased because the effect of elevated $CO_2$ was dominating over the effect of $eO_3$. (Kumari et al. 2013; Bhatia et al. 2013; Mishra et al. 2013). As of Leaf area and LAI, Chlorophyll content also affected (Bhatia et al. 2021a, b; Phothi et al. 2016; Kumari et al. 2015a, b; Heagle et al. 1998). Photosynthesis also increases with elevated $CO_2$ and decreases with $eO_3$ (Bhatia et al. 2021a, b; Bhatia et al. 2013; Phothi et al. 2016). The above-mentioned traits together result in the biomass change. Biomass increases when elevated $CO_2$ dominated over $eO_3$ and vice-versa (Kumari et al. 2013; Reinert et al. 1997; Phothi et al. 2016; Mishra et al. 2013). One important trait, i.e. pod initiation is also observed to be affected (Bhatia et al. 2021a, b) in chickpea. Improved PSII efficiency and water use efficiency were also significantly affected under elevated atmospheric concentration of $CO_2$ and $O_3$. Increase in carotenoid starch, ascorbic and malondialdehyde content and reduction in protein content, membrane damage soluble sugar under combination of elevated $CO_2$ and elevated $O_3$ (Kumari et al. 2013, 2015a, b). Table 2.3. Significant research reports assessing role of interactive effect of elevated $CO_2$ and $O_3$ on crop growth and yield has been mentioned in Table 2.3.

**Table 2.3** Significant research reports assessing role of interactive effect of elevated $CO_2$ and $O_3$ on crop growth and yield

| Location | Crop/varieties | $CO_2$ and $O_3$ Concentration | Experimental condition | Effects | Mechanism/mode of action | References |
|---|---|---|---|---|---|---|
| ICAR-Indian Agricultural Research Institute, New Delhi, India 28° 40′ N, 77° 12′ E | Chickpea variety Pusa 5023 | $eCO_2$: 550 ppm<br>$eO_3$: 60 ppb | FAOCE | • Early Pod initiation by 2–3 days Declined $P_n$, $g_s$, chlorophyll content | • Early translocation of carbon assimilates to leaves could attribute pod initiation in advance<br>• $eCO_2$-induced partial stomatal closure reduces negative impacts of elevated ozone by controlling their uptake through stomata and also decreasing transpiration loss and enhancing WUE<br>• $eCO_2$ shows protective effect against ozone damage and can increase availability of photosynthates enabling plants to maintain growth | Bhatia et al. (2021a, b) |

(continued)

**Table 2.3** (continued)

| Location | Crop/varieties | $CO_2$ and $O_3$ Concentration | Experimental condition | Effects | Mechanism/mode of action | References |
|---|---|---|---|---|---|---|
| Botanical Garden, Banaras Hindu University, India 24° 14′ N, 82° 03′ E | Potato cultivar Kufri chandramukhi | $eCO_2$: 570 ppm<br>$eO_3$: 70 ppb | OTC | • No visible injury<br>• Increased $P_n$, $C_i$, WUE, $F_v/F_m$, chlorophyll and carotenoid contents, ascorbic acid content and APX enzyme activity<br>• Declined carotenoid content at 60 days of exposure<br>• Decreased ozone flux, $g_s$, $T_r$, lipid peroxidation, solute leakage, protein content, antioxidant enzymes activities (SOD, POX, CAT and GR) | • Lipid peroxidation and solute leakage is diminished under $eCO_2 + eO_3$ treatment with high level of $H_2O_2$ indicating that high ozone flux may increase $H_2O_2$ content after 30 day of exposure<br>• Reduced ozone flux after 60 day of exposure suggests that $eCO_2$ can decrease ROS formation and basal rate of oxygen activation by increasing ratio of $CO_2/O_2$ in photosynthesis apparatus reducing detrimental effects of ozone on the stability of cell membrane by decreasing MDA content and solute leakage<br>• Enhanced ascorbic acid indicates excess production of carbohydrates due to $eCO_2$. Low level of other antioxidative enzymes suggests lower oxidative stress induced by $eO_3$. Reduced RuBp oxygenation rate and enhanced RuBp carboxylation rate may increase $P_n$ and $C_i$ in potato leaves. $C_i$ can act as signal linking mesophyll $CO_2$ demand to stomatal aperture that usually decreases under $eCO_2$<br>• Increment in WUE was due to increased $P_n$<br>• $eCO_2$ enhances $F_v/F_m$ ratio in combined treatment of $eCO_2 + eO_3$. Increment in chlorophyll content may facilitate primary photochemical reaction leading to significant enhancement in $P_n$ | Kumari et al. (2015a, b) |

(continued)

**Table 2.3** (continued)

| Location | Crop/varieties | $CO_2$ and $O_3$ Concentration | Experimental condition | Effects | Mechanism/mode of action | References |
|---|---|---|---|---|---|---|
| Air Pollution Laboratory, Faculty of Agriculture, Natural Resources and Environmen, Naresuan University, Phitsanulok, Thailand | Rice cultivar Khao Dawk Mali 105 | $eCO_2$: 700 ppm<br>$eO_3$: 40 and 70 ppb | Indoor climate control chamber | • High photosynthesis,<br>• Low chlorophyll content and soluble sugar content, biomass of shoot and root | • $eCO_2$ directly supplies adequate $CO_2$ to carboxylation reaction of photosynthesis. $eCO_2$ also increases $C_i$ that reduces oxygenation of Rubisco and at the same time enhances carboxylation of Rubisco leading to enhance photosynthesis<br>• Reduction in stomatal conductance decreases ozone entry and ozone-induced damages<br>• Increased chlorophyll content in combined $eCO_2$ and $eO_3$ treatment in comparison to ozone treatment suggests that $eCO_2$ can increase number of chloroplasts, starch grain number and size but decreases grana thylakoid membrane number<br>• $eCO_2$ also enhances antioxidative defence system to scavenge ROS produced by $eO_3$. Higher soluble sugar content in comparison to $eO_3$ treatment indicates $eCO_2$-induced photosynthetic acclimatization to long-term growth<br>• Shoot and root biomass is higher than ozone treatment alone because $eCO_2$ enhanced photosynthesis increases biomass in combined treatment of $eCO_2$ + $eO_3$ | Phothi et al. (2016) |

(continued)

**Table 2.3** (continued)

| Location | Crop/varieties | $CO_2$ and $O_3$ Concentration | Experimental condition | Effects | Mechanism/mode of action | References |
|---|---|---|---|---|---|---|
| Botanical Garden, Banaras Hindu University, India 25° 81′ N, 83° 1′ E | Wheat cultivars HUW-37 and K-9170 | $eCO_2$: 700 ppm $eO_3$: ambient + 10 ppb | OTC | • Foliar injury<br>• (chlorotic stippling and interveinal chlorosis)<br>• Increased plant height, number of leaves, leaf area, number of tillers, biomass, leaf weight ratio, root and shoot ratio, number and weight of ears per plant, grain number and weight, total soluble sugar and net assimilation rate<br>• Increment in reducing sugar content in K-9107<br>• No significant changes in protein and total free amino acid contents<br>• Reduced leaf area ratio | • Reduction in foliar injury symptoms such as chlorotic stippling and interveinal chlorosis in the combined treatment of $eCO_2 + eO_3$ as compared to ozone treatment indicates $eCO_2$ may provide protection against ozone by decreasing stomatal conductance declining entry of ozone leading to enhance availability of substrates that may detoxify the effects of ozone by the repair mechanisms<br>• In comparison to ozone treatment, increased growth and yield of wheat cultivars suggest that carbon enrichment due to $eCO_2$ in leaves could increase the number of branches and tillers resulting enhancement in leaf area. $eCO_2$ also helps plant to get more biomass to overcome the negative effects of ozone due to higher photosynthetic rate<br>• K-9107 have higher biomass as compared to HUW-37. Higher net assimilation rate indicates carbon gain by increasing photosynthetic rate | Mishra et al. (2013) |

(continued)

**Table 2.3** (continued)

| Location | Crop/varieties | $CO_2$ and $O_3$ Concentration | Experimental condition | Effects | Mechanism/mode of action | References |
|---|---|---|---|---|---|---|
| Indian Agricultural Research Institute, New Delhi<br>28° 37′ N, 77° 12′ E | Wheat cultivars HD-2967 and C-306 | $eCO_2$: 550 ppm<br>$eO_3$: 70 ppb | FAOCE | • No significant effect on $P_n$, leaf area index, spike, spike length, number of spikelets and grain, grain weight, biological yield, grain yield and harvest percent | • $eCO_2$ ameliorates the negative impact of $eO_3$ on growth and yield attributes and improve them to at par with ambient by reducing stomatal conductance and stomatal density due to high concentration of $CO_2$ | Yadav et al. (2019) |

Table covers the targeted crops, effects on the plant growth and possible mechanism/mode of action along the study location, experimental setup with exposed $CO_2$ and $O_3$ concentrations in individual research works

*FAOCE* free air $O_3$ and $CO_2$ enrichment, $P_n$ net photosynthetic rate, $g_s$ stomatal conductance, $C_i$ intercellular $CO_2$ concentration, $T_r$ transpiration rate, $F_v/F_m$ maximum quantum efficiency of PSII

## 2.8 Interactive Effects of Ozone and Carbon Dioxide on Plant Production

Currently, terrestrial ecosystems absorbed almost one-third of the $CO_2$ released by human activities and one-fifth of the $O_3$ produced in the troposphere by chemical production, Ainsworth et al. 2020. In general, the interaction between $CO_2$ and $O_3$ suggested that higher $CO_2$ would lessen the ozone-induced reactions. Initially reported that higher $CO_2$ levels lessened the impact of $O_3$ on photosynthesis. However, this advantage disappeared later on, possibly due to the fact that prolonged exposure to $O_3$ limits the overall response of plants to $CO_2$. $CO_2$ and $O_3$ appeared to have additive effects on grasslands (Payer et al. 1993). Hasan et al. 2021 found that in wheat photosynthesis being down-regulated in response to prolonged exposure to elevated $CO_2$ this was especially noticeable in the winter cultivars compared to spring cultivars. A reduction in the light-saturated rate of $CO_2$ assimilation, partial stomatal closure, and an accumulation of starch and water-soluble carbohydrates in leaves were the outcomes of prolonged exposure to $O_3$. Therefore, led to a reduction in growth, maximum affected with root growth compared to shoot growth. $O_3$ decreased the enhancement effects of $CO_2$ on photosynthesis and growth, while effects on stomatal conductance were less than additive in plants exposed to elevated $CO_2$ and $O_3$.

In ambient field conditions, in bean plants, increased $O_3$ concentration counteract the effects of increased $CO_2$, which leads to reduced seed set, weight, and floral attractiveness as well as a decrease in pollinator visits to the flowers, whereas higher $CO_2$ exposure has a beneficial effect on the physiology of field bean plants and enhances the attractiveness of their blooms, which increases pollinator visits, seed set, and seed weight (Ainsworth et al. 2020). Therefore, any benefit that $CO_2$ may have had for plant reproduction was outweighed by the detrimental effects of $O_3$. Ozone can change the volatile organic compounds (VOCs) that flowers release, which can have an impact on the odour plume or map that pollinators use to find flowers (Blande 2021; Ryalls et al. 2022). This gas can influence the scent of flowers by changing the physiology of the plant, which consequently changes in the emission of flowers, or by reacting with the flowers to change their chemical structure (Holopainen et al. 2013).

In addition to the ways that $CO_2$ and $O_3$ alter the scent of flowers, other factors that affect flower visitation include altered nectar and pollen quality and quantity as a result of changes in C:N ratios at higher $CO_2$ concentrations (Hoover et al. 2012) or direct $O_3$ damage to nectars (Stabler 2016). According to Berner et al. (2015), the interaction of $CO_2$ and $O_3$ in brassica did not affect the economic yield, but it did make the plants more vulnerable to pathogen and insect attack. Variations in the physiological response of plants to elevated levels of $CO_2$ and $O_3$ may impact the interaction between plants and pathogens concerning the availability and quality of nutrients for pathogen feeding. Variations in the concentrations of antioxidant enzymes and the content of stress compounds, such as reactive oxygen species (ROS), may stimulate defence-like reactions (Fiscus et al. 2005).

### 2.8.1 Effect of Elevated Carbon Dioxide ($CO_2$) and Ozone ($O_3$) on Crop Phenology and Growth Parameters

Bhatia et al. (2021a, b) revealed that, days to pod initiation of chickpea decreased by 3 days under elevated $O_3$ (EO) and elevated carbon dioxide ($ECO_2$) levels compared to ambient suggesting that photosynthates may have moved to the leaves early, which would explained the early pod initiation. Because of the source/sink imbalance, the pod maturity under EO occurred eight to ten days earlier than the ambient during the two years studied (Andersen 2003). Increased $O_3$may prevent sugar export from leaves, which could cause early pod development and leaf senescence (Grantz and Farrar 2000). When comparing the interaction treatment of EO and $ECO_2$ to the EO alone, a remarkable rise in all the growth parameters was seen at pod formation. Plants with higher photosynthetic rates were able to use larger quantities of photosynthates for growth, which was accompanied by a rise in LAI. As a leguminous crop, chickpeas may be able to fix more nitrogen in environments with greater $CO_2$, which the plants can then use to assist the process of growth improvement (Gamper et al. 2005). In a $CO_2$-enriched treatment, a higher leaf area index and a reduced stomatal aperture may enhance water usage efficiency at the leaf and canopy level, leading to improved growth (Mills and Harmens 2011). The number of secondary branches increased significantly, which in turn affects the total number of leaves and, consequently, the total photosynthetic area.

### 2.8.2 Effect of Elevated $O_3$ and $CO_2$ on the Photosynthetic Rate, Stomatal Conductance and Total Chlorophyll

There was a notable decrease in the net photosynthetic rate (Pn) between 14 and 21%. Similarly, 25 and 19% decrease in stomatal conductance (gs) in the EO treatment compared to the ambient treatment. This drops in gs in the EO treatment may have led to a fall in Pn via lowering internal leaf $CO_2$ levels. (Bhatia et al. 2021a, b). Since the stomata do not need to open as widely to allow enough $CO_2$ for photosynthesis to occur in the leaf, changes in stomatal aperture under elevated $CO_2$levels may be the cause of the lower stomatal conductance (Ainsworth and Long 2005). On the other hand, 13 and 16% higher photosynthesis rates (Pn) were observed when EO + $CO_2$interaction treatment was compared to EO alone. This was due to the fact that the treatment had greater $CO_2$ levels. Because greater $CO_2$caused partial stomatal closure, which limited $O_3$ uptake by stomata while also decreasing transpiration losses and improving water use efficiency, the effect of $O_3$ was lessened. Increased availability of photosynthates allowed plants to sustain growth that could be utilized for damage repair and detoxifying activities and the enhanced $CO_2$ had a protective impact against $O_3$ injury (Booker et al. 2007).

At the flowering stage, $ECO_2$ had a higher total chlorophyll content (12 and 11%) than the ambient. Chickpeas, being a legume, may obtain nitrogen through symbiotic nitrogen fixation, and increased atmospheric $CO_2$ may have expedited plant intake of nitrogen. The amount of chlorophyll in leaves and the concentration of foliar N may have increased as a result. Under EO, the total chlorophyll was less than that of the ambient. The formation of $O_3$-mediated ROS under EO conditions may cause a decrease in chlorophyll because of insufficient leaf antioxidant capability (Caregnato et al. 2013) and leaf chlorophyll content.

### 2.8.3 *Effect of Elevated $O_3$ and $CO_2$ on Growth and Biochemical Parameters*

Ozone is a strong oxidizing agent and induces negative impact on vegetation. High $O_3$ levels can cause injury in various plant species and reduce economic and biological output in agriculturally important crops (Cardoso et al. 2004). Additionally, ambient $CO_2$ is a limiting factor for photosynthetic output in $C_3$ plants, thus, increased $CO_2$ levels have positive impact on plant growth and biological output. However, responses to elevated $CO_2$ were varied in various plants due to their genomic difference. The effect of combined exposure ($eCO_2 + eO_3$) is need to understand fully in various crops considering their genomic differences (Baier et al. 2005).

Exposure to $O_3$ levels affected the rate of chlorosis of leaves and its progression into necrosis. Ozone affects significantly the crop/plant photosynthetic, decrease in leaf area index and N content (Noormets et al. 2001; Al-Rawahy et al. (2013). Toxicological symptoms increase with increase in the $O_3$ levels in the affects plant/crop species. Nodulation retardation is significant with enhanced $O_3$ exposure (duration and quantity), but not significant in combined treatment of $O_3 + CO_2$ (Al-Rawahy et al. (2013). Reduction in photosynthetic pigment, i.e. chlorophylls content is common biochemical damage symptom under $O_3$ stress, which can be protected by the elevated $CO_2$ as observed in sugar beet (Temmerman et al. 2007) and potato (Hacour et al. 2002). Toxicity levels increase with increase in the $O_3$ levels.

### 2.8.4 *Effect of Elevated $O_3$ Elevated $CO_2$ on the Seed Quality and Yield*

According to Jawaharjothi et al. (2023), elevated $CO_2$ mitigates the negative impacts of elevated $O_3$ on Indian mustard growth and yield, as well as its seed yield, under interaction treatment of $O_3$ and $CO_2$, which showed that elevated $O_3$ has a negative impact on seed yield. The research revealed that under elevated $CO_2$, the number of siliquae/plants, number of seeds/siliquae and harvest index were greater, whereas the lowest values were recorded under e $O_3$. The qualities of yield are directly correlated with the reduction in growth. Plant height, SLA, SLW and other growth parameters as well as yield attributes under $O_3$-exposed treatments are among the growth factors that are associated with the economic yield decline (Singh et al. 2017).

A similar reduction in seed yield was seen in Brassica campestris var. Kranti by Singh et al. (2009) and Bhatia et al. (2011), and in maize and soybean by McGrath et al. (2015), garlic by Gayathri et al. (2019), rice by Kumar et al. (2021), maize by Yadav et al. (2021) and wheat by Hansen et al. (2019). Higher concentration of $O_3$ may lead to reduced dry matter production and crop productivity, which can also hasten leaf senescence, modify photosynthetic activity and change stomatal conductance in leaves.

### *2.8.5 Effect of Elevated $O_3$ and $CO_2$ on Disease*

Rising of tropospheric $O_3$ concentrations could result in considerable losses in crop yield in India. $O_3$ may have an impact on a plant's reaction to external stresses like pathogens and diseases in addition to its direct effects on the plant. Paoletti et al. (2020), suggested that along with the other environmental stresses, the effects of $O_3$ on crops must be quantified. Higher $O_3$ levels may make plants more vulnerable by severely harming their cuticle layer, leaving them expose to pathogen and insect attack (Berner et al. 2015) and a rise in plant diseases.

Although elevated $O_3$ exposure has no direct effect on fungal infections, it may make necrotrophic pathogens more likely to colonize plants that are already weakened by. Pathogens classified as hemi biotrophic/necrotrophic prefer stressed, damaged, or weak plants. Numerous abiotic stressors have been shown to impair plants' defence systems and increase their vulnerability to pathogenic infection (Atkinson and Urwin 2012). Berner et al. (2015) found that while brassica's economic yield remained unaffected by the interaction of $CO_2$ and $O_3$, however the plants' susceptibility to disease and insect attack increased.

### *2.8.6 Effect of Elevated $O_3$ and $CO_2$ on Microbiome*

According to Smith and Paul (1990), soil microorganisms and the ecosystem-level processes can be significantly impacted by climate change factors including increased atmospheric $CO_2$ and $O_3$. Still unclear, nevertheless, are the fundamental processes by which soil microbes react to these environmental changes (Zak et al. 2000). Relying on the assumption that soil microbes are generally C limited, the current prevailing hypothesis states that while $O_3$ reduces soil microbial biomass and activities because of lower C allocation belowground, elevated $CO_2$ increases them due to enhanced soil C availability (Blagodatskaya et al. 2010).

Under $O_3$ enrichment, the average reduction in soil respiration is 15%. Because $CO_2$ dilutes litter, it inhibits the involvement of microbes in litter decomposition. This effect varies depending on the habitat and is particularly noticeable in tropical forests (Finzi et al. 2002). Elevated $CO_2$ alters the distribution (number and variety) of omnivorous insects, earthworms and enchytraeids, microarthropods, collembolans, omnivorous insects, and nematode communities (particularly those comprising omnivores, saprophagous feeders, and predators). It also affects the proportion of edaphic groups by affecting the quality and moisture of plant biomass. Changes in soil moisture, animal grazing activity, and soil nutrient availability caused by $CO_2$ can also affect how soil microbes react to increased $CO_2$.

According to Fiscus et al. (2005), many crop plants, especially $C_3$ crops, are often responsive to elevated $CO_2$ and $O_3$. Feng et al. (2008), reported that, elevated $CO_2$ alone increased the shoot biomass of wheat and soybean by 16% and 48%, respectively, whereas elevated $O_3$ reduced them by 21 and 18% respectively. Furthermore, increased $CO_2$ has been demonstrated to lessen the impacts of $O_3$ on plants by boosting rates of C assimilation and decreasing $O_3$ uptake (Rogers et al. 2009). The reactions of soil microbial activities, however, will primarily determine whether the $CO_2$- and $O_3$-induced changes in plant biomass can potentially translate into changes in soil C sequestration. Therefore, changes in the availability of C and N for microorganisms brought on by increased $CO_2$ and $O_3$ could potentially have an impact on microbial biomass.

As a result, changes in C and N availability for microorganisms brought on by increased $CO_2$ and $O_3$ may eventually have an additional impact on microbial biomass and activities as well as potentially cause a change in the composition of microbial communities.

### 2.8.7 *Effect of $O_3$, $CO_2$ and Disease on Pathogenesis-Related (PR) Proteins and Yield Attributes*

The importance of antioxidant enzymes, ROS, and oxidative burst as a defence mechanism against *F. oxysporum* was documented by Swarupa et al. (2014). Plant defence mechanisms against pathogenic infection are also linked to proteins relevant to plant pathogenesis (PR). According to Ebraham et al. (2011), the production of PR proteins in the far-off, uninfected sections of plants can result in systemic acquired resistance, which shields the afflicted plants from additional infection. As a response to combat elevated $O_3$, plants upregulate antioxidative defence system by higher enzyme activities such as, superoxide dismutase (SOD), catalase (CAT), and peroxidase (POX) to protect the plant from oxidative damage.

The highest rise in SOD was seen in plants treated with EO, while CAT and POX activities increased less. Showing that SOD's activity rose more than that of CAT and POX, it was more important in detoxifying the generated reactive oxygen species (ROS). While CAT directly breaks down $H_2O_2$ into water and oxygen, superoxide dismutase acts as the first line of defence by detoxifying superoxide radicals, protecting plant tissue membranes (Sairam and Saxena 2000). Under the interaction (EO + $CO_2$) treatment, the activity of SOD was substantially higher than that of ambient and elevated $CO_2$. Reduced antioxidant levels and reduced nutritional quality in the seed may arise from increased ROS generation in response to $O_3$ stress in the plant (Daripa et al. 2016).

## 2.9 Influencing Factors

The combination of elevated $CO_2$ and tropospheric $eO_3$ can have complex and interacting effects on crop physiology, and these effects can be influenced by a variety of environmental factors and crop genetics. Influencing factors in which environmental factors and genetics can interact with $CO_2$ and $O_3$ to affect crops.

### *2.9.1 Temperature*

Elevated temperatures can exacerbate the negative effects of $O_3$ on crop physiology, and can also affect the response of crops to elevated $CO_2$. For example, high temperatures can reduce the beneficial effects of elevated $CO_2$ on photosynthesis and growth in some crops.

### *2.9.2 Relative Humidity*

High relative humidity can reduce the negative effects of $O_3$ on plant physiology, while low relative humidity can exacerbate these effects. elevated $CO_2$ can also reduce transpiration rates in crops, which can impact relative humidity and water use efficiency.

### *2.9.3 Vapour Pressure Deficit*

Elevated $CO_2$ can reduce the transpiration rates of crops, which can lead to a decrease in the vapour pressure deficit (VPD) between the leaf surface and the surrounding air. This can reduce the negative effects of high VPD on crop growth and physiology, and can also impact water use efficiency.

### 2.9.4 *Crop Genetics*

Different crop species and varieties can have varying responses to elevated $CO_2$ and elevated $O_3$, depending on their genetic makeup. Some crops may be more resistant to $O_3$ damage or more responsive to elevated $CO_2$, while others may be more sensitive to these environmental stresses.

In addition to these factors, other environmental stresses, such as drought or nutrient deficiencies, can also interact with elevated $CO_2$ and $O_3$ to impact crop physiology. It is important for researchers to consider these interacting factors when studying the effects of air pollution on crops, in order to develop effective strategies for mitigating the negative impacts of these pollutants on food security and ecosystem health (Fig. 2.1).

## 2.10 Conclusion and Future Prospect

The synergistic effects of elevated $CO_2$ and elevated $O_3$ on agricultural crops are influenced by multiple factors that interact with each other. Here are some of the key factors responsible for these synergistic effects: Environmental conditions: Other biotic and abiotic factors can affect the interaction between elevated $CO_2$ and $O_3$ on agricultural crops in the field conditions. For example, high temperatures can exacerbate the negative effects of $O_3$, while high relative humidity can mitigate these effects. Plant physiology: The physiological response of crops to elevated $CO_2$ and $O_3$ can also influence their synergistic effects. For example, elevated $CO_2$ can stimulate plant growth and biomass accumulation, while $O_3$ can reduce photosynthetic rates and increase oxidative stress. Crop genetics: Different crop species and cultivars can have varying responses to elevated $CO_2$ and $O_3$, depending on their genetic makeup. Some crops may be more resistant to $O_3$ damage or more responsive to elevated $CO_2$, while others may be more sensitive to these environmental stresses. Interaction with other environmental stresses: The interaction between elevated $CO_2$ and $O_3$ can also be influenced by other environmental stresses, such as drought or nutrient deficiencies. These stresses can interact with elevated $CO_2$ and $O_3$ to impact crop physiology and yield. Duration and concentration of exposure: The duration and concentration of exposure to elevated $CO_2$ and $O_3$ can also influence their synergistic effects. For example, prolonged exposure to elevated concentration of $O_3$ can have more severe impacts on crop physiology and yield than short-term exposure.

Overall, the interaction between elevated $CO_2$ and $O_3$ on agricultural crops is complex and influenced by multiple factors. It is important for researchers to consider these interacting factors when studying the effects of air pollution under changing environmental conditions especially climate change on agricultural crops, in order to develop effective strategies for mitigating the negative impacts of these pollutants on food security and agroecosystem health.

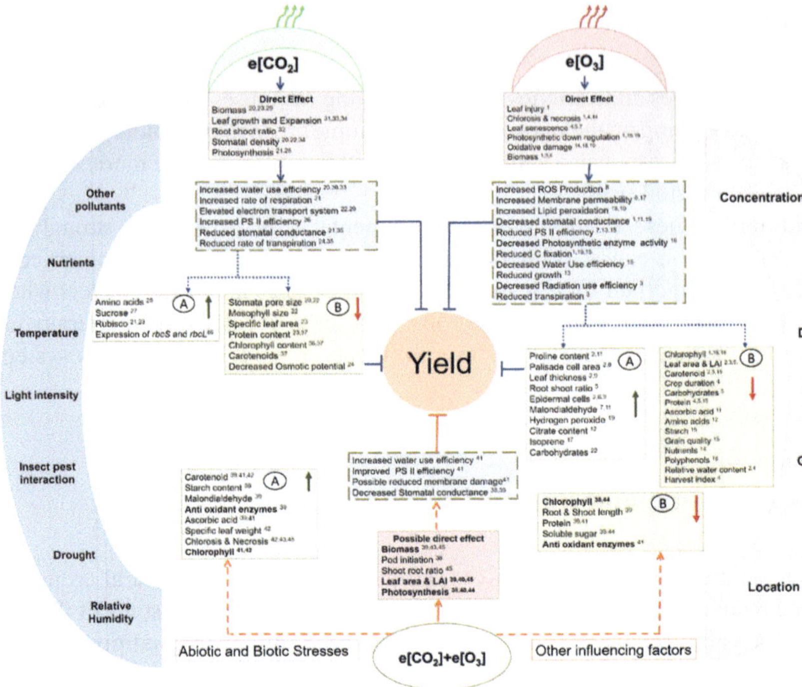


**Fig. 2.1** Illustration of direct effect of e[$CO_2$] and e[$O_3$] alone, mechanism (dashed square), traits affected (increased A/ decreased B), Solid arrow indicates the consequences of effects and mechanism on yield while dotted arrow indicates traits affected owing to involved mechanism or mode of action on different traits. For e[$CO_2$] and e[$O_3$] individually it is shown in blue colour arrow. Increase and decrease for affected traits is shown as upright green arrow and downwards red arrow, respectively. Similarly possible direct effect of e[$CO_2$] + e[$O_3$] interaction, plausible mechanism (dashed square), plausible traits affected (increased A/ decreased B) has been shown in lower part of figure. For e[$CO_2$] + e[$O_3$] interaction, it is shown in orange colour arrows. Dashed arrow indicates plausible effect or mechanism while solid arrow indicates known effect or mechanism. An unknown interaction with other abiotic and biotic stress factors with e[$CO_2$] + e[$O_3$] interaction and interplay with other influencing factors is also depicted in the figure. The traits shown in bold font explains that in e[$CO_2$] + e[$O_3$] interaction, if there is more influence of e[$CO_2$] mechanism over e[$O_3$], the individual trait is increased and vice-versa when e[$O_3$] mechanism is predominant. ([1]Dhevagi et al. 2022; [2]Al-Muwayhi et al. 2014; [3]Singh et al. 2021; [4]Chaudhary and Rathore 2022; [5]Chaudhary and Rathore 2021a; [6]Chaudhary and Rathore 2021b; [7]Peng et al. 2020; [8]Chaudhary and Agrawal 2015; [9]Al Sahli et al. 2013; [10]Ramya et al. 2021; [11]Sun et al. 2014a, 2014b; [12]Borowiak et al. 2010; [13]Kovilpillai et al. 2023; [14]Yadav et al. 2020; [15]Li et al. 2021; [16]Mina et al. 2010; [17]Bortolin et al. 2014; [18]Sahoo et al. 2023; [19]Chen et al. 2018; [20]Zhang et al. 2021; [21]Maroco et al. 1999; [22]Zheng et al. 2019; [23]Pal et al. 2005; [24]Mamatha et al. 2014; [25]Noguchi et al. 2018; [26]De Souza et al. 2008; [27]Vardhan et al. 2013; [28]Zhao et al. 2004; [29]Li et al. 2019; [30]Lee et al. 2020; [31]Usuda 2006; [32]Wu et al. 2017; [33]Wei et al. 2022; [34]Jain et al. 2007; [35]Ji et al. 2015; [36]Ruhil et al. 2014; [37]de la Mata et al. 2012; [38]Bhatia et al. 2021; [39]Kumari et al. 2013; [40]Bhatia et al. 2013; [41]Kumari et al. 2015a, 2015b; [42]Heagle et al. 1998; [43]Reinert et al. 1997; [44]Phothi et al. 2016; [45]Mishra et al. 2013; [46]Rai et al. 2016)

## References

Agathokleous E, Kitao M, Kinose Y (2018) A review study on ozone phytotoxicity metrics for setting critical levels in Asia. Asian J Atmos Environ 12(1):1–16

Ainsworth EA, Davey PA, Bernacchi CJ, Dermody OC, Heaton EA, Moore DJ, Morgan PB, Naidu SL, Yoo Ra HS, Zhu XG, Curtis PS, Long SP (2002) A meta-analysis of elevated $CO_2eCO_2$ effects on soybean (Glycine max) physiology, growth and yield. Glob Change Biol 8:695–709

Ainsworth EA, Lemonnier P, Wedow JM (2020) The influence of rising tropospheric carbon dioxide and ozone on plant productivity. Plant Biol J 22:5–11

Ainsworth EA, Long SP (2005) What have we learned from 15 years of free-air $CO_2$ enrichment (FACE)? A meta-analytic review of the responses of photosynthesis, canopy. New Phytol 165:351–371

Ainsworth EA (2017a) Understanding and improving global crop response to ozone. Plant J 90:886–897

Ainsworth EA, Rogers A (2007) The response of photosynthesis and stomatal conductance to rising [$CO_2$]: mechanisms and environmental interactions. Plant Cell Environ 30:258–262

Ainsworth EA (2008) Rice production in a changing climate: a meta- analysis of responses to elevated carbon dioxide and elevated ozone concentration: meta-analysis of rice responses to global change. Glob Change Biol 14:1642–1650

Ainsworth EA (2017b) Understanding and improving global crop response to ozone pollution. Plant J 90:886–897

Ainsworth EA, Leakey ADB, Ort DR, Long SP (2008) FACE-ing the facts: inconsistencies and interdependence among field, chamber and modeling studies of elevated [$CO_2$] impacts on crop yield and food supply. New Phytol 179:5–9

Al Sahli AA, Al-Muwayhi MA, Doaigey AR, Basalah MO, Ali HM, El-Zaidy M, Sakran AM (2013) Effect of ozone and ascorbic acid on the anatomical, physiological and biochemical parameters of pepper (*Capsicum frutescens* L.). J Pure Appl Microbiol 7:159–168

Al-Rawahy S, Sulaiman H, Farooq SA, Karam MF, Sherwani N (2013) Effect of $O_3$ and $CO_2$ levels on growth, biochemical and nutrient parameters of alfalfa (*Medicago sativa*). APCBEE Proc 5:288–295

Al-Muwayhi MA, Al Sahli AA, Doaigey AR, Basalah MO, Ali HM, El-Zaidy ME, Sakran AM (2014) Role of ascorbic acid in alleviating air pollutants in eggplant seedlings. J Pure Appl Microbiol 8(1):783–792

Allen LH, Baker JT, Boote KJ (1996) The $CO_2$ fertilization effect: higher carbohydrate production and retention as biomass and seed yield. In: Bazzaz F, Sombroek W (eds) Global climate change and agricultural production. FAO, Rome and Wiley, Chichester, pp 65–100

Allen LH, Pan D, Boote KJ, Pickering NB, Jones JW (2003) Carbon dioxide and temperature effects on evapotranspiration and water-use efficiency of soybean. Agron J 95(4):1071–1081

Atkinson NJ, Urwin PE (2012) The interaction of plant biotic and abiotic stresses: from genes to the field. J Exp Bot 63:3523–3543

Bhatia A, Mina U, Kumar V, Tomer R, Kumar A, Bidisha Chakrabarti RN, Singh BS (2021a) Effect of elevated ozone and carbon dioxide interaction on growth, yield, nutrient content and wilt disease severity in chickpea grown in Northern India. Heliyon 7(1):e06049

Badiani M, D'Annibale A, Paolacci AR, Miglietta F, Rashchi A (1993) The antioxidant status of soybean (Glycine max) leaves grown under nature $CO_2$ enrichment in the field. Aust J Plant Physiol 20:275–284

Bagard M, Jolivet Y, Hasenfratz-Sauder M-P, Gérard J, Dizengremel P, Le Thiec D (2015) Ozone exposure and flux-based response functions for photosynthetic traits in wheat, maize and poplar. Environ Pollut 206:411–420

Baier M, Kandlbinder A, Golldack D, Dietz KJ (2005). Oxidative stress and ozone: perception, signaling and response. Plant Cell Environ 28:1012–1020

Baker JT, Allen LH Jr, Boote KJ (1992) Temperature effects on rice at elevated $CO_2$ concentration. J Exp Bot 43:959–964

Barnes JD, Bannister G, Ollerenshaw J (1993) Interactive effects of $CO_2$ and/or $O_3$ on wheat (*Triticum aestivum* L.). UK Committee for Air Pollution Effects Research Annual Meeting, Bangor, Wales (Abstract)

Berner JM, Maliba B, Inbaraj P (2015) Impact of elevated levels of $CO_2$ and $O_3$ on the yield and photosynthetic capabilities of *Brassica napus*. Procedia Environ Sci 29:255–261

Bhatia A, Tomer R, Kumar V, Singh SD, Pathak H (2012) Impact of tropospheric ozone on crop growth and productivity—a review. J Sci Ind Res 71:97–112

Bhatia A, Kumar V, Kumar A, Tomer R, Singh B, Singh S (2013) Effect of elevated ozone and carbon dioxide interaction on growth and yield of maize. Maydica 58(3–4):291–298

Bhatia A, Mina U, Kumar V, Tomer R, Kumar A, Chakrabarti B, Singh B et al (2021) Effect of elevated ozone and carbon dioxide interaction on growth, yield, nutrient content and wilt disease severity in chickpea grown in Northern India. Heliyon 7(1)

Bilgin DD, Zavala JA, Zhu J, Clough SJ, Ort DR, DeLucia EH (2010) Biotic stress globally downregulates photosynthesis genes. Plant Cell Environ 33:1597–1613

Bindi M, Hacour A, Vandermeiren K, Craigon J, Ojanpera K, Sellden G, Hogy P, Finnan J, Fibbi L (2002) Chlorophyll concentration of potatoes grown under elevated carbon dioxide and/or ozone concentrations. Eur J Agron 17:319–335

Black VJ, Black CR, Roberts JA, Stewart CA (2000) Tansley review no. 115: impact of ozone on the reproductive development of plants. New Phytol 147:421–447

Blagodatskaya E, Blagodatsky S, Dorodnikov M, Kuzyakov Y (2010) Elevated atmospheric $CO_2$ increases microbial growth rates in soil: results of three $CO_2$ enrichment experiments. Glob Change Biol 16:836–848

Blande JD (2021) Effects of air pollution on plant–insect interactions mediated by olfactory and visual cues, Curr Opin Environ Sci Health 19:Article 100228

Booker FL, Burkey KO, Pursley WA, Heagle AS (2007) Elevated carbon dioxide and ozone effects on peanut: I. gas-exchange, biomass, and leaf chemistry. Crop Sci 47(4):1475–1487

Borowiak K, Zbierska J, Drapikowska M (2010) Differences in morpho-anatomical structure of ozone-sensitive and ozone-resistant tobacco cultivars. Acta Biologia Hungarica 61(1):90–100

Bortolin RC, Caregnato FF, Divan AM Jr, Reginatto FH, Gelain DP, Moreira JCF (2014) Effects of chronic elevated ozone concentration on the redox state and fruit yield of red pepper plant Capsicum baccatum. Ecotoxicol Environ Saf 100:114–121

Bowes G (1991) Growth at elevated $CO_2$e$CO_2$: photosynthesis response mediated through Rubisco. Plant, Cell Environ 14:795–806

Brajesh KS, Delgado-Baquerizo M, Egidi E, Guirado E, Leach JE, Liu H, Trivedi P (2023) Climate change impacts on plant pathogens, food security and paths forward. Nat Rev Microbiol 21(10):640–656

Bunce JA (1996) Stomatal responses to carbon dioxide enrichment and water stress in soybean. Environ Exp Bot 36(1):11–19

Bunce JA (2020) Growth, yield and stomatal responses of potato to carbon dioxide enrichment. Am J Potato Res 97(5):454–464

Burkey KO, Booker FL, Pursley WA, Heagle AS (2007) Elevated carbon dioxide and ozone effects on peanut: II. Seed yield and quality. Crop Sci 47:1488–1497

Cardoso-Vilhena J, Balaguer L, Eamus D, Ollerenshaw J, Barnes J (2004) Mechanisms underlying the amelioration of $O_3$-induced damage by elevated atmospheric concentrations of $CO_2$. J Experi Bot 55:771–781

Caregnato FF, Bortolin RC, Junior AMD, Moreira JCF (2013) Exposure to elevated ozone levels differentially affects the antioxidant capacity and the redox homeostasis of two subtropical *Phaseolus vulgaris L.* varieties. Chemosphere 93(2):320–330

Carrera FP, Noceda C, Maridueña-Zavala MG, Cevallos-Cevallos JM (2021) Metabolomics, a powerful tool for understanding plant abiotic stress. Agronomy 11:824

Cen YP, Layzell DB (2004) Does oxygen limit nitrogenase activity in soybean exposed to elevated $CO_2$e$CO_2$? Plant Cell Environ 27:1229–1238

Centritto M, Lucas ME, Jarvis PG (2002) Gas exchange, biomass, whole-plant water-use efficiency and water uptake of peach (Prunus persica) seedlings in response to elevated carbon dioxide concentration and water availability. Tree Physiol 22:699–706

Chaudhary IJ, Rathore D (2021a) Assessment of dose-response relationship between ozone dose and groundnut (Arachis hypogaea L) cultivars using OpenTop Chamber (OTC) and ethyenediurea (EDU). Environ Technol Innov 22:101494

Chaudhary IJ, Rathore D (2021b) Micro-morphological and anatomical response of groundnut (*Arachis hypogaea* L.) cultivars to ground-level ozone. J Appl Biol Biotechnol 9(04):137–150

Chaudhary IJ, Rathore D (2022) Effects of ambient and elevated ozone on morphophysiology of cotton (*Gossypium hirsutum* L.) and its correlation with yield traits. Eviron Technol Innov 25:102146

Chaudhary N, Agrawal SB (2015) The role of elevated ozone on growth, yield and seed quality amongst six cultivars of mung bean. Ecotoxicol Environ Saf 111:286–294

Chen Z, Cao J, Yu H, Shang H (2018) Effects of elevated ozone levelsconcentration on photosynthesis biomass and non-structural carbohydrates of Phoebe bournei and Phoebe zhennan in Subtropical China. Front Plant Sci 9:1764

Cook BI, Smerdon JE, Seager R, Coats S (2014) Global warming and 21st century drying

Cooper OR et al (2020) Multi-decadal surface ozone trends at globally distributed, remote locations. Elementa-Sci Anthrop

Cotrufo MF, Ineson P, Scott A (1998) Elevated $CO_2$e$CO_2$ reduces the nitrogen concentration of plant tissues. Glob Change Biol 4(1):43–54

Cure JD, Acock B (1986) Crop response to carbon dioxide doubling: a literature survey. Agric For Clim Dyn 43:2607–2627

Cushman JC, Denby K, Mittler R (2022) Plant responses and adaptations to a changing climate. Plant J 109:319–322

Daripa A, Bhatia A, Ojha S, Tomer R, Chattaraj S, Singh KP, Singh SD (2016) Chemical and natural plant extract in ameliorating negative impact of tropospheric ozone on wheat crop: a case study in a part of semiarid north west India. Aerosol Air Qual Res 16:1742–1756

Das M, Zaidi PH, Pal M, Sengupta UK (2002) Stage sensitivity of mung bean (*Vigna radiata* L. Wilczek) to an elevated level of carbon dioxide. J Agron Crop Sci 188:219–224

David T, Patrick MJ, Thomas RB (1997) Nitrogenase activity and $N_2$ fixation are stimulated by elevated $CO_2$e$CO_2$ in a tropical $N_2$-fixing tree & misc. Oecologia 109:28–33

De la Mata L, Cabello P, de la Haba P, Agüera E (2012) Growth under elevated atmospheric $CO_2$ concentration accelerates leaf senescence in sunflower (*Helianthus annus* L.) plants. J Plant Physiol 164(14):1392–1400

De Souza AP, Gaspar M, Da Silva EA, Ulian EC, Waclawovsky AJ, Nishiyama MY Jr, Dos Santos RV, Teixeira MM, Souza GM, Buckeridge MS (2008) Elevated $CO_2$ increases photosynthesis, biomass and productivity, and modifies gene expression in sugarcane. Plant, Cell Environ 31:1116–1127

De Souza AP, Massad KF, Zhang C, Maroco JP (2020) Elevated $CO_2$e$CO_2$ increases leaf growth and delays leaf senescence in soybean. Plant, Cell Environ 43(7):1582–1595. https://doi.org/10.1111/pce.13766

Dhevagi P, Ramya A, Priyatharshini S, Poornima R (2022) Effect of elevated trophospheric ozone on *Vigna mungo* L. Ozone Sci Eng 44(6):566–586

Ding L, Wang K, Jiang G, Liu M, Cheng X, Huang D (2021) Responses of photosynthesis and yield of winter wheat to elevated $CO_2$e$CO_2$ and temperature. Field Crop Res 265:108048. https://doi.org/10.1016/j.fcr.2021.108048

Docherty M, Wade F, Hurst D, Whittaker J, Lea P (1997) Responses of tree sap-feeding herbivores to elevated CO2. Glob Change Biol 3(1):51–59

Drake BG (1992) A field study of the effects of elevated $CO_2$e$CO_2$ on ecosystem processes in a Chesapeake Bay wetland. Aust J Bot 40:579–595

Drake BG, Gonzàlez-Meler MA (1997) More efficient plants: a consequence of rising atmospheric $CO_2$? Plant Mol Biol 48:609–639

Duque L, Poelman EH, Stefan-Dewenter I (2021) Plant age at the time of ozone exposure affects flowering patterns, biotic interactions and reproduction of wild mustard. Sci Rep 11:23448

Eastburn DM, McElrone AJ, Bilgin DD (2011) Influence of atmospheric and climatic change on plant-pathogen interactions: climatic change and host-pathogen interactions. Plant Pathol 60:54–69

Ebrahim S, Usha K, Singh B (2011) Pathogenesis related (PR) proteins in plant defense mechanism. In: Méndez-Vilas A (ed) Science against microbial pathogens: communicating current research and technological advances. Formatex microbiology series, vol 1, no 3, pp 1043–1054

Ebi KL, Anderson CL, Hess JJ, Kim SH, Loladze I, Neumann RB et al (2021) Nutritional quality of crops in a high $CO_2$ world: an agenda for research and technology development. Environ Res Lett 16:064045

Emberson LD, Pleijel H, Ainsworth EA, van den Berg M, Ren W, Osborne S, Mills G, Pandey D, Dentener F, Büker P, Ewert F, Koeble R, Van Dingenen R (2018) Ozone effects on crops and consideration in crop models. Eur J Agron 100:19–34

Esterbauer H (1982) Separation and characterization of the aldehydic products of lipid peroxidation stimulated by ADP-$Fe^{2+}$ in rat liver microsomes. Biochem J 208(1):129-40

Fajer ED, Bowers MD, Bazzaz FA (1989) The effects of enriched carbon dioxide atmospheres on plant—insect herbivore interactions. Science 243(4895):1198–200

Fang C, Fernie AR, Luo J (2019) Exploring the diversity of plant metabolism. Trends Plant Sci 24(1):83–98

Fangmeier A, Grüters U, Hertstein U, Sandhage-Hofmann A, Vermehren B, Jäger HJ (1996) Effects of elevated $CO_2eCO_2$, nitrogen supply and tropospheric ozone on spring wheat. I. Growth and Yield. Environ Pollut 91:381–390

Feng ZZ, Kobayashi K, Ainsworth EA (2008) Impact of elevated ozone concentration on growth, physiology, and yield of wheat (*Triticum aestivum* L.): a meta-analysis. Glob Change Biol 14:2696–2708

Feng Z, Kobayashi K, Ainsworth EA (2008) Impact of elevated ozone concentration on growth, physiology, and yield of wheat (*Triticum aestivuml*): a meta-analysis: meta-analysis of wheat responses to elevated ozone. Glob Change Biol 14:2696–2708

Fernández-Martínez M, Sardans J, Chevallier F et al (2019) Global trends in carbon sinks and their relationships with $CO_2$ and temperature. Nat Clim Chang 9:73–79

Fernie AR, Luo J (2019) Exploring the diversity of plant metabolism. Trends Plant Sci 24:83–98

Finzi AC, DeLucia EH, Hamilton JG, Richter DD, Schlesinger WH (2002) The nitrogen budget of a pine forest under free air $CO_2$ enrichment. Oecologia 132:567–578

Fiscus EL, Booker FL, Burkey KO (2005) Crop responses to ozone: uptake, modes of action, carbon assimilation and partitioning. Plant, Cell Environ 28:997–1011

Fiscus EL, Reid CD, Miller JE, Heagle AS (1997) Elevated $CO_2eCO_2$ reduces $O_3$ flux and $O_3$-induced yield losses in soybeans: possible implications for elevated $CO_2eCO_2$ studies. J Exp Bot 48:307–313

Foyer CH, Descourvières P, Kunert KJ (1994) Protection against oxygen radicals: an important defence mechanism studied in transgenic plants. Plant Cell Environ 17:507–523

Franks PJ, Beerling DJ (2009) Maximum leaf conductance driven by $CO_2$ effects on stomatal size and density over geologic time. Proc Natl Acad Sci 106(25):10343–10347

Franzaring J, Högy P, Erbs M, Fangmeier A (2010) Responses of soil and canopy climate in a six year $CO_2$ enrichment study with spring crops. Agric Meterol 150:354–360

Franzaring J, Weller S, Schmid I, Fangmeier A (2011) Growth, senescence and water use efficiency of spring oilseed rape (Brassica napus L. cv Mozart) grown in a factorial combination of nitrogen supply and elevated $CO_2eCO_2$. Environ Exp Bot 72:284–296

Friedlingstein P, Jones MW, O'Sullivan M et al (2019) Global carbon budget 2019. Earth Syst Sci Data 11:1783–1838

Gamper H, Hartwig UA, Leuchtmann A (2005) Mycorrhizas improve nitrogen nutrition of Trifolium repens after 8 yr of selection under elevated atmospheric $CO_2$ partial pressure. New Phytol 167(2):531–542

Garcia RL, Long SP, Wall GW, Osborne CP, Kimball BA, Nie GY, Pinter PJ Jr, Lamorte RL, Wechsung F (1998) Photosynthesis and conductance of spring-wheat leaves: field response to continuous free-air atmospheric $CO_2$ enrichment. Plant Cell Environ 21:659–669

Gaudel et al. (2020) Aircraft observations since the 1990s reveal increases of, tropospheric ozone at multiple locations across the Northern Hemisphere. Sci Adv

Gayathri J, Boomiraj K, Avudainayagam S, Maheswari M, Chandrasekhar CN, Karthikeyan S (2019) Impact of tropospheric ozone on growth and yield of garlic in high altitude region of Western Ghats. Int J Chem Stud 7(3):3099–3101

Gayathri J, Sharma DK, Kovilpillai B, Bhatia A, Kumar S, Prasad M, Suroshe SS, Kumar RR, Dunna V, Kumar SN (2023) Impacts of elevated ozone and $CO_2$ on growth and yield of double zero mustard (*Brassica juncea*). Indian J Agric Sci 93(7):743–749

Ghude SD, Jena C, Chate DM, Beig G, Pfister GG, Kumar R, Ramanathan V (2014) Reductions in India's crop yield due to ozone. Geophys Res Lett 41(15):5685–5691

Gousset-Dupont A, Huché-Thélier L, Reviron MP, Genet P (2021) Impact of high $CO_2$ concentration on grapevine morphology and anatomy: Implications for plant architecture and fruit quality. Sci Hortic 283:110115. https://doi.org/10.1016/j.scienta.2021.110115

Grantz DA, Farrar JF (2000) Ozone inhibits phloem loading from a transport pool: compartmental efflux analysis in Pima cotton. Funct Plant Biol 27(9):859–868

Grimmer MK, John Foulkes M, Paveley ND (2012) Foliar pathogenesis and plant water relations: a review. J Exp Bot 63:4321–4331

Gutiérrez-Rodríguez M, Resco P, Ferrero-Serrano A, García-Sánchez F (2021) Elevated $CO_2eCO_2$ concentration enhances biomass accumulation, morphological and anatomical changes in tomato seedlings. J Plant Physiol 263:153454. https://doi.org/10.1016/j.jplph.2021.153454

Hacour A, Craigon J, Vandermeiren K, Ojanpera K, Pleigel H, Danielsson H, Hogy P, Finnan J, Bindi M (2002) $CO_2$ and ozone effects on canopy development of potato crops across Europe. Europ J Agronomy 17:257–272

Halliwell B, Gutteridge JMC (1989) Free radicals in biology and medicine. Clarendon, Oxford, pp 277–289

Hansen EM, Hauggaard-Nielsen H, Launay M, Rose P, Mikkelsen TN (2019) The impact of ozone exposure, temperature and $CO_2$ on the growth and yield of three spring wheat varieties. Environ Exp Bot 168:103868

Hasegawa T, Sakai H, Tokida T, Nakamura H, Zhu C, Usui Y (2013) Rice cultivar responses to elevated $CO_2eCO_2$ at two free-air $CO_2$ enrichment (FACE) sites in Japan. Funct Plant Biol 40:148–159

Hasan MM, Rahman MA, Skalicky M, Alabdallah NM, Waseem M, Jahan MS, Ahammed GJ, El-Mogy MM, El-Yazied AA, Ibrahim MFM, Fang XW (2021) Ozone induced stomatal regulations, MAPK and phytohormone signaling in plants Int. J Mol Sci 22(12):6304

Hassan IA, Bell NB, Ashmore MR, Cotrozzi L, Haiba NS, Basahi JM, Summan A, Almeelbi T, Ismail IM (2018) Radish (*Raphanus sativus* L.) cultivar-specific response to $O_3$: patterns of biochemical and plant growth characteristics. Clean-Soil Air Water 46(6):1800124

Hoover SER, Ladley JJ, Shchepetkina AA, Tisch M, Gieseg SP, Tylianakis JM (2012) Warming, CO2, and nitrogen deposition interactively affect plant-pollinator mutualism. Ecol Lett 15(2012):227–234

Heagle AS, Miller JE, Booker FL (1998) Influence of ozone stress on soybean response to carbon dioxide enrichment: I Foliar properties. Crop Sci 38:113–121

Hetherington AM, Woodward I (2003a) The role of stomata in sensing and driving environmental change. Nature 52:215–220

Hetherington AM, Woodward FI (2003b) The role of stomata in sensing and driving environmental change. Nature 424(6951):901–908

Högy P, Keck M, Niehaus K, Franzaring J, Fangmeier A (2010) Effect of atmospheric $CO_2$ enrichment on biomass, yield and low molecular weight metabolites in wheat grain. J Cereal Sci 50:248–254

Holopainen JK, Nerg AM, Blande JD (2013) Multitrophic signaling in polluted atmospheres. In: Niinemets Ü, Monson RK (eds) Biology, controls and models of tree volatile organic compound emissions. Tree Physiology, Springer

Hou X, Wu X, Ma C, Tian D, Yan Z, Li P (2022) Effect of the elevated ozone on greening tree species of urban: alterations in CNP stoichiometry and nutrient stock allocation to leaves and fine roots. Urb Fores Urb Green 76:127735

Hu L, Wang Z, Huang B, Cao F (2021) Elevated $CO_2eCO_2$ enhanced leaf photosynthesis and antioxidant capacity of maize seedlings under low temperature. Plant Physiol Biochem 159:131–138. https://doi.org/10.1016/j.plaphy.2021.04.003

Iglesias DJ, Calatayud A, Barreno E, Primo-Millo E, Talon M (2006) Responses of citrus plants to ozone: leaf biochemistry, antioxidant mechanisms and lipid peroxidation. Plant Physiol Biochem 44:125–131

Imai K, Cleman DF, Yanagisawa T (1985) Increase in atmospheric partial pressure of carbon dioxide and growth and yield of rice (*Oryza sativa* L.). Jpn J Crop Sci 54:413–418

IPCC (2013) Climate Change 2013: the physical science basis. In: Stocker TF, Qin D, Plattner G-K, Tignor M, Allen SK, Boschung J, Nauels A, Xia Y, Bex V, Midgley PM (eds) Contribution of Working Group I to the fifth assessment report of the intergovernmental panel on climate change. Cambridge University Press, Cambridge

IPCC (2021) Climate change 2021: The physical science basis. In: Contribution of working group I to the sixth assessment report of the intergovernmental panel on climate change. Cambridge University Press

Jablonski LM, Wang X, Curtis PS (2002) Plant reproduction under elevated $CO_2eCO_2$ conditions: a meta-analysis of reports on 79 crop and wild species. New Phytol 156:9–26

Jain V, Pal M, Raj A, Khetarpal S (2007) Photosynthesis and nutrient composition of spinach and fenugreek grown under elevated carbon dioxide concentration. Biol Plant 51(3):559–562

Ji G, Xue H, Seneweera S, Ping L, Yu-Zheng Z, Qi D, Er-da L, Xing-yu H (2015) Leaf photosynthesis and yield components of mung bean under fully open-air elevated [$CO_2$]. J Integr Agric 14(5):977–983

Kang S, Zhang F, Hu X, Zhang J (2002) Benefits of $CO_2$ enrichment on crop plants are modified by soil water status. Plant Soil 238(1):69–77

Kangasjarvi J, Talvinen J, Utriainen M, Karjalainen R (1994) Plant defence systems induced by ozone. Plant, Cell Environ 17:783–794

Karnosky DF, Percy KE, Xiang B, Callan B, Noormets A, Mankovska B et al (2002) Interacting elevated $CO_2eCO_2$ and tropospheric $O_3$ predisposes aspen (*Populus tremuloides* Michx.) to infection by rust (*Melampsora medusaef* sp. tremuloidae): interaction of $CO_2$ and $O_3$ on aspen leaf rust. Glob Change Biol 8:329–338

Kee LS, Deli C, Rob N, Roger A (2012) Does phosphorus stimulate the effect of elevated [$CO_2$] on growth and symbiotic nitrogen fixation of grain and pasture legumes? Crop Pasture Sci 63(1):53–62

Keeling RF, Piper SC, Bollenbacher AF, Walker JS (2009) Atmospheric $CO_2$ records from sites in the SIO air sampling network. In: Trends: a compendium of data on global change. Carbon dioxide information analysis center, Oak Ridge National Laboratory, U.S. Department of Energy, Oak Ridge

Kim HY, Lieffering M, Kobayshi K, Okada M, Mitchell MW, Gumpertz M (2003) Effects of free-air $CO_2$ enrichment and nitrogen supply on the yield of temperate paddy rice crops. Field Crops Res 83:261–270

Kim HY, Lieffering M, Miura S, Kobayashi K, Okada M (2001) Growth and nitrogen uptake of $CO_2$-enriched rice under field conditions. New Phytol 150:223–229

Kimball BA, Kobayashi K, Bindi M (2002) Responses of agricultural crops to free-air $CO_2$ enrichment. Adv Agron 77:293–368

Kinney KK, Lindroth RL, Jung SM, Nordheim EV (1997) Effects of $CO_2$ and $NO_3-$availability on deciduous trees: phytochemistry and insect performance. Ecology 78(1):215–230

Kissoudis C, van de Wiel C, Visser RGF, van der Linden G (2014) Enhancing crop resilience to combined abiotic and biotic stress through the dissection of physiological and molecular crosstalk. Front Plant Sci 5:1–20

Köhler IH, Huber SC, Bernacchi CJ, Baxter IR (2019) Increased temperatures may safeguard the nutritional quality of crops under future elevated $CO_2eCO_2$ concentrations. Plant J 97:872–886

Kovilpillai B, Nedumaran S, Mani S, Mani JR, Natarajan S, Ramasamy J (2023) Impacts of elevated ozone and ozone protectants on plant growth. Nutrients, biochemical and yield properties of Turnip (*Brassica rapa* L.). Science & Engineering, Ozone.https://doi.org/10.1080/01919512.2023.2165475

Kumar P, Rai P, Chaturvedi Ashish K, Khetarpal S, Pal M (2012) High atmospheric $CO_2$ delays leaf senescence and crop maturity in chickpea (*Cicer arietinum* L.) Ind J Plant Physiol 17:254–258

Kumar R, Bhatia A, Chakrabarti B, Kumar V, Tomer R, Sharma DK, Kumar SN (2021) Effect of elevated ozone and carbon dioxide on growth and yield of rice (*Oryza sativa*). Ind J Agric Sci 91:1607–1611

Kumari S, Agrawal M, Tiwari S (2013) Impact of elevated $CO_2eCO_2$ and $O_3$ on Beta vulgaris L.: Pigments, metabolites, antioxidants, growth and yield. Environ Pollut 174:279–288

Kumari S, Agrawal M, Singh A (2015a) Effects of ambient and elevated $CO_2eCO_2$ and ozone on physiological characteristics, antioxidative defense system and metabolites of potato in relation to ozone flux. Environ Exp Bot 109:276–287

Kumari S, Agrawal M, Singh A (2015b) Effects of ambient and elevated $CO_2eCO_2$ and ozone on physiological characteristics, antioxidative defense system and metabolites of potato in relation to ozone flux. Environ Exp Botany 109:276–287

Lake JA, Quick WP (2006) Ecological impacts of excessive atmospheric $CO_2$ on terrestrial ecosystems. Ann Appl Biol 149(3):315–327. The authors of this review article discuss how elevated $CO_2eCO_2$ levelsconcentration can lead to a reduction in stomatal density

Lal DM et al (2012) Tropospheric ozone and aerosol long-term trends over the Indo, Gangetic Plain (IGP), India. Atmos Res

Leakey ADB, Ainsworth EA, Bernacchi CJ, Rogers A, Long SP, Ort DR (2009a) Elevated $CO_2$ effects on plant carbon, nitrogen, and water relations: six important lessons from FACE. J Exp Bot 60:2859–2876

Leakey ADB (2009) Rising atmospheric carbon dioxide concentration and the future of C4 crops for food and fuel. Proc R Soc B 276:2333–2343

Leakey ADB, Uribelarrea M, Ainsworth EA, Naidu SL, Rogers A, Ort DR, Long SP (2006) Photosynthesis, productivity, and yield of maize are not affected by open-air elevated of $CO_2$ concentration in the absence of drought. Plant Physiol 140:779–790

Leakey ADB, Ainsworth EA, Bernacchi CJ, Rogers A, Long SP, Ort DR (2009b) Elevated $CO_2eCO_2$ effects on plant carbon, nitrogen, and water relations: six important lessons from FACE. J Exp Bot 60:2859–2876

Lee Y-H, Sang W-G, Baek J-K, Kim J-H, Shin P, Seo M-C, Cho J-I (2020) The effect of concurrent elevation in $CO_2$ and temperature on the growth, photosynthesis, and yield of potato crops. PLoS ONE 15(10):e0241081

Leisner CP, Ainsworth EA (2012) Quantifying the effects of ozone on plant reproductive growth and development. Glob Change Biol 18:606–616

Lenka S, Lenka NK (2012) Impact of tropospheric ozone on agroecosystem: an assessment. J Agric Phys, 12(1):1–11

Li JH, Dugas WA, Hymus GJ, Johnson DP, Hinkle CR, Drake BG, Hungate BA (2003) Direct and indirect effects of elevated $CO_2eCO_2$ on transpiration from Quercus myrtifolia in a scrub-oak ecosystem. Global Change Biol 9(1):96–105

Li P, Li B, Seneweera S, Zong Y, Li FY, Han Y, Hao X (2019a) Photosynthesis and yield response to elevated $CO_2eCO_2$, $C_4$ plant foxtail millet behaves similarly to $C_3$ species. Plant Sci 285:239–247

Li S, Moller CA, Mitchell NG, Lee D, Ainsworth EA (2021a) Bioenergy sorghum maintains photosynthetic capacity in elevated ozone concentrations. Plant, Cell Environ 44(3):729–746

Li F, Zhou Y, Shi P, Wu X (2021b) A meta-analysis of the effects of elevated $CO_2eCO_2$ on leaf area index in crops. Field Crop Res 263:108082. https://doi.org/10.1016/j.fcr.2021.108082
Li X, Sun Z, Shao S, Zhang S, Ahammed GJ, Zhang G et al (2015) Tomato-Pseudomonas syringae interactions under elevated $CO_2eCO_2$ concentration: the role of stomata. J Exp Bot 66:307–316
Lincoln DE, Couvet D, Sionit N (1986) Response of an insect herbivore to host plants grown in carbon dioxide enriched atmospheres. Oecologia 69(4):556–560
Lincoln DE, Sionit N, Strain BR (1984) Growth and feeding response of Pseudoplusia includens (Lepidoptera: Noctuidae) to host plants grown in controlled carbon dioxide atmospheres. Environ Entomol 13(6):1527–1530
Lincoln DE, Fajer ED, Johnson RH (1993) Plant-insect herbivore interactions in elevated CO2 environments. Trends Ecol Evol 8(2):64–68
Lindroth RL, Kinney KK, Platz CL (1993) Responses of diciduous trees to elevated atmospheric $CO_2$: productivity, phytochemistry, and insect performance. Ecology 74(3):763–777
Lindroth RL, Arteel GE, Kinney KK (1995) Responses of three saturniid species to paper birch grown under enriched $CO_2$ atmospheres. Funct Ecol 306–311
Leishman MR, Sanbrooke KJ, Woodfin RM (1999) The effects of elevated CO# and light environment on growth and reproductive performance of four annual species. New Phytol 144:455–462
Li Y, Li C, Bradshaw HD, Stover EW (2019b) High atmospheric $CO_2$ affects stomatal density and anatomy of young apple leaves. J Plant Physiol 233:112–117
Li Y, Li X, Zhang Y (2019) Effects of elevated $CO_2eCO_2$ concentration on stomatal density and pore index of cotton (*Gossypium hirsutum* L.) leaves under drought stress. J Integr Agric 18(5):1111–1117
Lin M, Horowitz LW, Payton R, Fiore AM, Tonnesen G (2017) US surface ozone trends and extremes from 1980 to 2014: quantifying the roles of rising Asian emissions, domestic controls, wildfires, and climate. Atmos Chem Phys 17:2943–2970
Liu W, Lu J, Liu J, Huang Y, Yang W, Zhang Y, Gao H et al (2021) Elevated $CO_2eCO_2$ enhances rice leaf growth and nitrogen metabolism through promoting nitrogen uptake and assimilation. Plant Soil 466(1–2):447–462. https://doi.org/10.1007/s11104-020-04824-4
Liu Y, Guan H, Wang Q, Zhang Z (2015) Elevated atmospheric $CO_2$ reduces stomatal and metabolic limitations on photosynthesis caused by salinity in the C4 crop sorghum. Environ Exp Bot 119:78–87
Lobell DB, Gourdji SM (2012) The influence of climate change on global crop productivity. Plant Physiol 160(4):1686–1697
Loladze I, Nolan JM, Ziska LH, Knobbe AR (2019) Rising atmospheric $CO_2$ lowers concentrations of plant carotenoids essential to human health: a meta-analysis. Mol Nutr Food Res 63:1801047
Lombardozzi D, Sparks JP, Bonan G, Levis S (2012) Ozone exposure causes a decoupling of conductance and photosynthesis: Implications for the Ball-Berry stomatal conductance model. Oecologia 169:651–659
Long SP, Naidu SL (2002) Effects of oxidants at the biochemical, cell and physiological levels concentration, with particular reference to ozone. In: Bell JNB, Treshow M (eds) Air pollution and plant life. Wiley, Chichester
Lu et al (2018) Lower tropospheric ozone over India and its linkage to the South Asian monsoon. Atmos Chem Phys
Mamatha H, Rao NKS, Laxman RH, Shivashankara KS, Bhatt RM, Pavithra KC (2014) Impact of elevated $CO_2eCO_2$ on growth, physiology, yield and quality of tomato (*Lycopersicon esculentum* Mill) cv, Arka Ashish. Photosynthetica 52(4):519–528
Manderscheid R, Pacholski A, Frühauf C, Weigel HJ (2009) Effects of free air carbon dioxide enrichment and nitrogen supply on growth and yield of winter barley cultivated in a crop rotation. Field Crops Res 110:185–196
Maroco JP, Edwards GE, Ku MSB (1999) Photosynthetic acclimation of maize to growth under elevated levels concentration of carbon dioxide. Planta 210:115–125

Marshall DL, Tyler AP, Abrahamson NJ, Avritt JJ, Barnes MG, Larkin LL, Medeiros JS, Reynolds J, Shaner MGM, Simpson HL, Maliakal-Witt S (2010) Pollen performance of *Raphanus sativus* (Brassicaceae) declines in response to elevated [$CO_2$]. Sex Plant Reprod 23(4):325–336

Martínez-Lüscher J, Kizildeniz T, Vučetić V, Dai Z, Luedeling E (2019) Sensitivity of grapevine phenology to water availability, temperature and $CO_2$ concentration. Front Plant Sci 10:1084

Mavrogianopoulos GN, Spanakis J, Tsikalas P (1998) Effect of carbon dioxide enrichment and salainity on photosynthesis and yield in melon. Sci Hortic 79:51–63

Mbikayi NT, Hileman DR, Bhattacharya NC, Ghosh PP, Biswas PK (1983) Effect of $CO_2$ enrichment on physiology and biomass production in cowpea (*Vigna unguiculata* L.) grown in open top chambers. In: Proceedings of international congress of plant physiology, New Delhi, pp 640–645

McGrath JM, Betzelberger AM, Wang S, Shook E, Zhu XG, Long SP, Ainsworth EA (2015) An analysis of ozone damage to historical maize and soybean yields in the United States. Proc Natl Acad Sci U S A 112:14390–14395

Miglietta F, Giuntoli J, Bindi M, Vaccari FP (2005) Free-air $CO_2$ enrichment (FACE) of grapevine (*Vitis vinifera* L.): II. Growth and quality of grape and wine in response to elevated $CO_2eCO_2$ concentrations. Europ J Agron 22(4):369–390

Mills G et al (2011) Evidence of widespread effects of ozone on crops and (semi-natural vegetation in Europe (1990–2006) in relation to AOT40-and flux, based risk maps. Global Change Biol

Mina U, Kumar P, Varshney CK (2010) Effect of ozone exposure on growth, yield and isoprene emission from tomato (*Lycopersicon esculentum* L.) plants. Veget Crops Res Bull 72(1):35–48

Mishra AK, Agrawal SB (2015) Biochemical and physiological characteristics of tropical mungbean (*Vigna radiata* L.) cultivars against chronic ozone stress: an insight to cultivar-specific response. Protoplasma 252:797–811

Mishra AK, Rai R, Agrawal SB (2013) Differential response of dwarf and tall tropical wheat cultivars to elevated ozone with and without carbon dioxide enrichment: Growth, yield and grain quality. Field Crop Res 145:21–32

Mitchell RAC, Black CR, Burkart S, Burke JI, Donnelly A, de Temmmerman L, Fangmeier A, Mulholland BJ, Theobald JC, Van Oijen M (1999) Photosynthetic responses in spring wheat grown under elevated $CO_2eCO_2$ concentrations and stress conditions in the European, multiple-site experiment 'ESPACE-wheat.' Eur J Agron 10:205–214

Miyagi KM, Kinugasa T, Hikosaka K, Hirose T (2007) Elevated $CO_2eCO_2$ concentration, nitrogen use, and seed production in annual plants. Glob Change Biol 13:2161–2170

Mo XG, Lin ZH, Liu SX (2007) Climate change impacts on the ecohydrological processes in the Wuding River basin. Acta Ecol Sin 27:4999–5007

Monks PS (2015) A review of the observations and origins of the spring ozone maximum. Atmos Environ@@

Morgan PB, Ainsworth EA, Long SP (2003) How does elevated ozone impact soybean? A meta-analysis of photosynthesis, growth and yield: impact of elevated ozone on soybean. Plant, Cell Environ 26:1317–1328

Morison JIL (1987) Intercellular $CO_2$ concentration and stomatal response to $CO_2$. In: Zeiger E, Farquhar GD, Cowan IR (eds) Stomatal Function. Stanford University Press, Stanford, CA, pp 229–251

Mulholland BJ, Craigon J, Black CR, Colls JJ, Atherton J, Landon G (1997) Effects of elevated carbon dioxide and ozone on the growth and yield of spring wheat (*Triticum aestivum* L.). J Exp Bot 48:113–122

Myers SS, Zanobetti A, Kloog I, Huybers P, Leakey AD, Bloom AJ, Carlisle E, Dietterich LH, Fitzgerald G, Hasegawa T, Holbrook NM (2014) Increasing $CO_2$ threatens human nutrition. Nature 510(7503):139–142

Nasser RR, Fuller MP, Jellings AJ (2008) Effect of elevated $CO_2$ and nitrogen levels on lentil growth and nodulation. Agron Sustain Dev 28(2):175–180

Nie GY, Long SP, Garcia RL, Kimball BA, Lamorte RL, Pinter PJ, Wall GW, Webber AN (1995) Effects of free-air $CO_2$, enrichment on the development of the photosynthetic apparatus in wheat, as indicated by changes in leaf proteins. Plant Cell Environ 18:855–864

Noguchi K, Tsunoda T, Miyagi A, Kawai-Yamada M, Sugiura D, Miyazawa S-I, Tokida T, Usui Y, Nakamura H, Sakai H, Hasegawa T (2018) Effects of elevated atmospheric $CO_2$ on respiratory rates in mature leaves of two rice cultivars grown at a free-air $CO_2$ enrichment site and analyses of the underlying mechanisms. Plant Cell Physiol 59(3):637–649

Noormets A, Sober A, Pell EJ, Dickson RE, Podila GK, Sober J, Isebrands JG, Karnosky DF (2001) Stomatal and non-stomatal limitation to photosynthesis in two trembling aspen (*Populus tremuloides* Michx.) clones exposed to elevated $CO_2$ and/or $O_3$. Plant Cell and Environ 24:327–336

Nowroz F, Hasanuzzaman M, Siddika A, Parvin K, Garcia P, Nahar K, Prasad PV (2024) Elevated tropospheric ozone and crop production: potential negative effects and plant defence mechanisms. Front Plant Sci 14:1244515

Pal M, Rao LS, Jain V, Srivastava AC, Pandey R, Raj A, Singh KP (2005) Effects of elevated $CO_2$ and nitrogen on wheat growth and photosynthesis. Biol Plant 49(3):467–470

Pal M, Karthikeyapandian V, Jain V, Srivastava AC, Raj A, Sengupta UK (2004) Biomass production and nutritional levels concentration of berseem (*Trifolium alexandrium*) grown under elevated $CO_2eCO_2$. Agric Ecosyst Environ 101:31–38

Pal M, Chaturvedi AK, Pandey SK, Bahuguna RN, Khetarpal S, Anand A (2014) Rising atmospheric $CO_2$ may affect oil quality and seed yield of sunflower (*Helianthus annus* L.). Acta Physiol Plant 36:2853–2861

Pal M, Sharma P, Bhadoria PBS (2019) Elevated carbon dioxide and water stress altered root: shoot ratio, anatomy and morphology of sorghum. J Plant Growth Regul 38(3):927–936

Paoletti E, Feng Z, Marco AD, Hoshika Y, Harmens H et al (2020) Challenges, gaps and opportunities in investigating the interactions of ozone pollution and plant ecosystems. Sci Total Environ 709:136188@@

Payer H, Firsching K, Steiner K, Djelassi K (1993) Factorial effects of $CO_2$, $O_3$ and irrigation on the yield and water-use efficiency of experimental grass swards grown in closed chambers. UK Committee for Air Pollution Effects Research Annual Meeting, Bangor, Wales (Abstract)

Pell EJ, Dann MS (1991) Multiple stress-induced foliar senescence and implications for whole plant longevity. In: Monney HA, Winner WE, Pell EJ (eds) Responses of plants to multiple stresses. Academic, San Diego, CA, pp 389–403

Peng J, Shang B, Xu Y, Feng Z, Calatayud V (2020) Effects of ozone on maize (*Zea mays* L.) photosynthetic physiology, biomass and yield components based on exposure- and flux-response relationships. Environ Pollut 256:113466

Phothi R, Umponstira C, Sarin C, Siriwong W, Nabheerong N (2016) Combining effects of ozone and carbon dioxide application on photosynthesis of Thai jasmine rice (*Oryza sativa* L.) cultivar Khao Dawk Mali 105. Austr J Crop Sci 10(4):591–597

Piercy and Bjorkman (1983) $CO_2$ and plants: the response of plants to rising levels concentration of atmospheric $CO_2$. Physiol Effects 65–105

Pinter PJ, Kimball BA, Garcia RL, Wall GW, Hunsacker DJ, Lamorte RL (1996) Free-air $CO_2$ enrichment—responses of cotton and wheat crops. In: Koch G, Mooney HA (eds) Carbon dioxide and terrestrial ecosystems. Academic Press, San Diego, pp 215–249

Polley HW, Johnson HB, Tischler CR, Torbert HA (1999) Links between transpiration and plant nitrogen: variation with atmospheric $CO_2$ concentration and nitrogen availability. Int J Plant Sci 160:535–542

Poorter H, Nagel O (2000) The role of biomass allocation in the growth response of plants to different levels concentration of light, CO2, nutrients and water: a quantitative review. Aust J Plant Physiol 27(6):595–607

Poorter H (1993) Interspecific variation in the growth response of plants to an elevated ambient $CO_2$ concentration. In: Rozema J, Lambers H, Van de Geijn SC, Cambridge ML (eds) $CO_2$ and biosphere. Kluwer Academic Publishers, Dordrecht, pp 77/97

Prasad PVV, Boote KJ, Allen LH Jr (2006) Adverse high temperature effects on pollen viability, seed-set, seed yield and harvest index of grain sorghum [*Sorghum bicolor* (L.) Moench] are more severe at elevated carbon dioxide due to higher tissue temperatures. Agric Meteorol 139:237–251

Prasad PVV, Boote KJ, Allen LH Jr, Thomas JMG (2002) Effects of elevated temperature and carbon dioxide on seed-set and yield of kidney bean (*Phaseolus vulgaris* L.). Global Change Biol 8:710–721

Prior SA, Runion GB, Rogers HH, Arriaga FJ (2010) Elevated atmospheric carbon dioxide effects on soybean and sorghum gas exchange in conventional and no-tillage systems. J Environ Qual 39(2):596–608

Qian H, Chen Y, Fang X, Wang X, Xu G, Chen Y (2020) Elevated $CO_2eCO_2$ promotes leaf photosynthesis and anatomical changes in tobacco plants grown under different nitrogen conditions. Front Plant Sci 11:768. https://doi.org/10.3389/fpls.2020.00768

Rai P, Chaturvedi AK, Shah D, Viswanathan C, Pal M (2016) Elevated CO 2 enhances carbohydrate assimilation at flowering stage and seed yield in chickpea (*Cicer arietinum*). Indian J Plant Physiol 21:114–121

Rabah NR, Fuller MP, Jellings AJ (2008) Effect of elevated $CO_2eCO_2$ and nitrogen levels concentration on lentil growth and nodulation. Agron Sustain Dev 28(2):175–180

Rai R, Agrawal M, Agrawal SB (2011) Effects of ambient $O_3$ on wheat during reproductive development: gas exchange, photosynthetic pigments, chlorophyll flouresence, and carbohydrates. Photosynthetica 49:285–294

Ramya A, Dhevagi P, Priyatharshini S, Saraswathi R, Avudainayagam S, Venkataramani S (2021) Response of rice (*Oryza sativa* L.) cultivars to elevated ozone stress. Environ Monit Assess 193:808

Rao MV, Hale BA, Ormrod DP (1995) Amelioration of ozone induced oxidative damage in wheat plants grown under high carbon dioxide. Plant Physiol 109:421–432

Rathore A, Gopikrishnan GS, Kuttippurath J (2023) Changes in tropospheric ozone over India: variability, long-term trends and climate forcing. Atmos Environ 309

Raza A, Hussain M, Li Y, Wu Y (2020) Effects of elevated $CO_2eCO_2$ concentration on plant morphology, anatomy and yield of wheat (*Triticum aestivum* L.) under field conditions. Environ Exp Botany 171:103983. https://doi.org/10.1016/j.envexpbot.2019.103983

Reddy VR, Reddy KR, Acock MC, Trent A (1994) Carbon dioxide enrichment and temperature effects on root growth in cotton. Biotronics 23:47–57

Reinert RA, Eason G, Barton J (1997) Growth and fruiting of tomato as influenced by carbon dioxide and ozone. New Phytol 137(3):411–420

Rivero RM, Mittler R, Blumwald E, Zandalinas SI (2022) Developing climate-resilient crops: Improving plant tolerance to stress combination. Plant J 109:373–389

Rogers A, Ainsworth EA, Leakey ADB (2009) Will elevated carbon dioxide concentration amplify the benefits of nitrogen fixation in legumes? Plant Physiol 151:1009–1016

Rogers HH, Dahlman RC (1993) Crop responses to $CO_2$ enrichment. In: Rozema J, Lambers H, Van de Geijn SC, Cambridge ML (eds) $CO_2$ and biosphere. Kluwer Academic Publishers, Dordrecht, pp 117/131

Roy S, Beig G, Jacob D (2008) Seasonal distribution of ozone and its precursors over the tropical Indian region using regional chemistry-transport model. J Geophys Res: Atmos 113(D21)

Ruhil K, Sheeba AA, Iqbal M, Tripathy BC (2014) Photosynthesis and growth responses of mustard (*Brassica juncea* L. cv Pusa Bold) plants to free air carbon dioxide enrichment (FACE). Protoplasma 252(4):935–946

Ryalls JMW, Langford B, Mullinger NJ, Bromfield LM, Nemitz E, Pfrang C et al (2022) Anthropogenic air pollutants reduce insect-mediated pollination services, Environ Pollut 297, Article 118847

Sage RF, Sharkey TD, Seemann JR (1989) Acclimation of photosynthesis to elevated $CO_2eCO_2$ in five C3 species. Plant Physiol 89:590/596

Saha S, Sehgal VK, Nagarajan S, Pal M (2012) Impact of elevated atmospheric $CO_2$ on radiation utilization and related plant biophysical properties in pigeon pea (*Cajanus cajan* L.). Agric for Meteorol 158:63–70

Sairam RK, Saxena DC (2000) Oxidative stress and antioxidants in wheat genotypes: possible mechanism of water stress tolerance. J Agron Crop Sci 184:55–61

Salt DT, Brooks GL, Whittaker JB (1995) Elevated carbon dioxide affects leaf-miner performance and plant growth in docks (Rumex spp.). Glob Change Biol 1(2):153–156

Saurav S, Kumar SV, Debasish C, Madan P (2015) Atmospheric carbon dioxide enrichment induced modifications in canopy radiation utilization, growth and yield of chickpea (*Cicer arietinum* L.). Agric for Meteorol 202:102–111

Sahoo A, Madheshiya P, Mishra AK, Tiwari S (2023) Combating ozone stress through N fertilization: a case study of Indian bean (*Dolichos lablab* L.). Front Plant Sci 14:1125529. https://doi.org/10.3389/fpls.2023.1125529

Sandermann H (2000) Ozone/biotic disease interactions: molecular biomarkers as a new experimental tool. Environ Pollut 108:327–332

Sarkar A, Agrawal SB (2010) Elevated ozone and two modern wheat cultivars: an assessment of dose dependent sensitivity with respect to growth, reproductive and yield parameters. Environ Exp Bot 69:328–337

Sarkar A, Rakwal R, Agrawal SB, Shibato J, Ogawa Y, Yoshida Y, Agrawal GK, Agrawal M (2010) Investigating the impact of elevated levels concentration of ozone on tropical wheat using integrated phenotypical, physiological, biochemical, and proteomics approaches. J Proteome Res 9:4565–4584

Schwalm CR, Glendon S, Duffy PB (2020) RCP8.5 tracks cumulative $CO_2$ emissions. In: Proceedings of the national academy of sciences, USA 117, 19656

Severino JF, Stich K, Soja G (2007) Ozone stress and antioxidant substances in *Trifolium repens* and *Centaurea jacea* leaves. Environ Pollut 146:707–714

Sharma V, Gupta P, Priscilla K et al (2021a) Metabolomics intervention towards better understanding of plant traits. Cells 10:346

Sharma P, Dubey RS, Agrawal SB (2021b) Elevated $CO_2eCO_2$ alters photosynthetic electron transport chain and thylakoid membrane structure in Indian mustard (*Brassica juncea* L.). J Environ Sci 107:227–237. https://doi.org/10.1016/j.jes.2021.03.023

Singh P, Agrawal M, Agrawal SB (2009) Evaluation of physiological, growth and yield responses of a tropical oil crop (*Brassica campestris* L. var. Kranti) under ambient ozone pollution at varying NPK levels. Environ Pollut 157(3):871–80

Singh F, Diwakar B (1995) Chickpea botany and production practices, skill development series no. 16. In: International crops research institute for semi-arid tropics, Patancheru, Andhra Pradesh, India

Singh RN, Mukherjee J, Sehgal VK, Bhatia A (2017) Effect of elevated ozone, carbon dioxide and their interaction on growth, biomass and water use efficiency of chickpea (*Cicer arietinum* L.). J Agrometeorol 19:301–305

Singh RN, Mukherjee J, Sehgal VK, Krishnan P, Das DK, Dhakar RK, Bhatia A (2021) Interactive effect of elevated trophosheric ozone and carbon dioxide on radiation utilization, growth and yield of chickpea (*Cicer arietinum* L.). Int J Biometeorol 65(11):1939–1952

Singh A, Pandey R, Yadav RK, Singh B (2020) Response of wheat to elevated carbon dioxide concentration: leaf growth, photosynthesis, and yield. J Plant Nutr 43(15):2272–2283. https://doi.org/10.1080/01904167.2020.1839282

Slattery RA, Ort DR (2019) Carbon assimilation in crops at high temperatures. Plant, Cell Environ 42:2750–2758

Smith JL, Paul EA (1990) The significance of soil microbial biomass estimations. In: Bollag JM, Stotzky G (eds) Soil biochemistry. Marcel Dekker, Inc., New York, pp 357–396

Soares JC, Santos CS, Carvalho SMP, Pintado MM, Vasconcelos MW (2019) Preserving the nutritional quality of crop plants under a changing climate: importance and strategies. Plant Soil 443:1–26

Srivastava AC, Tiwari LD, Pal M, Sengupta UK (2002) $CO_2$—mediated changes in mungbean chemistry: impact on plant-herbivore interactions. Curr Sci 82(9):1148–1151

Srivastava GC, Pal M, Das M, Sengupta UK (2001) Growth, $CO_2$ exchange rate and dry matter partitioning in mungbean (*Vigna radiata* L.) grown under elevated $CO_2eCO_2$. Indian J Exp Biol 39:572–577

Stabler D (2016) Impacts of ozone pollution on nectar and pollen quality and their significance for pollinators. School of Biology, Newcastle University, PhD thesis, Newcastle, UK
Sun J, Feng Z, Ort DR (2014a) Impacts of rising trophospheric ozone on photosynthesis and metabolite levels concentration on field grown soybean. Plant Sci 226:147–161
Sun J, Feng Z, Ort DR (2014b) Impacts of rising trophospheric ozone on photosynthesis and metabolite levels on field grown soybean. Plant Sci 226:147–161
Swarupa V, Ravishankar KV, Rekha A (2014) Plant defense response against Fusarium oxysporum and strategies to develop tolerant genotypes in banana. Planta 239:735–751
Temmerman L, Legrand G, Vandermeiren K (2007) Effects of ozone on sugar beet grown in open top chambers. Europ J Agronomy 26:1–9
Tissue DT, Thomas RB, Strain BR (1997) Atmospheric $CO_2$ enrichment increases growth and photosynthesis of Pinus taeda: a 4 year experiment in the field. Plant Cell Environ 20(9):1123–1134
Twine TE, Bryant JJ, Richter K, Bernacchi CJ, McConnaughay KD, Morris SJ, Leakey AD (2013) Impacts of elevated $CO_2eCO_2$ concentration on the productivity and surface energy budget of the soybean and maize agroecosystem in the Midwest USA, Global Change Biol 19(9):2838–285
Uprety DC, Mishra RS, Abrol YP (1995) Effect of elevated $CO_2eCO_2$ on the photosynthesis, growth and water relation of Brassica species under moisture stress. J Agron Crop Sci 175:231–237
Usuda H (2006) Effects of elevated $CO_2$ on the capacity for photosynthesis of a single leaf and a whole plant, and on growth in a radish. Plant Cell Physiology 47(2):262–269
Vanaja M, Raghuram RP, Jyothi LN, Maheshwari M, Vagheera P, Ratnakumar P, Jyothi M, Yadav SK, Venkateswarlu B (2007) Effect of elevated $CO_2eCO_2$ concentrations on growth and yield of blackgram (*Vigna mungo* L. Hepper)—a rainfed pulse crop. Plant Soil Environ 53:81–88
Vanaja M, Vagheera P, Ratnakumar P, Jyothi LN, Raghuram RP, Yadav SK, Maheswari M, Venkateswarulu B (2006) Evaluation of certain rainfed food and oil seed crops for their response to elevated $CO_2eCO_2$ at vegetative stage. Plant Soil Environ 52:164–170
Vandermeiren K, Black C, Pleijel H, De Temmerman L (2005) Impact of rising tropospheric ozone on potato: effects on photosynthesis, growth, productivity and yield quality. Plant Cell Environ 28:982–996
Vardhan DA, VAanaja M, Srinivas K, Lal M, Satish T, Reddy P (2013) Effect of elevated $CO_2$ concentration on groundnut (*Arachis hypogaea*) ($C_3$) and pearl millet (*Pennisetum glaucum*) ($C_4$) crop plants and some implication on growth and photosynthetic activity. Ind J Agric Sci 83(7):721–723
Walker BJ, South PF, Ort DR (2016) Physiological evidence for plasticity in glycolate/glycerate transport during photorespiration. Photosynth Res 129:93–103
Wang W, Cai C, He J, Gu J, Zhu G, Zhang W, Zhu J, Liu G (2020) Yield, dry matter distribution and photosynthetic characteristics of rice under elevated $CO_2eCO_2$ and increased temperature conditions. Field Crop Res 248:107605
Wang H, Li J, Zhang X, Liu B, Chen G (2021) Elevated $CO_2eCO_2$ enhances maize leaf growth through cell number expansion rather than cell size expansion. Plant, Cell Environ 44(1):202–215. https://doi.org/10.1111/pce.14019
Ward JK, Strain BR (1997) Effects of low and elevated $CO_2$ partial pressure on growth and reproduction of Arabidopsis thaliana from different elevations. Plant Cell Environ 20(2):254–260
Webber AN, Nie GY, Long SP (1994) Acclimation of photosynthetic proteins to rising atmospheric $CO_2$. Photosynth Res 39:413/4
Wei Z, Abdelhakim LOA, Fang L, Peng X, Liu J, Liu F (2022) Elevated $CO_2$ effect on the response of stomatal control and water use efficiency in amaranth and maize plants to progressive drought stress. Agric Water Manag 266:107609
Williams RS, Thomas RB, Strain BR, Lincoln DE (1997) Effects of elevated $CO_2$, soil nutrient levels, and foliage age on the performance of two generations of Neodiprion lecontei (Hymenoptera: Diprionidae) feeding on loblolly pine. Environ Entomol 26(6):1312–1322
Wu DX, Wang GX (2000) Interaction of $CO_2$ enrichment and drought on growth, water use, and yield of broad bean (*Vicia faba*). Environ Exp Bot 43(2):131–139

Wu DX, Wang GX, Bai YF, Liao JX (2004) Effects of elevated $CO_2$ concentration on growth, water use, yield and grain quality of wheat under two soil water levels concentration. Agric Ecosyst Environ 104:493–507

Wu X-J, Sun S, Xing G-M, Wang G-L, Wang F, Xu Z-S, Tian Y-S, Hou X-L, Xiong A-S (2017) Elevated carbon dioxide altered morphological and anatomical characteristics, ascorbic acid accumulation and related gene expression during taproot development in carrots. Front Plant Sci 7:2026

Wu W, Ye Z, Chen R (2021) Morphological and anatomical changes of maize under elevated $CO_2eCO_2$: a meta-analysis. Photosynthetica 59(1):101–112. https://doi.org/10.32615/ps.2021.032

Xu J, Li J, Yan P, Li H, Liang Y (2021) Responses of photosynthesis and nitrogen metabolism to elevated $CO_2eCO_2$ and warming in soybean. J Agron Crop Sci 207(5):559–569. https://doi.org/10.1111/jac.12477

Yadav A, Bhatia A, Yadav S, Singh A, Tomer R, Kumar V, Singh B (2021) Growth, yield and quality of maize under ozone and carbon dioxide interaction in North West India. Aerosol Air Qual Res 20

Yadav A, Bhatia A, Yadav S, Kumar V, Singh B (2019) The effects of elevated $CO_2$ and elevated $O_3$ exposure on plant growth, yield and quality of grains of two wheat cultivars grown in north India. Heliyon 5:e02317

Yadav DS, Mishra AK, Rai R, Chaudhary N, Mukherjee A, Agrawal SB, Agrawal M (2020) Responses of an old and a modern Indian wheat cultivar to future $O_3$ level: physiological, yield and grain quality parameters. Environ Pollut 259:113939

Yendrek CR, Tomaz T, Montes CM, Cao Y, Morse AM, Brown PJ, Ainsworth EA et al (2017) High-throughput phenotyping of maize leaf physiological and biochemical traits using hyperspectral reflectance. Plant Physiol 173:614–626

Zak DR, Pregitzer KS, King JS, Holmes WE (2000) Elevated atmospheric $CO_2$, fine roots and the response of soil microorganisms: a review and hypothesis. New Phytol 147:201–222

Zandalinas SI, Sengupta S, Fritschi FB, Azad RK, Nechushtai R, Mittler R (2021) The impact of multifactorial stress combination on plant growth and survival. New Phytol 230:1034–1048

Zhang L, Zhou X, Li G, Liu F (2021) Elevated $CO_2$ modulates carbon assimilation and leaf water use efficiency of *Nicotiana tabacum* L. (tobacco) under patchy soil nutrient deficiency. Ind Crops Prod 166:113500

Zhang H, Zhu J, Gong Z, Zhu J-K (2022) Abiotic stress responses in plants. Nat Rev Genet 23:104–119

Zhao D, Reddy KR, Kakani VG, Mohammed AR, Read JJ, Gao W (2004) Leaf and canopy photosynthetic characteristics of cotton (*Gossypium hirsutum*) under elevated $CO_2$ concentration and UV-B radiation. J Plant Physiol 161:581–590

Zheng Y, Li F, Hao L, Yu J, Guo L, Zhou H, Ma C, Zhang X, Xu M (2019) Elevated $CO_2$ concentration induces photosynthetic down-regulation with changes in leaf structure, non-structural carbohydrates and nitrogen content of soybean. BMC Plant Biol 19:255

Zhu C, Zhu J, Cao Y, Sun Y, Zhang H, Zheng S (2021) Elevated $CO_2eCO_2$ alleviates the adverse effects of salt stress on photosynthesis and biomass production in tomato plants. Environ Exp Bot 187:104462. https://doi.org/10.1016/j.envexpbot.2021.104462

Ziska LH, Blowsky R (2007) A quantitative and qualitative assessment of mungbean (Vigna mungo wilczek) seed in response to elevated atmospheric carbon dioxide: potential changes in fatty acid composition. J Agric Food Chem 87:920–923

Ziska LH, Bunce JA (2007) Predicting the impact of changing $CO_2$ on crop yields: some thoughts on food. New Phytol 175:607–618

Ziska LH, Namuco O, Moya T, Quilang J (1997) Growth and yield responses of field-grown tropical rice to increasing carbon dioxide and air temperature. Agron J 89:45–53

# Chapter 3
# The Role of Nitric Oxides (NO) in Alleviating Effects of High Ozone Conditions on Crop Yield

**Bhavna Jaiswal, Durgesh Singh Yadav, and Madhoolika Agrawal**

**Abstract** Tropospheric ozone ($O_3$) is a pollutant formed through complex chemical reactions involving precursor pollutants emitted from anthropogenic activities. Ozone pollution has been shown to have detrimental effects on various crops and other $O_3$-sensitive plants, and there are substantial research studies available that investigate the $O_3$ damage mechanisms in crops, primarily wheat. Nitric oxide (NO) is a signaling molecule involved in multiple physiological processes and enhances the tolerance of crops to various abiotic stresses. The application of NO on crops has shown promise in mitigating the harmful impacts of $O_3$ pollution. However, the application of NO on crops is still an area of active research. The review underscores the detrimental impact of high $O_3$ pollution on crop yields, synthesis and role of NO, and highlights the potential of NO as a protective agent against $O_3$ stress. Additionally, the role of NO in activation of defense mechanisms, regulating stomatal behavior, and promoting overall plant resilience for maintaining crop productivity in $O_3$-polluted environments is emphasized.

**Keywords** Ozone · Nitric oxide · Stresses · Crops · Yield

## 3.1 Introduction

Since the beginning of industrialization, the concentration of tropospheric $O_3$, a potent greenhouse gas in the lower atmosphere, has indeed increased (Young et al. 2013). The rising concentration of tropospheric $O_3$ contributes to global average

B. Jaiswal
Department of Botany, Mahila Vidyalaya Degree College, Lucknow, Uttar Pradesh, India

D. S. Yadav
Department of Botany, Government Raza P.G. College, Rampur, Uttar Pradesh, India

M. Agrawal (✉)
Laboratory of Air Pollution and Global Climate Change, Department of Botany, Institute of Science, Banaras Hindu University, Varanasi, Uttar Pradesh, India
e-mail: madhoo.agrawal@gmail.com



A. K. Pandey and E. Oksanen (eds.), *Tropospheric Ozone and Food Security*,
https://doi.org/10.1007/978-3-032-03608-7_3

radiative forcing (the ability of a substance to alter the balance of incoming and outgoing energy in the Earth's atmosphere) of + 0.40 W $m^{-2}$, and predicted to increase by + 0.07 W $m^{-2}$ from 2010 to 2050 (Turnock et al. 2018). The increase of $O_3$ concentration is primarily attributed to anthropogenic activities. Anthropogenic emissions play a significant role in the release of $O_3$ precursor pollutants. The important sources of anthropogenic emissions include fossil fuel combustion, industrial processes, agricultural practices, residential and commercial activities (Cooper et al. 2014; Mills et al. 2018a, b, c). These activities contribute to the release of nitrogen oxides ($NO_x$), carbon monoxide (CO), and volatile organic compounds (VOCs), which are important $O_3$ precursors (Monks et al. 2015). Natural sources also contribute to the emission of $O_3$ precursors such as lightning NOx emissions (Schumann and Huntrieser 2007) and biogenic emissions, mainly isoprene (Guenther 1995). Biogenic emissions refer to the release of gases and volatile organic compounds (VOCs) from natural sources such as plants, trees, and other vegetation (Young et al. 2013). Isoprene is one of the most abundant VOCs which are commonly emitted by plants, particularly in response to environmental factors such as temperature and light. The higher temperatures and increased light exposure can lead to higher isoprene emissions, while higher $CO_2$ concentrations tend to suppress isoprene emission rates (Arneth et al. 2011). Natural emissions of $O_3$ precursors have existed for a long time, but the increase in anthropogenic emissions since the Industrial Revolution has significantly contributed to higher $O_3$ precursor levels (Stevenson et al. 2013). The interplay between natural and anthropogenic emissions, along with atmospheric chemistry and meteorological conditions, determines the formation and concentrations of $O_3$ in the troposphere (Perring et al. 2022).

High $O_3$ concentration has detrimental effects on plants, particularly on their photosynthetic process and reproductive capabilities (Sun et al. 2014; Yadav et al. 2021a, b). It interferes with chloroplasts, the cellular organelles responsible for photosynthesis. This reduction in photosynthesis leading to a decline in the plant's ability to produce carbohydrates and limiting its growth and biomass accumulation in plants (Chen et al. 2018). Furthermore, $O_3$ exposure during the flowering stage can have adverse effects on pollen development. It can reduce pollen viability and impair fertilization, ultimately leading to reduced crop yields (Gupta et al. 2022b). Different crop species and cultivars exhibit varying degrees of sensitivity to $O_3$ pollution (Gupta et al. 2022a). Some crops, such as wheat, soybeans, and potatoes, are more sensitive to $O_3$ damage than others (Mukherjee et al. 2020). Therefore, the impact of rising tropospheric $O_3$ pollution on crop yield loss can vary depending on the specific crop being grown. While the damage mechanisms of $O_3$ to these crops have been extensively studied (Feng et al. 2015; Mills et al. 2018a, b, c; Yadav and Agrawal 2021; Gupta et al. 2022a, b, c), there is limited knowledge on how to mitigate these negative impacts and increase grain production under elevated $O_3$ conditions (Harmens et al. 2019; Mukherjee et al. 2020).

Unlike ground level of $O_3$, nitric oxide (NO) is a bioactive gas that has been found to play a regulatory role in various plant growth and developmental processes (Asgher et al. 2017). It is considered a signaling molecule in plants and regulates physiological responses. NO is synthesized by plants through the action of specific enzymes, such as nitric oxide synthase-like enzymes (NOS-like enzymes) (Santolini et al. 2017). It is involved in multiple aspects of plant biology, including seed germination, root development, flowering, fruit ripening, and responses to environmental stresses (Wimalasekera et al. 2011). NO has been found to positively impact the quality features and nutritional contents of fruit plants (Imran et al. 2023). One of the key functions of NO is to promote plant resilience and survival under challenging conditions. It has been observed that NO can enhance plant tolerance to abiotic stresses such as ozone, drought, salinity, temperature extremes, and heavy metal toxicity (Ahlfors et al. 2009; Gill et al. 2013; Khan et al. 2019; Xu et al. 2021; Mata-Pérez et al. 2023). NO acts as a signaling molecule in stress signaling pathways, helping plants to adapt and cope with adverse conditions (Malik et al. 2020; Hancock and Veal 2021). Furthermore, NO also induces defense responses and activate the plant's immune system, thereby enhancing resistance to various diseases and pests (Kumar and Ohri 2023). NO can regulate the expression of defense associated genes and stimulate the production of secondary metabolites with antimicrobial properties (Huang et al. 2002). Furthermore, it regulates stomatal closure, which helps plants conserve water during drought stress (Sidana et al. 2015). The aim of this study is to compile the available literature on potential role of NO in mitigating $O_3$ impact.

## 3.2 Yield Loss by $O_3$

The fortuitous occurrence of elevated $O_3$ levels during the crop growing season hastens the decline in crop yield. Many studies have been conducted to measure the current and future extent of crop yield loss attributed to prevalent $O_3$ concentrations (Lal et al. 2017; Mills et al. 2018a, b, c; Yadav et al. 2021a, b). The findings consistently affirm the perspective that the current $O_3$ levels are adequately high to cause a reduction in crop productivity. According to (Mills et al. 2018a, b, c), the global production of soybean, wheat, maize and rice experienced yearly reductions of 12.4%, 7.1%, 6.1%, and 4.4%, respectively, attributed to ambient $O_3$ levels.

Some recent studies showing crop yield loss due to surface $O_3$ are presented in Table 3.1. A study conducted by Li et al. (2018) examined the impact of $O_3$ on a wheat cultivar Liangxing99. The results indicated a yield loss of 22.6% when exposed to ambient $O_3$ concentration with an additional 80 ppb. A meta-analysis of studies comparing the impacts of current ambient $O_3$ levels to pre-industrial levels on wheat crops revealed an average yield loss of 8.4% (Pleijel et al. 2018). This finding indicates a noteworthy and consistent crop response to the existing levels of $O_3$. According to Yadav et al. (2019a, b), the use of the FAOCE facility resulted in a 17.3% decline in grain yield for HD2967 and an 18.1% decline for C306 when exposed to 70 ppb of $O_3$. Similarly, yield losses in Western wheat ranged from 0 to 27%, while in Asian wheat, the yield losses ranged from 4 to 39% due to ambient $O_3$ levels (Schauberger et al. 2019). These findings indicate that both Western wheat and Asian wheat are susceptible to the detrimental effects of ambient $O_3$ conditions. Furthermore, extent of damage caused by $O_3$ exposure can vary across different regions and wheat varieties. A study by Fatima et al. (2019) evaluated the sensitivity of three wheat cultivars to $O_3$: HD 2987 (sensitive), PBW 502 (moderately sensitive), and Kharchiya 65 (tolerant). The study found that HD 2987 experienced a yield loss of 32.9%, PBW 502 experienced a yield loss of 13.3%, and Kharchiya 65 experienced a yield loss of 8.8% under ambient $O_3$ conditions. These findings demonstrate the varying degrees of sensitivity to $O_3$ among the different wheat cultivars. The wheat cultivar HD 2967 was studied under various field conditions and levels of $O_3$ treatment, which resulted in a decrease in yield ranging from 11 to 16.2%. (Ghosh et al. 2020; Mina et al. 2021). Similar to the above, Chaudhary et al. (2021) also investigated the impact of elevated $O_3$ on three different wheat cultivars: Zahir, Gedera, and Ruta. Specifically, Zahir exhibited a yield loss ranging from 7 to 16%, Gedera experienced a yield loss between 22.4 and 30%, and Ruta showed a yield loss ranging from 18.9 to 30%. According to a study by Yadav et al. (2021a), four wheat cultivars, namely HUW468, HD3086, HUW234, and HD3118, exhibited yield losses of 29.3%, 26.9%, 22.5%, and 25.2%, respectively, when subjected to elevated $O_3$ treatment. The above results indicated that understanding the specific sensitivities of different wheat cultivars to $O_3$ exposure can aid in the selection of appropriate cultivars and the development of agricultural practices that can help to minimize yield losses in $O_3$-polluted environments.

Gupta et al. (2022a) focused on examining the impact of ambient (45.57 ppb) + 20 ppb elevated $O_3$ on the yield of eight tomato (*Solanum lycopersicum*) cultivars. According to the findings of the study, the tomato plants experienced a yield loss ranging from 11.5% to 58.7% due to the $O_3$ treatment. This indicates that the exposure to elevated $O_3$ levels had a detrimental effect on the yield of the tomato cultivars. A recent research of Singh et al. (2023) considered three different species of *Abelmoschus* genus for their yield under elevated $O_3$ and observed that yield loss of 29.2, 22.9, and 22.4% in *A. moschatus*, *A. esculentus*, and *A. caillei*, respectively, compared to their ambient ones. In a study, Singh et al. (2019) investigated the effect of $O_3$ exposure on the weight of kernels in two different maize cultivars, which were found to be reduced in the elevated $O_3$ condition with an AOT40 of 22.2 ppm h. The reduction in kernel weight was reported to be 14.8% in DHM117 cultivar and 11.1% in HQPM1. However, findings of Schauberger et al. (2019) highlight the sensitivity of *Glycine max* to ambient $O_3$ levels. The range of yield losses observed from 2 to 10% suggests that even moderate levels of $O_3$ can have a noticeable impact on soybean productivity. Similarly, Shao et al. (2021) showed that under elevated $O_3$ treatment of 100 ppb, the Yongyou 538 cultivar of rice experienced a significant yield loss. The reported range of yield loss was between 24 and 35.2%.

## 3.3 Synthesis of Nitric Oxide

Multiple paths and multiple precursors have been recognized in the plant system for the biosynthesis of NO through extensive research. These pathways include reduction of nitrite ($NO_2^-$) or the oxidation of aminated molecules (Astier et al. 2018; Sun

**Table 3.1** Variations in crop yield losses in the field experiments under different $O_3$ concentrations

| Crop plant | Cultivar | $O_3$ concentration (ppb) | Yield loss (%) | Reference |
|---|---|---|---|---|
| Wheat (*Triticum aestivum* L.) | Liangxing99 | Ambient + 80 ppb (7 h day$^{-1}$) | 22.6 | Li et al. (2018) |
| | 17 cultivars | Ambient | 8.4 | Pleijel et al. (2018) |
| | HD-2967 | Ambient + 40 ppb (7 h day$^{-1}$) | 17.3 | Yadav et al. (2019a; b) |
| | | | 18.1 | |
| | C-306 | | | |
| | Western wheat | Ambient | 0–27 | Schauberger et al. (2019) |
| | Asian wheat | | 4–39 | |
| | HD 2987 | Ambient /EDU | 32.9 | Fatima et al. (2019) |
| | | | 13.3 | |
| | PBW 502 | | 8.8 | |
| | Kharchiya 65 | | | |
| | HD 2967 | Ambient + 20 ppb (8 h day$^{-1}$) | 16.2 | Ghosh et al. (2020) |
| | HD 2967 | Ambient + 140 ppb (11 h day$^{-1}$) | 11 | Mina et al. (2021) |
| | Zahir | Ambient + 102.8 ppb (7 h for 10 day) | 7–16 | Chaudhary et al. (2021) |
| | | | 22.4–30 | |
| | Gedera | | 18.9–30 | |
| | Ruta | | | |
| | HUW468 | Ambient + 20 ppb (8 h day$^{-1}$) | 29.3 | Yadav et al. (2021a, b) |
| | HD3086 | | 26.9 | |
| | HUW234 | | 22.5 | |
| | HD3118 | | 25.2 | |
| Tomato (*Solanum lycopersicum* L.) | 8 cultivars | Ambient + 20 ppb (6 h day$^{-1}$) | 11.5–58.7 | Gupta et al. (2022a, b, c) |
| Lady Finger (*Abelmoschus* Spp.) | *A. moschatus* | Ambient + 20 ppb (8 h day$^{-1}$) | 29.2 | Singh et al. (2023) |
| | *A. esculentus* | | 22.9 | |
| | *A. caillei* | | 22.4 | |
| Maize (*Zea mays* L.) | DHM117 | Ambient + 30 ppb (8 h day$^{-1}$) | 14.2 | Singh et al. (2019) |
| | HQPM1 | | 11.1 | |
| Soybeans (*Glycine max* L. Merr.) | – | Ambient | 2–10 | Schauberger et al. (2019) |
| Rice (*Oryza sativa* L.) | Yongyou 538 | Ambient + 100 ppb (8 h day$^{-1}$) | 24–35.2 | Shao et al. (2021) |

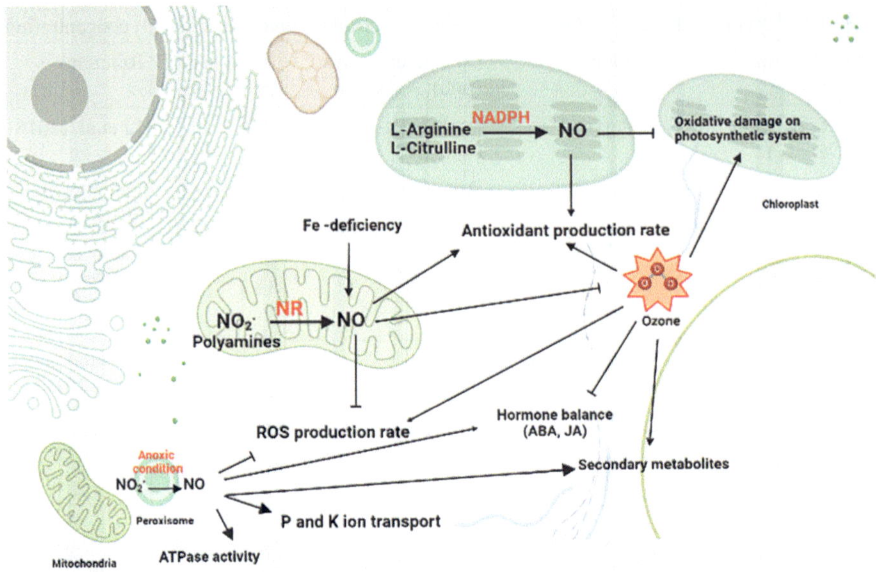


**Fig. 3.1** Cellular illustration of nitric oxide (NO) production and regulatory functions during $O_3$ stress. NR denotes nitrite reductase, ROS: Reactive oxygen species; $NO_2^-$: Nitrite, ABA: Abscisic Acid; JA: Jasmonic acid, P: Phosphorus, K: Potassium

et al. 2021). The site for the synthesis of NO, reported are protoplasts, chloroplasts, mitochondria, peroxisomes, cytoplasm, cell membrane, endoplasmic reticulum, and apoplast (Rőszer 2014; Igamberdiev et al. 2010; Fig. 3.1). Majorly, the pathways for NO synthesis are categorized as reductive and oxidative pathways, which are as follows:

### 3.3.1 *Reductive Pathway*

The reduction of $NO_2^-$ to NO can occur enzymatically or non-enzymatically under specific cases. The non-enzymatic reduction occurs under low pH or reducing environment and in high concentration of $NO_3^-$. The enzymatic reduction of $NO_2^-$ occurs by the enzymes nitrate reductase (NR), NO forming nitrite reductase (NOFNiR) and by mitochondrial nitrite reduction. Among the above stated routes, the major being the NR-mediated pathway for the synthesis of NO. NR is a multifunctional cytoplasmic enzyme with two active homodimeric constituents which require molybdopterin, heme, and FAD as cofactors (Astier et al. 2018). Though NR constitute only 1% activity for the reduction of $NO_2^-$, the major activity done is the reduction of $NO_3^-$ to $NO_2^-$. Thus, under optimum condition, very low level of NO is produced by NR-mediated route, while under high concentrations of nitrite and low nitrate, and acidic conditions reduction of $NO_2^-$ to NO is enhanced (Sun et al. 2021).

Apart from cytoplasmic NR, a membrane bound nitrite reductase is also reported for the synthesis of NO (Astier et al. 2018; Stöhr and Stremlau 2006). The enzyme is found in apoplasm of the root cells which uses NAD(P)H as an electron donor. This membrane bound nitrite reductase function under low oxygen pressure and along with the apoplastic membrane-bound NR that provide $NO_2^-$ from $NO_3^-$ (Stöhr and Stremlau 2006). Along with other enzymes, xanthine oxidase, aldehyde oxidases, and sulfite oxidases, which uses molybdenum as a cofactor reduce $NO_2^-$ to NO (Godber et al. 2000; Astier et al. 2018).

Mitochondrial electron transport chain also generates NO as under $O_2$ deprived condition $NO_2^-$is used as an alternative electron acceptor for ATP synthesis. Reduction of $NO_2^-$ to NO occur at the cytochrome bc1 (complex III) and the cytochrome-c oxidase (complex IV) under $O_2$ deprived condition ((hypoxia) (Igamberdiev et al. 2010). As plant lack specific $O_2$ transport system, rapidly growing cells of the photosynthetic tissues and stem etc., face $O_2$ deficiency, and hence, the mitochondrial reduction of $NO_2^-$ become a possible pathway for NO synthesis (Sun et al. 2021). Production of NO through mitochondrial electron transport chain has been reported in many plant species such as pea (*Pisum sativum*), tobacco (*Nicotiana tobaccum*) and barley (*Hordeum vulgare*) (Gupta and Kaiser 2010).

Synthesis of NO by $NO_2^-$ reduction is also reported in plant peroxisomes under hypoxic or anoxic conditions, which is due to the presence of deoxygenated heme-containing proteins in the matrix (Igamberdiev et al. 2010).

### 3.3.2 Oxidative Pathway

The oxidation of L-Arginine to NO and L-Citrulline has been spotted in the chloroplasts and peroxisomes of leaf tissue through double mono-oxygenation (Rőszer and Rőszer 2012). This oxidation process requires NADPH in chloroplast while $Ca^{2+}$, calmodulin, FAD, FMN and NADPH in leaf peroxisomes (Sun et al. 2021). Guo and Crawford (2005) reported that plant mitochondria are also involved in the oxidation of L-Arginine to NO, while the enzyme responsible is not known. It is supposed that it may be the enzyme of mitochondrial matrix or intermembrane space enzyme.

NO synthase (NOS) commonly not occur in plants but Guo et al. (2003) recognized a plant specific NOS in the mitochondria of *Arabidopsis thaliana* which was renamed as *A. thaliana* NOS-associated protein-1 (AtNOA1). This AtNOA1 oxidize L-Arginine with NADPH and $Ca^{2+}$ to NO and its activity was found sensitive to mammalian NOS inhibitors (Guo et al. 2003).

Polyamines and hydroxylamines are also known to enhance the NO synthesis by oxidation in plant cells (Rumer et al. 2009; Rőszer 2014). The probable route suggested is the interaction of polyamines with the nitrate reductase for the synthesis of NO and the effect of polyamine on the metabolism of L-Arginine (Rőszer 2014). Rumer et al. (2009) reported that in tobacco cell culture hydroxylamine can also be oxidized to NO as it is the intermediate in the nitrification process.

Non-enzymatically NO is released from a protonated form of $NO_2^-$, nitrous acid ($HNO_2$) under acidic condition which occurs in the apoplast of germinating or hypoxic seeds (Bethke et al. 2004). Bethke et al. (2004) showed NO synthesis from $NO_2^-$ in the apoplast of the aleurone layer of the barley (*Hordeum vulgare*) which involved the phenolic compounds occurring in the apoplast and seed coat.

## 3.4 Functions of NO

NO is a paramagnetic molecule carrying an unpaired electron and so diffuse easily across the membranes. Nitrosonium ion ($NO^+$) is yielded upon oxidation of NO, which takes part in nitrosation reactions when added to a thiol, amine or hydroxyl aromatic group (Simontacchi et al. 2015). The interconversion of NO, $NO^+$ and $NO^-$ occur in cellular environments. NO modulates responses in plant by reacting with metal centers, thiols, oxygen molecule, and free radicals (Simontacchi et al. 2015). The ionic environment in the cell created by NO help in basic cellular functions and it also regulate the gene expression patterns and protein activities, thereby alleviating the stress responses (Simontacchi et al. 2015).

Role of NO in germination of dormant seeds is widely reported and accepted (Rőszer 2014; Sun et al. 2021). Mitochondrial reduction of $NO_2^-$ to NO and enzymatically produced NO may control the initiation of seed germination (Rőszer 2014). Seed dormancy in many plants is interrupted by imbibition where water penetrates the seed coat and creates a temporal hypoxic condition. The process of imbibition is coupled with increased NO levels and that may lead to the mitochondrial NO synthesis (Liu et al. 2009; Gupta and Igamberdiev 2011). The NO produced within the mitochondria inhibits cytochrome-c oxidase, which further stimulates the seed germination (Gniazdowska et al. 2010). NO might offer an antimicrobial protection to the germinating seed in the soil (Bethke et al. 2004). Programmed cell death induced by NO has also been reported during germination, where the aleuron cells are being abolished (Lombardi et al. 2010).

A distinct and complex forms of dormancy is observed in many fruit tree species, e.g., apple and pomegranate (*Punica granatum*). This dormancy is broken by several months of cold or warm stratification. This period of dormancy can be reduced by pre-treatment with an NO donor such as sodium nitroprusside (SNP, 5 mM) which promote the embryo germination by almost 60% after 8 days of treatment, while in nontreated seeds, it promoted the germination by only 16%. This effect in the embryos was found to be associated with the modulation of ROS metabolism by NO during the early germination stage (Gniazdowska et al. 2010). Further, NO accumulation has been observed in the axes of embryos of seeds (*Malus domestica*) and in the endosperm (*Sechium deule*) after seed dormancy is released (Gniazdowska et al. 2010; Lombardi et al. 2012). Through seed dissection, it has been suggested that NO is perceived by the aleurone layer during seed embryogenesis and it also respond to NO (Sun et al. 2021).

The seed-to-seedling transition involves various phytohormones; for instance, seed dormancy is regulated by abscisic acid (ABA), whereas seed germination requires gibberellic acid (GA). An extensive cross-talk between NO and phytohormones during seed dormancy release has been reported (Arc et al. 2013). In connection with this, increased NO production has been observed after ABA treatment, which may ameliorate the repressive effect of ABA. However, NO accumulation in seeds decreases the ABA contents (Liu et al. 2009).

Liu et al. (2021) reported the involvement of NO in seed oil content and fatty acid composition. Thus, regulating the NO signaling can be a promising approach to improve the seed oil content from commercially significant crops. Along with the seed, NO regulate the root growth and development mediated through auxin (Yu et al. 2014; Sun et al. 2021). Lateral root formation was lacking under NO depleted condition (Sun et al. 2021). Regulation of cell cycle occurs at gene level by NO in the pericycle cells which ultimately form the lateral roots (Correa-Aragunde et al. 2006). Root hair formation is also found to be influenced by NO as it acts on Golgi vesicle formation and trafficking. In the mutant of NO synthesis, alteration in vesicle transfer and reduction in root hair length was observed (Sun et al. 2021). It was also observed that NO act upon cytoskeletal organization along with the ABA in order to alter the growth of root hairs (Lombardo and Lamattina 2018; Sun et al. 2021).

## 3.5 Role of NO in Stress Tolerance

NO has been reported to play a beneficial role in many plants under various abiotic stresses such as salinity, drought, $O_3$, and extreme temperature (Li et al. 2018; Sun et al. 2021) (Table 3.2). Under stress conditions, NO protect the seed germination of various horticultural species. Pre-soaking treatment attenuated the inhibition of seed germination and early growth of *Brassica chinensis* under salt stress (Ren et al. 2020).

Role of NO in modulation of mineral nutrition was first observed in maize plant (*Zea mays*) against tolerance of Fe-deficiency (Graziano et al. 2002). Increased NO production has been observed under Fe-deficiency in many herbaceous and fruit tree plants which stimulate the LeFER, LeFRO1, and LeIRT1 genes which are associated with the uptake of Fe (Jin et al. 2011; Zhai et al. 2016).

Uptake and homeostasis of many macronutrients such as N, P and K, which are the essential constituents of various macromolecules (nucleotides and proteins), is regulated by NO (Sun et al. 2021). It has been also reported to play a significant role in N-fixation by legumes. Buet et al. (2019) reported the contribution of NO in remodeling the root system and escalating the transport of P and K under their deficiency condition. Similar result in white lupin (*Lupinus albus*) was observed under P deficiency (Wang et al. 2010). NO was found to be involved in cluster root formation and citrate exudation, and cell wall P reutilization under P deficiency in white lupin and rice, respectively (Wang et al. 2010; Zhu et al. 2017). The K-homeostasis in the cell has also been found to be influenced by NO, as the increased

**Table 3.2** Protective role of NO against various abiotic stresses

| Stress | Level | Plant | NO source | Effects | References |
|---|---|---|---|---|---|
| Salt stress | NaCl (150 and 300 mM) | *Triticum aestivum* (wheat) | SNP (1 μM) | Enhanced antioxidant activity and decreased oxidative damage | Hasanuzzaman et al. (2011) |
| Salt stress | NaCl (100 mM) | *Brassica chinensis* (pakchoi) | SNP (5–100 μM) | Enhanced seed germination and growth of radicles and plumules by maintenance of ion homeostasis, improvement of osmolytes accumulation and antioxidant system | Ren et al. (2020) |
| Salt stress | NaCl (50–200 mM) | *Lycopersicum esculentum* (tomato) | GSNO (10–200 μM) | Promoted photosynthesis, osmotic balance, antioxidant enzymes, and increased levels of protein S-nitrosylation | Wang et al. (2023) |
| Salt stress | NaCl (200 mM) | *Lycopersicum esculentum* (tomato) | SNAP (50 μM) | Improved activity of antioxidants and accumulation of metabolites | Ahmad et al. (2018) |
| Heavy metal | As (25 and 50 μM) | *Oryza sativa* (rice) | SNP (50 μM) | Altered the root and coleoptile architecture and enhanced antioxidant enzymes | Singh et al. (2009) |
| Heavy metal | Pb (1.2 mM) | *Oryza sativa* (rice) | SNP (100 μM) | Enhanced antioxidants and transcript accumulation of heavy metals stress-related genes, and lowered oxidative stress markers | Rahim et al. (2022) |

(continued)

**Table 3.2** (continued)

| Stress | Level | Plant | NO source | Effects | References |
|---|---|---|---|---|---|
| Heavy metal | Cd (10 μM) | *Oryza sativa* (rice) | SNP (30 μM) | Improved biomass, antioxidant enzyme activity and maintained ionic balance | Yang et al. (2016) |
| Drought | PEG (0.1%) | *Tagetes erecta* (marigold) | SNP (5–500 μM) | Encouraged the adventitious root development | Liao et al. (2012) |
| Ozone | $O_3$ (80 ppb) | *Triticum aestivum* (wheat) | SNP (200 mM) | Improved antioxidant enzyme, net photosynthetic rate and yield | Li et al. (2018) |
| Heat | 40 °C | *Triticum aestivum* (wheat) | SNP (100 μM) | Improved photosynthesis and growth | Iqbal et al. (2022) |
| Osmotic stress | PEG (15%) | *Triticum aestivum* (wheat) | GSNO (0.2 mM) | Enhanced osmolytes and antioxidant enzymes | Khan et al. (2017) |
| Water-logging | 10 days | *Gossypium hirsutum* (cotton) | SNP (50 μM) | Rebalanced the plant growth promoting hormones and enhanced compensatory growth | Zhang et al. (2022) |

SNP—sodium nitroprusside, GSNO—S-nitrosoglutathione, SNAP—S-nitroso-N-acetylpenicillamine and PEG—Polyethylene glycol level of it adversely controls Shaker-like $K^+$ channel (AKT1) after K-restriction (Xia et al. 2014). These roles of NO are beneficial for the plants under various abiotic (drought and heat) stresses. Along with these, NO also showed association with the deficiencies of some micronutrients such as Cu, Zn, Mg and B, etc. (Farag et al. 2017; Liu et al. 2016; Zhang et al. 2009). These micronutrients are required in low quantity but the excess of these cause toxicity to the plants; thus, NO has been reported to alleviate the stress of these elements. Liu et al. (2016) and Zhang et al. (2009) showed the advantageous role of NO under Cu stress in *Catharanthus roseus* (periwinkle) and *Lycopersicon esculentum* (tomato) seedlings, respectively. The activity of ATPase and production of secondary metabolites were enhanced under the influence of NO in Cu stress (Liu et al. 2016). Watermelon seedlings have been protected by B-induced injury as NO reduced the B-accumulation and protected the seedling from oxidative damage (Farag et al. 2017). Liu et al. (2017) observed the regulation of root hair morphogenesis by NO under Mg-deficiency in *Arabidopsis*.

Under salt stress, NO regulates the water status and maintain the ion homeostasis in plants (Simontacchi et al. 2015; Nabi et al. 2019). NO has been reported to reduce the oxidative damage and regulate the secondary metabolites in pak choi (*Brassica chinensis*) and sunflower (*Helianthus annuus*) grown under saline condition (Tailor et al. 2019; Ren et al. 2020). Further, studies have suggested that protein S-nitrosation is responsible to mediate the acclimation against salt stress after the NO treatment of citrus (*Citrus aurantium*) plants (Tanou et al. 2009). Wang et al. (2023) found that NO promoted the photosynthetic rate, osmotic balance, antioxidative capability, and protein S-nitrosylation levels in tomato (*Lycopersicum esculentum*) grown under NaCl treatment.

In drought stress, NO has been observed to alleviate the stress by directly acting upon root morphology, stomatal closure and antioxidant system (Simontacchi et al. 2015; Sun et al. 2021). In marigold (*Tagetes erecta*), the ultrastructure of mesophyll cells was protected and the adventitious root formation was encouraged by NO under drought stress (Liao et al. 2012). NO helps the plants in drought stress by stimulating the antioxidant and osmolytes synthesis. It has also been reported to be synthesized and accumulated in the guard cells under drought stress and help in stomatal closure via ABA and calcium ($Ca^{2+}$) dependent pathway to minimize the water loss (Mata and Lamattina 2001; Simontacchi et al. 2015). The expression of proteins that enhance NO level improves the plant performance in long-term drought stress by improving the plant biomass, height, and less damage to membranes (Simontacchi et al. 2015).

## 3.6 NO Against $O_3$ Stress Mechanism and Yield Maintenance

Tropospheric $O_3$ increases oxidative stress levels, leading to lipid peroxidation, protein degradation, pigment fading, enzyme inactivation, and membrane damage (Ansari et al. 2023; Pazarlar et al. 2017). When enzymes are inactivated by $O_3$, it disrupts the normal metabolic processes in the plant, hindering its growth and development (Ansari et al. 2020). Numerous physiological and biochemical alterations occur as a consequence, disrupting the homeostasis of ROS in plants (Yadav et al. 2019a, b; Fig. 3.1). NO is a molecule that contains an unpaired electron, making it highly reactive. This reactivity allows NO to interact with ROS in plants (Nabi et al. 2019; Fig. 3.1). NO acts as a signaling molecule that triggers a series of defense mechanisms in plants when they encounter $O_3$ stress (Astier et al. 2018). Ozone exposure can upregulate the expression of NR1, thereby stimulate the activation of nitrate reductase (NR) in plant leaves (Li et al. 2018). The activation of NR enzyme in plant leaves under $O_3$ exposure is believed to be a protective response. By increasing the activity of NR, the plant enhances its ability to convert nitrate into nitrite. This response may help in detoxifying the excess ROS generated due to $O_3$ stress and maintaining the balance of nitrogen metabolism in the plant (Fig. 3.1).

The mechanism by which NO counters the effects of $O_3$ stress involves multiple pathways. Firstly, NO acts as a signaling molecule that triggers various defense mechanisms in plants, such as the activation of antioxidant systems (Basit et al. 2023; Li et al. 2018). This helps to scavenge ROS generated during oxidative stress caused by $O_3$ exposure. NO protects the photosynthetic system from oxidative damage by reducing ROS levels (Basit et al. 2023; Wani et al. 2023). Furthermore, it regulates the expression and activity of antioxidant enzymes, such as superoxide dismutase (SOD), catalase (CAT), and peroxidase (POD) and minimizes $O_3$-induced damage (Li et al. 2018; Fig. 3.1). NO also involves in overexpression of genes related to stress-related proteins that protect plants against $O_3$-induced injury. High $O_3$ levels can induce stomatal closure, limiting the entry of $O_3$ but also reducing $CO_2$ uptake for photosynthesis. NO has been shown to modulate stomatal movement, promoting stomatal opening and improving $CO_2$ diffusion into the leaves (Banerjee and Roychoudhury, 2021; Sun et al. 2019). This helps maintain photosynthetic activity and minimize the negative impact of $O_3$ on crop yield (Li et al. 2018). Moreover, NO can counteract the negative effects of $O_3$ on pollen development and fertilization, contributing to enhanced crop yields (Wong et al. 2020). NO modulates the balance between hormones including abscisic acid (ABA) and jasmonic acid (JA), leading to a more favorable response to $O_3$ stress Fig. 3.1.

In wheat plants, the tolerance to abiotic stress is mediated by the involvement of NO and abscisic acid (Iqbal et al. 2022). This regulatory mechanism can enhance plants' ability to adapt to elevated $O_3$ levels and alleviate decreases in crop yield. By modulating the hormonal pathways of ethylene and abscisic acid which regulate stomatal closure, leaf senescence, and other processes, NO can potentially help plants cope with $O_3$ stress and maintain their productivity (Banerjee and Roychoudhury, 2021). However, high levels of NO production in some cases might have negative consequences, such as impaired photosynthesis or disruption of other signaling pathways. Li et al. (2018) reported that endogenous NO accumulates heavily in response to $O_3$, which contributes to mitigating the adverse effects and promoting a higher grain yield in wheat.

The role of NO in maintaining crop yield under high $O_3$ exposure is still an active area of research. While NO has the potential to mitigate $O_3$-induced stress and protect plants, its effects can be context-dependent and influenced by multiple factors. Further studies are needed to better understand the specific mechanisms by which NO interacts with $O_3$ stress and its implications for crop yield maintenance.

## 3.7 Methods for the Application of Nitric Oxide

NO in plants can be modulated or provided by various methods such as by some exogenous synthetic NO donors, genetic manipulations, interaction with microorganisms, changing the endogenous synthesis and scavenger mechanisms (Simontacchi et al. 2015).

Li et al. (2018) used sodium nitroprusside (SNP) of 200 μM concentration as NO donor against the $O_3$ stress. SNP was sprayed to the leaves of wheat plants. While, Dong et al. (2014) applied SNP (0.1 mM and 0.25 mM) in Hoagland nutrient solution against salt stress for growing cotton (*Gossypium hirsutum*) seedlings. Wang et al. (2023) added nitrosoglutathione (GSNO) as NO donor at 10, 50, 100, and 200 μM concentrations to Hoagland solution for growing tomato (*Lycopersicum esculentum)* against salt stress.

S-nitroso-N-acetylpenicillamine (SNAP), S-nitrosoglutathione (GSNO), and diethylamine NONOate sodium (DEA-NONOate) are the other NO donor used for exogenous application (Ahmad et al. 2018; Nabi et al. 2019; Wang et al. 2023).

## 3.8 Detection of NO in Plants

Quantification of NO has been done by various methods and the methods still need further investigations in order to appropriately measure and localize the NO in tissues.

The most widely used detection method of NO is the fluorescent dyes such as DAF-2 (Diaminofluorescein-2), DAF-FM, DAR-4M, which detect the NO inside and outside the cell (Ruemer et al. 2016; Vishwakarma et al. 2019). The fluorescent dye reacts with an oxidation product of NO by nitrosation and produce highly fluorescing triazol (DAF-2T) which is measured either in extracts by a fluorimeter or in samples by confocal laser scanning microscopy (LSM).

Another method is gas phase NO measurement where ozone-based chemiluminescence technique is used which is highly specific and sensitive (Ruemer et al. 2016; Vishwakarma et al. 2019). The photons are emitted by excited $NO_2$ which is counted by photomultiplier tube.

Griess reaction is a method which indirectly measure NO by oxidizing the available NO in the sample to $NO_2$ and followed by reaction with Griess reagent (Vishwakarma et al. 2019). The reaction gives an azo dye of magenta color which is further measured spectrophotometrically.

Electron spin resonance (ESR) or electron paramagnetic resonance (EPR) is also used for detecting NO in plants. The instrument is based on the observation of unpaired electrons in magnetic fields and measure the strength of magnetic field at a frequency when a signal is received (Mur et al. 2011; Vishwakarma et al. 2019). The magnetic field exhibits a "resonance" in the microwave region between parallel and antiparallel electron spin orientations.

NO electrodes are also used which is made up of teflon/platinum and Ag/$AgCl_2$ (Mur et al. 2011; Vishwakarma et al. 2019). The former serves as the main electrode and the later as the reference electrode dipped in a solution of 30 mM NaCl/0.3 nM HCl. NO is detected by its oxidation at + 0.8 V to + 0.9 V as compared to the reference electrode.

Mass spectrometry and laser based infrared spectroscopy are the other methods for the detecting NO (Vishwakarma et al. 2019).

## 3.9 Conclusions

The scarcity of existing literature and research on the role of nitric oxide (NO) in mitigating high $O_3$ effects on crop yield leaves several critical aspects of the mechanism unclear. While some preliminary findings suggest that NO play a significant role in mitigating the negative effects of high $O_3$ conditions on crop yield. Studies reported its involvement in controlling metabolic processes, growth and developmental processes such as germination of seed, root development, flower transition, ripening of fruits, etc. Through various experiments and analyses, it has been observed that the application of NO in $O_3$-stressed environments led to a noticeable improvement in crop productivity and overall yield of tomato and wheat. The protective effects of NO were attributed to its ability to scavenge harmful ROS generated under high $O_3$ stress, thereby reducing oxidative damage and enhancing the resilience of crops. This book chapter highlights the potential of using NO as a practical and eco-friendly approach to safeguard crop productivity in regions experiencing high $O_3$ levels due to pollution or climate change. Further research and field trials are warranted to better understand the underlying mechanisms and optimize the application of NO for agricultural practices under $O_3$-stressed conditions.

**Acknowledgements** Prof. Madhoolika Agrawal expresses sincere gratitude to the Science and Engineering Research Board (SERB), New Delhi, for awarding the J.C. Bose National Fellowship (Sanction No. JCB/2021/000040).

## References

Ahlfors R, Brosché M, Kollist H, Kangasjärvi J (2009) Nitric oxide modulates ozone-induced cell death, hormone biosynthesis and gene expression in Arabidopsis thaliana. Plant J 58:1–12. https://doi.org/10.1111/j.1365-313X.2008.03756.x

Ahmad P, Abass Ahanger M, Nasser Alyemeni M, Wijaya L, Alam P, Ashraf M (2018) Mitigation of sodium chloride toxicity in Solanum lycopersicum L. by supplementation of jasmonic acid and nitric oxide. J Plant Interact 13:64–72. https://doi.org/10.1080/17429145.2017.1420830

Ansari N, Agrawal M, Agrawal SB (2020) An assessment of growth, floral morphology, and metabolites of a medicinal plant *Sida cordifolia* L. under the influence of elevated ozone. Environ Sci Pollut Res 1–14. https://doi.org/10.1007/s11356-020-10340-y

Ansari N, Yadav D, Singh P, Agrawal M, Agrawal S (2023) Ozone exposure response on physiological and biochemical parameters vis-a-vis secondary metabolites in a traditional medicinal plant *Sida cordifolia* L. Ind Crop Prod 194:11626. https://doi.org/10.1016/j.indcrop.2023.116267

Arc E, Galland M, Godin B, Cueff G, Rajjou L (2013) Nitric oxide implication in the control of seed dormancy and germination. Front Plant Sci 4:346. https://doi.org/10.3389/fpls.2013.00346

Arneth A, Schurgers G, Lathiere J, Duhl T, Beerling DJ, Hewitt CN, Martin M, Guenther A (2011) Global terrestrial isoprene emission models: sensitivity to variability in climate and vegetation. Atmos Chem Phys 11:8037–8052. https://doi.org/10.5194/acp-11-8037-2011

Asgher M, Per TS, Masood A, Fatma M, Freschi L, Corpas FJ, Khan NA (2017) Nitric oxide signaling and its crosstalk with other plant growth regulators in plant responses to abiotic stress. Environ Sci Pollut Res 24:2273–2285. https://doi.org/10.1007/s11356-016-7947-8

Astier J, Gross I, Durner J (2018) Nitric oxide production in plants: an update. J Exp Bot. https://doi.org/10.1093/jxb/erx420

Banerjee A, Roychoudhury A (2021) Nitric oxide mediated regulation of stomatal movement during desiccation. In: Nitric oxide in plant biology: an ancient molecule with emerging roles. Academic Press, pp 493–502. https://doi.org/10.1016/B978-0-12-818797-5.00031-5

Basit F, Bhat JA, Alyemeni MN, Shah T, Ahmad P (2023) Nitric oxide mitigates vanadium toxicity in soybean (*Glycine max* L.) by modulating reactive oxygen species (ROS) and antioxidant system. J Hazard Mater 451:131085. https://doi.org/10.1016/j.jhazmat.2023.131085

Bethke PC, Badger MR, Jones RL (2004) Apoplastic synthesis of nitric oxide by plant tissues. Plant Cell 16:332–341. https://doi.org/10.1105/tpc.017822

Buet A, Galatro A, Ramos-Artuso F, Simontacchi M (2019) Nitric oxide and plant mineral nutrition: current knowledge. J Exp Bot 70:4461–4476. https://doi.org/10.1093/jxb/erz129

Chaudhary N, Bonfil DJ, Tas E (2021) Physiological and yield responses of spring wheat cultivars under realistic and acute levels of ozone 1–21

Chen Z, Cao J, Yu H, Shang H (2018) Effects of elevated ozone levels on photosynthesis, biomass and non-structural carbohydrates of *Phoebe bournei* and *Phoebe zhennan* in subtropical China. Front Plant Sci 871. https://doi.org/10.3389/fpls.2018.01764

Cooper OR, Parrish DD, Ziemke J, Balashov NV, Cupeiro M, Galbally IE, Gilge S, Horowitz L, Jensen NR, Lamarque JF, Naik V, Oltmans SJ, Schwab J, Shindell DT, Thompson AM, Thouret V, Wang Y, Zbinden RM (2014) Global distribution and trends of tropospheric ozone: an observation-based review. Elementa 2. https://doi.org/10.12952/journal.elementa.000029

Correa-Aragunde N, Graziano M, Chevalier C, Lamattina L (2006) Nitric oxide modulates the expression of cell cycle regulatory genes during lateral root formation in tomato. J Exp Bot 57:581–588. https://doi.org/10.1093/jxb/erj045

Dong YJ, Jinc SS, Liu S, Xu LL, Kong J (2014) Effects of exogenous nitric oxide on growth of cotton seedlings under NaCl stress. J Soil Sci Plant Nutr 14:1–13

Farag M, Najeeb U, Yang J, Hu Z, Fang ZM (2017) Nitric oxide protects carbon assimilation process of watermelon from boron-induced oxidative injury. Plant Physiol Biochem 111:166–173. https://doi.org/10.1016/j.plaphy.2016.11.024

Fatima A, Singh AA, Mukherjee A, Dolker T, Agrawal M, Agrawal SB (2019) Assessment of ozone sensitivity in three wheat cultivars using ethylenediurea. Plants 8. https://doi.org/10.3390/plants8040080

Feng Z, Hu E, Wang X, Jiang L, Liu X (2015) Ground-level $O_3$ pollution and its impacts on food crops in China: a review. Environ Pollut 199:42–48. https://doi.org/10.1016/j.envpol.2015.01.016

Ghosh A, Pandey AK, Agrawal M, Agrawal SB (2020) Assessment of growth, physiological, and yield attributes of wheat cultivar HD 2967 under elevated ozone exposure adopting timely and delayed sowing conditions. Environ Sci Pollut Res 27:17205–17220. https://doi.org/10.1007/s11356-020-08325-y

Gill SS, Hasanuzzaman M, Nahar K, Macovei A, Tuteja N (2013) Importance of nitric oxide in cadmium stress tolerance in crop plants. Plant Physiol Biochem 63:254–261. https://doi.org/10.1016/j.plaphy.2012.12.001

Gniazdowska A, Krasuska U, Debska K, Andryka P, Bogatek R (2010) The beneficial effect of small toxic molecules on dormancy alleviation and germination of apple embryos is due to NO formation. Planta 232:999–1005. https://doi.org/10.1007/s00425-010-1214-x

Godber BLJ, Doel JJ, Sapkota GP, Blake DR, Stevens CR, Eisenthal R, Harrison R (2000) Reduction of nitrite to nitric oxide catalyzed by xanthine oxidoreductase. J Biol Chem 275:7757–7763

Graziano M, Verónica Beligni M, Lamattina L (2002) Nitric oxide improves internal iron availability in plants. Plant Physiol 130:1852–1859. https://doi.org/10.1104/pp.009076

Guenther A (1995) A global model of natural volatile organic compound emissions. J Geophys Res 100:8873–8892. https://doi.org/10.1029/94JD02950

Guo FQ, Crawford NM (2005) Arabidopsis nitric oxide synthase1 is targeted to mitochondria and protects against oxidative damage and dark-induced senescence. Plant Cell 17:3436–3450. https://doi.org/10.1105/tpc.105.037770

Guo FQ, Okamoto M, Crawford NM (2003) Identification of a plant nitric oxide synthase gene involved in hormonal signaling. Science 302:100–103. https://doi.org/10.1126/science.1086770

Gupta KJ, Igamberdiev AU (2011) The anoxic plant mitochondrion as a nitrite: NO reductase. Mitochondrion 11:537–543. https://doi.org/10.1016/j.mito.2011.03.005

Gupta KJ, Kaiser WM (2010) Production and scavenging of nitric oxide by barley root mitochondria. Plant Cell Physiol 51:576–584. https://doi.org/10.1093/pcp/pcq022

Gupta A, Yadav DS, Agrawal SB, Agrawal M (2022c) Sensitivity of agricultural crops to tropospheric ozone: a review of Indian researches. Environ Monit Assess 194:894. https://doi.org/10.1007/s10661-022-10526-6

Gupta A, Bhushan S, Madhoolika A (2022a) Evaluation of toxicity of tropospheric ozone on tomato (*Solanum lycopersicum* L.) cultivars: ROS production, defense strategies and intraspecific sensitivity. J Plant Growth Regul 7. https://doi.org/10.1007/s00344-022-10870-4

Gupta A, Yadav DS, Agrawal SB, Agrawal M (2022b) Individual effects of high temperature and tropospheric ozone on tomato: a review. J Plant Growth Regul. https://doi.org/10.1007/s00344-022-10678-2

Hancock JT, Veal D (2021) Nitric oxide, other reactive signalling compounds, redox, and reductive stress. J Exp Bot 72:819–829. https://doi.org/10.1093/jxb/eraa331

Harmens H, Hayes F, Sharps K, Radbourne A, Mills G (2019) Can reduced irrigation mitigate ozone impacts on an ozone-sensitive African wheat variety? Plants 8. https://doi.org/10.3390/plants8070220

Hasanuzzaman M, Hossain MA, Fujita M (2011) Nitric oxide modulates antioxidant defense and the methylglyoxal detoxification system and reduces salinity-induced damage of wheat seedlings. Plant Biotechnol Rep 5:353–365

Huang X, Kiefer E, Von Rad U, Ernst D, Foissner I, Durner J (2002) Nitric oxide burst and nitric oxide-dependent gene induction in plants. In: Plant physiology and biochemistry. Elsevier Masson, pp 625–631. https://doi.org/10.1016/S0981-9428(02)01401-8

Igamberdiev AU, Bykova NV, Shah JK, Hill RD (2010) Anoxic nitric oxide cycling in plants: participating reactions and possible mechanisms. Physiol Plant 138:393–404

Imran M, Hussain S, Iqbal A, Saleem MH, ur Rehman N, Mo Z, Chen X, Tang X (2023) Nitric oxide confers cadmium tolerance in fragrant rice by modulating physio-biochemical processes, yield attributes, and grain quality traits. Ecotoxicol. Environ Saf 261:115078. https://doi.org/10.1016/j.ecoenv.2023.115078

Iqbal N, Sehar Z, Fatma M, Umar S, Sofo A, Khan NA (2022) Nitric oxide and abscisic acid mediate heat stress tolerance through regulation of osmolytes and antioxidants to protect photosynthesis and growth in wheat plants. Antioxidants 11. https://doi.org/10.3390/antiox11020372

Jin CW, Du ST, Shamsi IH, Luo BF, Lin XY (2011) NO synthase-generated NO acts downstream of auxin in regulating Fe-deficiency-induced root branching that enhances Fe-deficiency tolerance in tomato plants. J Exp Bot 62:3875–3884. https://doi.org/10.1093/jxb/err078

Khan M, Imran QM, Shahid M, Mun BG, Lee SU, Khan MA, Hussain A, Lee IJ, Yun BW (2019) Nitric oxide—induced AtAO3 differentially regulates plant defense and drought tolerance in *Arabidopsis thaliana.* BMC Plant Biol 19:1–19. https://doi.org/10.1186/s12870-019-2210-3

Khan MN, Mobin M, Abbas ZK, Siddiqui MH (2017) Nitric oxide-induced synthesis of hydrogen sulfide alleviates osmotic stress in wheat seedlings through sustaining antioxidant enzymes, osmolyte accumulation and cysteine homeostasis. Nitric Oxide 68:91–102. https://doi.org/10.1016/j.niox.2017.01.001

Kumar D, Ohri P (2023) Say "NO" to plant stresses: unravelling the role of nitric oxide under abiotic and biotic stress. Nitric Oxide—Biol Chem. https://doi.org/10.1016/j.niox.2022.11.004

Lal S, Venkataramani S, Naja M, Kuniyal JC, Mandal TK, Bhuyan PK, Kumari KM, Tripathi SN, Sarkar U, Das T, Swamy YV, Gopal KR, Gadhavi H, Kumar MKS (2017) Loss of crop yields in India due to surface ozone: an estimation based on a network of observations. Environ Sci Pollut Res 24:20972–20981. https://doi.org/10.1007/s11356-017-9729-3

Li C, Song Y, Guo L, Gu X, Muminov MA, Wang T (2018) Nitric oxide alleviates wheat yield reduction by protecting photosynthetic system from oxidation of ozone pollution. Environ Pollut 236:296–303. https://doi.org/10.1016/j.envpol.2018.01.093

Liao WB, Huang GB, Yu JH, Zhang ML (2012) Nitric oxide and hydrogen peroxide alleviate drought stress in marigold explants and promote its adventitious root development. Plant Physiol Biochem 58:6–15. https://doi.org/10.1016/j.plaphy.2012.06.012

Liu Y, Shi L, Ye N, Liu R, Jia W, Zhang J (2009) Nitric oxide-induced rapid decrease of abscisic acid concentration is required in breaking seed dormancy in *Arabidopsis*. New Phytol 183:1030–1042. https://doi.org/10.1111/j.1469-8137.2009.02899.x

Liu S, Yang R, Pan Y, Ren B, Chen Q, Li X, Xiong X, Tao J, Cheng Q, Ma M (2016) Beneficial behavior of nitric oxide in copper-treated medicinal plants. J Hazard Mater 314:140–154. https://doi.org/10.1016/j.jhazmat.2016.04.042

Liu M, Liu XX, He XL, Liu LJ, Wu H, Tang CX, Zhang YS, Jin CW (2017) Ethylene and nitric oxide interact to regulate the magnesium deficiency-induced root hair development in Arabidopsis. New Phytol 213:1242–1256. https://doi.org/10.1111/nph.14259

Liu J, Zhu XY, Deng LB, Liu HF, Li J, Zhou XR, Wang HZ, Hua W (2021) Nitric oxide affects seed oil accumulation and fatty acid composition through protein S-nitrosation. J Exp Bot 72:385–397. https://doi.org/10.1093/jxb/eraa456

Lombardi L, Ceccarelli N, Picciarelli P, Sorce C, Lorenzi R (2010) Nitric oxide and hydrogen peroxide involvement during programmed cell death of *Sechium edule* nucellus. Physiol Plant 140:89–102. https://doi.org/10.1111/j.1399-3054.2010.01381.x

Lombardi L, Mariotti L, Picciarelli P, Ceccarelli N, Lorenzi R (2012) Ethylene produced by the endosperm is involved in the regulation of nucellus programmed cell death in *Sechium edule* Sw. Plant Sci 187:31–38. https://doi.org/10.1016/j.plantsci.2012.01.011

Lombardo MC, Lamattina L (2018) Abscisic acid and nitric oxide modulate cytoskeleton organization, root hair growth and ectopic hair formation in *Arabidopsis*. Nitric Oxide—Biol Chem 80:89–97. https://doi.org/10.1016/j.niox.2018.09.002

Malik Z, Afzal S, Danish M, Abbasi GH, Bukhari SAH, Khan MI, Dawood M, Kamran M, Soliman MH, Rizwan M, Alhaithloulf HAS, Ali S (2020) Role of nitric oxide and calcium signaling in abiotic stress tolerance in plants. In: Protective chemical agents in the amelioration of plant abiotic, pp 563–581. https://doi.org/10.1002/9781119552154.ch28

Mata CG, Lamattina L (2001) Nitric oxide induces stomatal closure and enhances the adaptive plant responses against drought stress. Plant Physiol 126:1196–1204. https://doi.org/10.1104/pp.126.3.1196

Mata-Pérez C, Sánchez-Vicente I, Arteaga N, Gómez-Jiménez S, Fuentes-Terrón A, Oulebsir CS, Calvo-Polanco M, Oliver C, Lorenzo Ó (2023) Functions of nitric oxide-mediated post-translational modifications under abiotic stress. Front Plant Sci 14:1–15. https://doi.org/10.3389/fpls.2023.1158184

Mills G, Sharps K, Simpson D, Pleijel H, Broberg M, Uddling J, Jaramillo F, Davies WJ, Dentener F, Van den Berg M, Agrawal M, Agrawal SB, Ainsworth EA, Büker P, Emberson L, Feng Z, Harmens H, Hayes F, Kobayashi K, Paoletti E, Van Dingenen R (2018b) Ozone pollution will compromise efforts to increase global wheat production. Glob Chang Biol 24:3560–3574. https://doi.org/10.1111/gcb.14157

Mills G, Sharps K, Simpson D, Pleijel H, Frei M, Burkey K, Emberson L, Uddling J, Broberg M, Feng Z, Kobayashi K, Agrawal M (2018c) Closing the global ozone yield gap: quantification and cobenefits for multistress tolerance. Glob Chang Biol 24:4869–4893. https://doi.org/10.1111/gcb.14381

Mills G, Pleijel H, Malley CS, Sinha B, Cooper OR, Schultz MG, Neufeld HS, Simpson D, Sharps K, Feng Z, Gerosa G, Harmens H, Kobayashi K, Saxena P, Paoletti E, Sinha V, Xu X (2018a) Tropospheric ozone assessment report: present-day tropospheric ozone distribution and trends relevant to vegetation. Elementa 6. https://doi.org/10.1525/elementa.302

Mina U, Kandpal A, Bhatia A, Ghude S, Bisht DS, Kumar P (2021) Wheat cultivar growth, biochemical, physiological and yield attributes response to combined exposure to tropospheric ozone, particulate matter deposition and ascorbic acid application. Bull Environ Contam Toxicol. https://doi.org/10.1007/s00128-021-03373-7

Monks PS, Archibald AT, Colette A, Cooper O, Coyle M, Derwent R, Fowler D, Granier C, Law KS, Mills GE, Stevenson DS, Tarasova O, Thouret V, Von Schneidemesser E, Sommariva R, Wild O, Williams ML (2015) Tropospheric ozone and its precursors from the urban to the global scale from air quality to short-lived climate forcer. Atmos Chem Phys. https://doi.org/10.5194/acp-15-8889-2015

Mukherjee A, Singh Yadav D, Agrawal SB, Agrawal M (2020) Ozone a persistent challenge to food security in India: Current status and policy implications. Curr Opin Environ Sci Heal 19:100220. https://doi.org/10.1016/j.coesh.2020.10.008

Mur LAJ, Mandon J, Cristescu SM, Harren FJM, Prats E (2011) Methods of nitric oxide detection in plants: a commentary. Plant Sci 181:509–519. https://doi.org/10.1016/j.plantsci.2011.04.003

Nabi RBS, Tayade R, Hussain A, Kulkarni KP, Imran QM, Mun BG, Yun BW (2019) Nitric oxide regulates plant responses to drought, salinity, and heavy metal stress. Environ Exp Bot 161:120–133. https://doi.org/10.1016/j.envexpbot.2019.02.003

Pazarlar S, Cetinkaya N, Bor M, Ozdemir F (2017) Ozone triggers different defence mechanisms against powdery mildew (*Blumeria graminis* DC. Speer f. sp. *tritici*) in susceptible and resistant wheat genotypes. Funct Plant Biol 44:1016–1028. https://doi.org/10.1071/FP17038

Perring MP, Bullock JM, Alison J, Holder AJ, Hayes F (2022) Out of sight, out of mind—but not out of scope: the need to consider ozone ($O_3$) in restoration science, policy, and practice. Restor Ecol 1–14. https://doi.org/10.1111/rec.13622

Pleijel H, Broberg MC, Uddling J, Mills G (2018) Current surface ozone concentrations significantly decrease wheat growth, yield and quality. Sci Total Environ 613–614:687–692. https://doi.org/10.1016/j.scitotenv.2017.09.111

Rahim W, Khan M, Al Azzawi TNI, Pande A, Methela NJ, Ali S, Imran M, Lee DS, Lee GM, Mun BG, Moon YS, Lee IJ, Yun BW (2022) Exogenously applied sodium nitroprusside mitigates lead toxicity in rice by regulating antioxidants and metal stress-related transcripts. Int J Mol Sci 23(17):9729. https://doi.org/10.3390/ijms23179729

Ren Y, Wang W, He J, Zhang L, Wei Y, Yang M (2020) Nitric oxide alleviates salt stress in seed germination and early seedling growth of pakchoi (*Brassica chinensis* L.) by enhancing physiological and biochemical parameters. Ecotoxicol Environ Saf 187:109785. https://doi.org/10.1016/j.ecoenv.2019.109785

Rőszer T, Rőszer T (2012) Nitric oxide synthesis in the chloroplast. In: The biology of subcellular nitric oxide, pp 49–66

Rőszer T (2014) Biosynthesis of nitric oxide in plants. In: Nitric oxide in plants: metabolism and role in stress physiology, pp 17–32

Ruemer S, Krischke M, Fekete A, Lesch M, Mueller MJ, Kaiser WM (2016) Methods to detect nitric oxide in plants: are DAFs really measuring NO? BT—Plant nitric oxide: methods and protocols. In: Gupta KJ (ed) Springer, New York, pp 57–68. https://doi.org/10.1007/978-1-4939-3600-7_6

Rümer S, Kapuganti JG, Kaiser WM (2009) Oxidation of hydroxylamines to NO by plant cells. Plant Signal Behav 4:853–855. https://doi.org/10.4161/psb.4.9.9378

Santolini J, André F, Jeandroz S, Wendehenne D (2017) Nitric oxide synthase in plants: where do we stand? Nitric Oxide—Biol Chem 63:30–38. https://doi.org/10.1016/j.niox.2016.09.005

Schauberger B, Rolinski S, Schaphoff S, Müller C (2019) Global historical soybean and wheat yield loss estimates from ozone pollution considering water and temperature as modifying effects. Agric for Meteorol 265:1–15. https://doi.org/10.1016/j.agrformet.2018.11.004

Schumann U, Huntrieser H (2007) The global lightning-induced nitrogen oxides source. Atmos Chem Phys 7:3823–3907. https://doi.org/10.5194/acp-7-3823-2007

Shao Z, Zhao Y, Zhang Y, Wang Y, Wang Y, Yang L (2021) Effect of ozone stress on yield characteristics of indica-japonica hybrid rice Yongyou 538 in two consecutive growing seasons. Environ Exp Bot 186:104447. https://doi.org/10.1016/j.envexpbot.2021.104447

Sidana S, Bose J, Shabala L, Shabala S (2015) Nitric oxide in drought stress signalling and tolerance in plants. In: Nitric oxide action in abiotic stress responses in plants. Springer, Berlin, pp 95–114. https://doi.org/10.1007/978-3-319-17804-2_6

Simontacchi M, Galatro A, Ramos-Artuso F, Santa-María GE (2015) Plant survival in a changing environment: the role of nitric oxide in plant responses to abiotic stress. Front Plant Sci 6:1–19. https://doi.org/10.3389/fpls.2015.00977

Singh HP, Kaur S, Batish DR, Sharma VP, Sharma N, Kohli RK (2009) Nitric oxide alleviates arsenic toxicity by reducing oxidative damage in the roots of *Oryza sativa* (rice). Nitric Oxide—Biol Chem 20:289–297. https://doi.org/10.1016/j.niox.2009.02.004

Singh AA, Agrawal SB, Shahi JP, Agrawal M (2019) Yield and kernel nutritional quality in normal maize and quality protein maize cultivars exposed to ozone. J Sci Food Agric 99:2205–2214. https://doi.org/10.1002/jsfa.9414

Singh P, Ansari N, Pandey S, Madhoolika R, Shashi A, Agrawal B (2023) Effect of elevated ozone on the antioxidant response, genomic stability, DNA methylation pattern and yield in three species of *Abelmoschus* having different ploidy levels. Environ Sci Pollut Res. https://doi.org/10.1007/s11356-023-26538-9

Stevenson DS, Young PJ, Naik V, Lamarque JF, Shindell DT, Voulgarakis A, Skeie RB, Dalsoren SB, Myhre G, Berntsen TK, Folberth GA, Rumbold ST, Collins WJ, MacKenzie IA, Doherty RM, Zeng G, Van Noije TPC, Strunk A, Bergmann D, Cameron-Smith P, Plummer DA, Strode SA, Horowitz L, Lee YH, Szopa S, Sudo K, Nagashima T, Josse B, Cionni I, Righi M, Eyring V, Conley A, Bowman KW, Wild O, Archibald A (2013) Tropospheric ozone changes, radiative forcing and attribution to emissions in the atmospheric chemistry and climate model intercomparison project (ACCMIP). Atmos Chem Phys 13:3063–3085. https://doi.org/10.5194/acp-13-3063-2013

Stöhr C, Stremlau S (2006) Formation and possible roles of nitric oxide in plant roots. J Exp Bot 57:463–470. https://doi.org/10.1093/jxb/erj058

Sun J, Feng Z, Ort DR (2014) Impacts of rising tropospheric ozone on photosynthesis and metabolite levels on field grown soybean. Plant Sci 226:147–161. https://doi.org/10.1016/j.plantsci.2014.06.012

Sun LR, Yue CM, Hao FS (2019) Update on roles of nitric oxide in regulating stomatal closure. Plant Signal Behav. https://doi.org/10.1080/15592324.2019.1649569

Sun C, Zhang Y, Liu L, Liu X, Li B, Jin C, Lin X (2021) Molecular functions of nitric oxide and its potential applications in horticultural crops. Hortic Res 8. https://doi.org/10.1038/s41438-021-00500-7

Tailor A, Tandon R, Bhatla SC (2019) Nitric oxide modulates polyamine homeostasis in sunflower seedling cotyledons under salt stress. Plant Signal Behav 14:1667730. https://doi.org/10.1080/15592324.2019.1667730

Tanou G, Job C, Rajjou L, Arc E, Belghazi M, Diamantidis G, Molassiotis A, Job D (2009) Proteomics reveals the overlapping roles of hydrogen peroxide and nitric oxide in the acclimation of citrus plants to salinity. Plant J 60:795–804. https://doi.org/10.1111/j.1365-313X.2009.04000.x

Turnock ST, Wild O, Dentener FJ, Davila Y, Emmons LK, Flemming J, Folberth GA, Henze DK, Jonson JE, Keating TJ, Kengo S, Lin M, Lund M, Tilmes S, O'Connor FM (2018) The impact of future emission policies on tropospheric ozone using a parameterised approach. Atmos Chem Phys 18:8953–8978. https://doi.org/10.5194/acp-18-8953-2018

Vishwakarma A, Wany A, Pandey S, Bulle M, Kumari A, Kishorekumar R, Igamberdiev AU, Mur LAJ, Gupta KJ (2019) Current approaches to measure nitric oxide in plants. J Exp Bot 70:4333–4343

Wang BL, Tang XY, Cheng LY, Zhang AZ, Zhang WH, Zhang FS, Liu JQ, Cao Y, Allan DL, Vance CP, Shen JB (2010) Nitric oxide is involved in phosphorus deficiency-induced cluster-root development and citrate exudation in white lupin. New Phytol 187:1112–1123. https://doi.org/10.1111/j.1469-8137.2010.03323.x

Wang C, Wei L, Zhang J, Hu D, Gao R, Liu Y, Feng L, Gong W, Liao W (2023) Nitric oxide enhances salt tolerance in tomato seedlings by regulating endogenous S-nitrosylation levels. J Plant Growth Regul 42(1):275–293. https://doi.org/10.1007/s00344-021-10546-5

Wani KI, Naeem M, Khan MMA, Aftab T (2023) Nitric oxide induces antioxidant machinery, PSII functioning and artemisinin biosynthesis in *Artemisia annua* under cadmium stress. Plant Sci 334:111754. https://doi.org/10.1016/j.plantsci.2023.111754

Wimalasekera R, Tebartz F, Scherer GFE (2011) Polyamines, polyamine oxidases and nitric oxide in development, abiotic and biotic stresses. Plant Sci 181:593–603. https://doi.org/10.1016/j.plantsci.2011.04.002

Wong A, Donaldson L, Portes MT, Eppinger J, Feijó JA, Gehring C (2020). *Arabidopsis* DIACYLGLYCEROL KINASE4 is involved in nitric oxide-dependent pollen tube guidance and fertilization. Development 147. https://doi.org/10.1242/DEV.183715

Xia J, Kong D, Xue S, Tian W, Li N, Bao F, Hu Y, Du J, Wang Y, Pan X, Wang L, Zhang X, Niu G, Feng X, Li L, He Y (2014) Nitric oxide negatively regulates AKT1-mediated potassium uptake through modulating vitamin B6 homeostasis in *Arabidopsis*. Proc Natl Acad Sci USA 111:16196–16201. https://doi.org/10.1073/pnas.1417473111

Xu J, Liu T, Qu F, Jin X, Huang N, Wang J, Hu X (2021) Nitric oxide mediates γ-aminobutyric acid-enhanced muskmelon tolerance to salinity–alkalinity stress conditions. Sci Hortic (Amsterdam) 286:110229. https://doi.org/10.1016/j.scienta.2021.110229

Yadav A, Bhatia A, Yadav S, Kumar V, Singh B (2019a) The effects of elevated $CO_2$ and elevated $O_3$ exposure on plant growth, yield and quality of grains of two wheat cultivars grown in north India. Heliyon 5:e02317. https://doi.org/10.1016/j.heliyon.2019.e02317

Yadav DS, Rai R, Mishra AK, Chaudhary N, Mukherjee A, Agrawal SB, Agrawal M (2019b) ROS production and its detoxification in early and late sown cultivars of wheat under future $O_3$ concentration. Sci Total Environ 659:200–210. https://doi.org/10.1016/j.scitotenv.2018.12.352

Yadav DS, Agrawal SB, Agrawal M (2021a) Ozone flux-effect relationship for early and late sown Indian wheat cultivars: growth, biomass, and yield. F Crop Res 263:108076. https://doi.org/10.1016/j.fcr.2021.108076

Yadav DS, Jaiswal B, Agrawal SB, Agrawal M (2021b) Diurnal variations in physiological characteristics, photoassimilates, and total ascorbate in early and late sown Indian wheat cultivars under exposure to elevated ozone. Atmosphere (Basel) 12:1568. https://doi.org/10.3390/atmos12121568

Yadav DS, Agrawal M (2021) Global crop yield losses due to tropospheric ozone: The present and future scenario. In: Agrawal SB, Agrawal M, Singh A (eds) Tropospheric ozone: a hazard for vegetation and human health. Cambridge Scholars Publishing, pp 515–550

Yang L, Ji J, Harris-Shultz KR, Wang H, Wang H, Abd-Allah EF, Luo Y, Hu X (2016) The dynamic changes of the plasma membrane proteins and the protective roles of nitric oxide in rice subjected to heavy metal cadmium stress. Front Plant Sci 7:190. https://doi.org/10.3389/fpls.2016.00190

Young PJ, Archibald AT, Bowman KW, Lamarque J-FF, Naik V, Stevenson DS, Tilmes S, Voulgarakis A, Wild O, Bergmann D, Cameron-Smith P, Cionni I, Collins WJ, Dalsøren SB, Doherty RM, Eyring V, Faluvegi G, Horowitz LW, Josse B, Lee YH, MacKenzie IA, Nagashima T, Plummer DA, Righi M, Rumbold ST, Skeie RB, Shindell DT, Strode SA, Sudo K, Szopa S, Zeng G (2013) Pre-industrial to end 21st century projections of tropospheric ozone from the atmospheric chemistry and climate model intercomparison project (ACCMIP). Atmos Chem Phys 13:2063–2090. https://doi.org/10.5194/acp-13-2063-2013

Yu M, Lamattina L, Spoel SH, Loake GJ (2014) Nitric oxide function in plant biology: a redox cue in deconvolution. New Phytol 202:1142–1156. https://doi.org/10.1111/nph.12739

Zhai L, Xiao D, Sun C, Wu T, Han Z, Zhang X, Xu X, Wang Y (2016) Nitric oxide signaling is involved in the response to iron deficiency in the woody plant *Malus xiaojinensis*. Plant Physiol Biochem 109:515–524. https://doi.org/10.1016/j.plaphy.2016.11.005

Zhang Y, Han X, Chen X, Jin H, Cui X (2009) Exogenous nitric oxide on antioxidative system and ATPase activities from tomato seedlings under copper stress. Sci. Hortic. (Amsterdam) 123:217–223. https://doi.org/10.1016/j.scienta.2009.08.015

Zhang Y, Liu G, Xu S, Dai J, Li W, Li Z, Zhang D, Cui Z, Li C, Dong H (2022) Nitric oxide reduces the yield loss of waterlogged cotton by enhancing post-stress compensatory growth. F Crop Res 283:108524. https://doi.org/10.1016/j.fcr.2022.108524

Zhu XF, Zhu CQ, Wang C, Dong XY, Shen RF (2017) Nitric oxide acts upstream of ethylene in cell wall phosphorus reutilization in phosphorus-deficient rice. J Exp Bot 68:753–760. https://doi.org/10.1093/jxb/erw480

# Chapter 4
# Role of Volatile Organic Compounds: Antioxidants or Plant Regulators

**Ashish Kumar Mishra, Parvati Madheshiya, Shivani Gupta, Ansuman Sahoo, Gereraj Sen Gupta, and Supriya Tiwari**

**Abstract** Volatile organic compounds (VOCs), such as isoprene, monoterpenes, and sesquiterpenes, encompass a wide range of chemicals emitted by plants, significantly influencing plant physiology and ecosystem dynamics. Among these VOCs, isoprene stands out as a particularly fascinating subject of research. This chapter delves into the dual role of isoprene as both a natural antioxidant and a plant regulator. Isoprene's ability to stabilize plant membranes and mitigate oxidative stress caused by ground-level ozone and elevated temperatures has sparked interest in its potential antioxidant functions. While the antioxidant activities of isoprene are increasingly recognized, the byproducts resulting from its interaction with reactive oxygen species remain under investigation. Notably, isoprene emission is restricted to certain plant species, indicating its selective role in modulating plant responses to various biotic and abiotic stresses. The interaction of isoprene with reactive oxygen and nitrogen species within leaves further suggests its involvement in complex plant signalling pathways and adaptive mechanisms. The intricate relationship between isoprene, its antioxidant properties, and its regulatory functions underscores the dynamic interplay between plants and their environment. Further research is essential to fully elucidate isoprene's role as a natural antioxidant, plant regulator, or both. This chapter emphasizes isoprene's potential as a dual-functioning compound that not only shields plants from environmental stresses but also influences their ecological responses and adaptations.

**Keywords** Tropospheric ozone · Volatile organic compounds · Isoprene · Antioxidants

A. K. Mishra · P. Madheshiya · S. Gupta · A. Sahoo · G. S. Gupta · S. Tiwari (✉)
Laboratory of Ecotoxicology, Department of Botany, Institute of Science, Banaras Hindu University, Varanasi, India
e-mail: stiwari@bhu.ac.in

 
A. K. Pandey and E. Oksanen (eds.), *Tropospheric Ozone and Food Security*,
https://doi.org/10.1007/978-3-032-03608-7_4

## 4.1 Introduction

Volatile organic compounds (VOCs) are a diverse group of chemicals emitted by both natural and anthropogenic sources. VOCs, including substances such as benzene, formaldehyde, toluene, and isoprene, play a significant role in atmospheric chemistry. These compounds can react with nitrogen oxides ($NO_x$) in the presence of sunlight to form ground-level ozone ($O_3$), a key component of photochemical smog. $O_3$ is a multifaceted gas with significant importance in the Earth's atmosphere. Its presence in the stratosphere acts as a protective shield by absorbing dangerous ultraviolet (UV) radiation. However, when found at ground level, $O_3$ can have negative impacts. Elevated levels of ground-level $O_3$ are linked to crop damage and negative effects on ecosystems (Manisalidis et al. 2020; Sahoo et al. 2023; Mishra et al. 2024a, Mishra, et al., 2024b). Due to their sedentary lives, plants need to possess the ability to adapt to different alterations in factors such as light levels, temperature fluctuations, moisture variations, and other abiotic conditions within their surroundings (Asensi-Fabado et al. 2017). When these parameters deviate from a specific range, plants face stress, resulting in a decline in growth rate, diminished reproductive capacity, and in some cases, mortality. These stress-induced effects extend beyond individual plants, influencing ecological traits on both immediate and extended timescales. This has the potential to impact factors like biodiversity, the functioning of ecosystems, and the storage of carbon. The convergence of swiftly escalating temperatures, drought conditions, and pollutants stands out as a notable occurrence linked to worldwide transformations. These occurrences pose a significant threat to plants that lack adaptation and the capacity for swift adjustment to cope with these elements (Kanakidou et al. 2022). In order to navigate through these challenges, plants employ a range of strategies, and one of these tactics involves the emission of volatile organic compounds (VOCs). Land-based ecosystems release a multitude of biogenic volatile organic compounds (BVOCs), playing a substantial role in shaping the atmospheric chemical composition (Guenther 2013). Among these, plants are notable producers of volatile isoprenoids, encompassing isoprene, monoterpenes (MTs), and sesquiterpenes. These compounds are particularly abundant and chemically reactive within the BVOC group. Isoprene ($C_5H_8$; 2-methyl-1,3-butadiene), although small, holds a crucial position among BVOCs, contributing to around 50% of the global BVOC emissions (Zenone et al. 2016). Every year, approximately 1000 teragrams (Tg) of biogenic volatile organic compounds (BVOCs) are released globally, and among these emissions, isoprene accounts for 53% (Guenther et al. 2012). Isoprene is discharged from diverse terrestrial ecosystems, encompassing elements such as leaves, blossoms, stems, trunks, roots, fallen leaves, and soil microorganisms. Nonetheless, the primary source of isoprene emission is foliage, as highlighted in studies by Loreto and Schnitzler (2010), Guenther et al. (2012), and Greenberg et al. (2012). The release of isoprene is not a widespread characteristic and is restricted to specific plant species termed "emitters". An analysis of research related to isoprene indicates that approximately 20% of the perennial vegetation in tropical and temperate regions worldwide are categorized as emitters (Loreto and Fineschi

2015). Emitters incur a relatively high energy cost for isoprene emission, consuming approximately 20 ATP and 14 NADPH for each isoprene molecule (Sharkey and Yeh 2001). In typical conditions, the release of isoprene accounts for 2–5% of the total carbon fixed through photosynthesis. However, during periods of stress, this proportion can rise to about 10–15%, or even surpass the amount fixed by photosynthesis itself (Brilli et al. 2014). BVOCs emitted by plants can be released as a regular process or in reaction to different environmental pressures, including biological and non-biological stresses (Loreto and Schnitzler 2010). For example, isoprene, a significant BVOC, has been proposed to act as a protective mechanism by stabilizing plant cell membranes and guarding against oxidative stress due to factors like $O_3$ or high temperatures (Loreto and Fares 2007; Velikova et al. 2011). BVOCs, including isoprene, can interact with reactive oxygen and nitrogen species within leaves, potentially mitigating oxidative harm (Velikova et al. 2008; Jardine et al. 2012). However, the existence of byproducts from the interaction of isoprene with reactive oxygen species remains uncertain (Cappellin et al. 2019). Isoprene, while produced by certain plant species termed "emitters", has shown the potential to serve as both a natural antioxidant and a plant regulator. This dual role has attracted significant attention due to its implications for plant physiology and ecosystem dynamics.

Isoprene emissions have been linked to its capacity to act as a protective mechanism against oxidative stress. Studies suggest that isoprene can stabilize plant membranes and mitigate oxidative damage caused by factors such as $O_3$ and high temperatures. This antioxidative property has raised the question of whether isoprene serves as a natural antioxidant within plants. In addition to its potential antioxidant role, isoprene has also been proposed as a regulator of various plant processes. Its emission is not uniform across all plant species but is instead confined to specific emitter species. This selective emission indicates that isoprene might play a role in regulating plant responses to both biological and environmental stresses. Moreover, isoprene's reactivity with reactive oxygen and nitrogen species within leaves suggests its involvement in plant signalling and adaptation. The intricate relationship between isoprene, its antioxidative properties, and its regulatory functions underscores the complexity of plant-environment interactions. Understanding whether isoprene primarily functions as a natural antioxidant, a plant regulator, or both, requires comprehensive research that delves into its impacts on plant physiology, responses to stressors, and its contribution to ecosystem dynamics. This article delves into the scientific exploration of isoprene's roles, shedding light on its potential to act as both a natural antioxidant and a plant regulator, thereby contributing to the broader understanding of plant-environment interactions.

## 4.2 $NO_x$-VOCs Interplay ($O_3$ Formation and Degradation)

Atmospheric isoprene assures the availability of OH and provides fuel for the $NO–NO_2–O_3$ cycle, which is responsible for $O_3$ creation and destruction in the troposphere (Jardine et al. 2012; Rivera-Rios et al. 2014). Much attention has been paid to isoprene chemistry recently (Fuchs et al. 2013; Jacobs et al. 2013; Peeters et al. 2014; Dunker et al. 2016). The intricate process of $O_3$ creation and the intricate chemistry of its precursor molecules are still under extensive investigation (Fittschen et al. 2017). The initial step in $O_3$ formation is typically the photolysis of nitrogen dioxide ($NO_2$). The regeneration of $NO_2$ from NO is a crucial step in the cycle and occurs when carbon monoxide (CO), methane, or volatile organic compounds (VOCs) react with hydroxyl radicals (·OH) to produce hydroxyl and peroxy radicals. These radicals then readily react with NO to regenerate $NO_2$. This cycle of $NO_x$ recycling is known as the $RO_x$–$NO_x$ cycle. The relationship between photochemical $O_3$ formation and its precursors, namely VOCs and $NO_x$, is nonlinear and depends on their respective concentrations (Liu et al. 2012; Ling et al. 2014).

In cases where isoprene is absent, the overall $O_3$ production is primarily influenced by the total levels of nitrogen oxides ($NO_x$) in the atmosphere. However, when isoprene is present, the reaction of isoprene with hydroxyl radicals in the atmosphere converts NO to $NO_2$, thereby promoting the generation of $O_3$ (Gilman et al. 2009; Sharkey et al. 2008a, b). Consequently, under conditions of low $NO_x$ concentrations, the formation of $O_3$ is constrained by $NO_x$ availability, whereas under conditions of high $NO_x$ concentrations coupled with the presence of VOCs like isoprene, the recycling of hydroxyl radicals becomes efficient, sustaining elevated levels of VOCs and shifting the limitation of $O_3$ formation to VOCs (Carter 1994).

Furthermore, during the oxidation of isoprene by hydroxyl radicals, the resulting $RO_2$ and NO can contribute to the creation of isoprene nitrate ($RONO_2$), a compound that can be further recycled to produce additional $NO_2$ and $O_3$ (Paulot et al. 2012). Nitric oxide (NO) reacts with the peroxy radical to create $O_3$ in the troposphere. Figure 4.1 depicts the NO conversion to $NO_2$, which is then photolyzed to $O_3$, and the NO is recycled. The amount of $O_3$ that may be generated by $NO_x$ pollution at a given place and time varies. For instance, $NO_x$ quickly interacts to generate $HNO_3$ in contaminated places near the Earth's surface, which functions as a repository for $NO_x$ (Ye et al. 2016). More $NO_x$ molecules react with peroxy radicals to produce $O_3$ in less polluted regions because $NO_2$ photolysis can more effectively compete with $HNO_3$ synthesis. $O_3$ generation is more effective in the tropics because $NO_x$ is more easily launched into the free troposphere there (Butler et al. 2020; Jacob et al. 1996). VOC: $NO_x$ ratio also controls $O_3$ concentration (Simon et al. 2015). Urban regions with excessive $NO_x$ owing to high emissions have locally suppressed $O_3$ concentrations because $O_3$ production is controlled by VOCs. Suburban locations may have higher $O_3$ concentrations due to the mixing of $NO_x$ and VOCs brought from metropolitan centres.

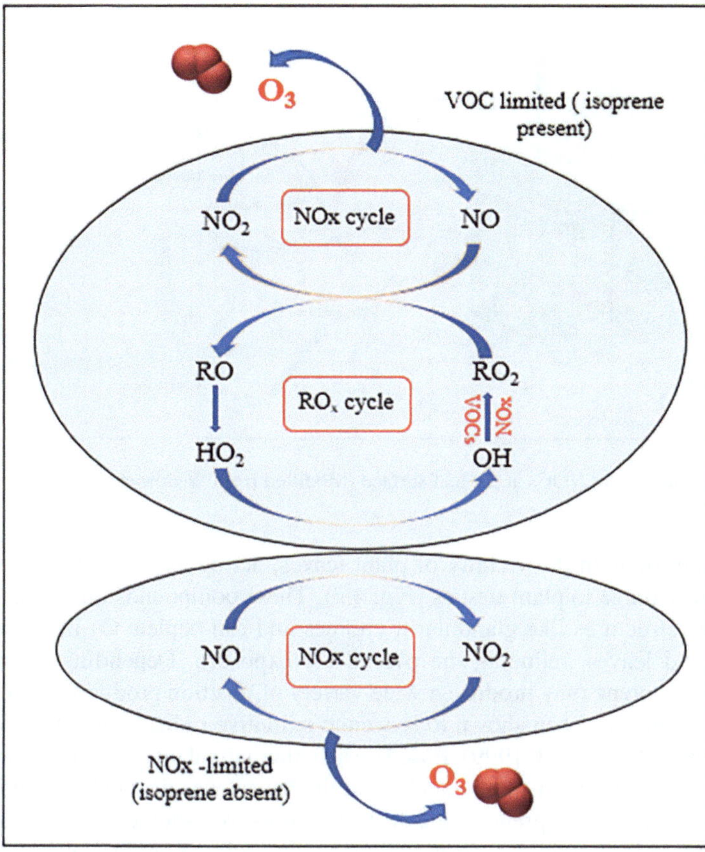


**Fig. 4.1** $RO_X$-$NO_X$ interplay in the formation of tropospheric $O_3$ (modified from Sahoo and Tiwari 2021)

## 4.3 Isoprene: An $O_3$ Quencher Entity?

Isoprene is not typically considered an "$O_3$ quencher" in the same way that some other compounds, like certain terpenes and diterpenoids found in plants, can act as $O_3$ quenchers or sinks. Isoprene is a volatile organic compound (VOC) emitted by many plants, and it has its own role in atmospheric chemistry.

Isoprene is known for its involvement in the formation of secondary organic aerosols and its role in atmospheric chemistry, particularly in the generation of atmospheric pollutants. While it can react with atmospheric oxidants like $O_3$, its primary impact on $O_3$ is not as a quencher to reduce $O_3$ concentrations. Instead, the interactions of isoprene with $O_3$ can lead to the formation of other compounds and particulate matter.

In ecosystems, certain plant-derived compounds like terpenes, sesquiterpenes, and diterpenoids are more specifically recognized for their ability to reduce or "quench"

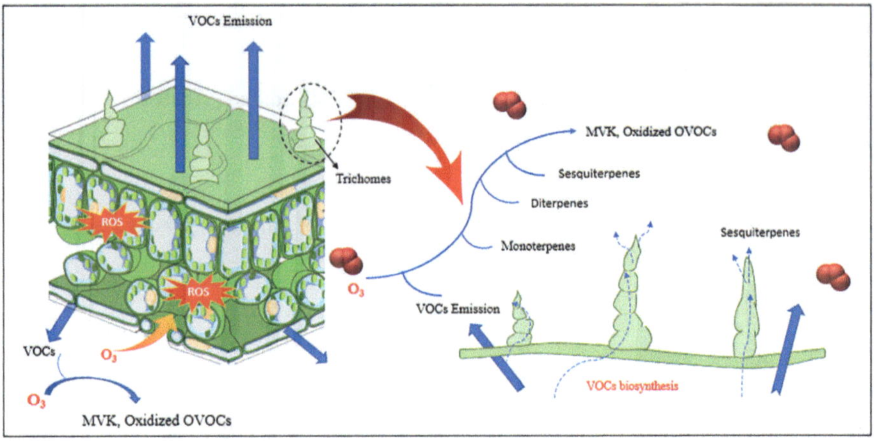


**Fig. 4.2** Ozonolysis of VOCs at the leaf surface (Modified from Wedow et al. 2021)

$O_3$ concentrations in the vicinity of plant leaves, acting as protective mechanisms against $O_3$ damage to plant tissues (Fig. 4.2). These compounds are often stored in specialized structures like glandular trichomes and can deplete $O_3$ in the boundary layer around leaves, reducing the overall $O_3$ exposure. Depending on the other substrates, isoprene may produce a wide variety of reaction products (Fuentes et al. 2000). Isoprene has been shown to scavenge oxidative radicals in highly oxidizing environments (Sauer et al. 1999). Based on a mathematical model (Chameides 1989), the interaction between isoprene and $O_3$ in the leaves was deemed to be negligible. The concentration of isoprene inside the leaves, however, should be much larger than what is shown in Chameides' formulation, as was pointed out by Salter and Hewitt (1992). The researchers came to the conclusion that isoprene, at physiological quantities, may efficiently react with $O_3$, generating hydroxymethyl hydroperoxide and exacerbating the $O_3$-induced damages.

### *4.3.1 Trichomes: A Storehouse of Volatile Isoprenoids*

Glandular trichomes are specialized structures on plant leaves that store and release non-isoprene volatile compounds, providing protection against $O_3$ damage. Isoprene, due to its high volatility and hydrophobic nature, is quickly released into the atmosphere and is not stored within leaves or on their surfaces. In contrast, glandular trichomes store and exude a range of metabolites, including monoterpenes, sesquiterpenes, diterpenes, and occasionally flavonoids and benzenoids (Wagner 1991; Tissier et al. 2017; Jud et al. 2016).

Plants may take varying amounts of time, from days to months, to fill their glandular trichomes with these protective metabolites. Once stress arises, these compounds are essential in bolstering resilience against both biological and environmental challenges. For instance, there is evidence supporting the role of a diterpenoid called cis-abienol, secreted by glandular trichomes on Nicotiana tabacum leaves, in reducing stomatal uptake of $O_3$. This reduction occurs by depleting $O_3$ in the boundary layer around the leaf, effectively reducing the overall $O_3$ dose (Jud et al. 2016).

Studies by Li et al. (2018) have provided further confirmation that plant sensitivity to $O_3$, the uptake of $O_3$ by stomata, and $O_3$-induced leaf damage are strongly influenced by the density of glandular trichomes, including both capitate and peltate types. These findings have expanded our understanding of the protective function of glandular trichomes across various plant species with different trichome densities, types, and exudates.

Additionally, some plant species, such as eucalypts, possess oil glands within their leaf mesophyll that contain significant amounts of secondary metabolites, including mono- and sesquiterpenes. These compounds have the potential to act as efficient $O_3$ sinks, substantially reducing $O_3$ concentrations within leaves after $O_3$ has entered the leaf interior (Goodger et al. 2018).

It's worth noting that the density and distribution of glandular trichomes can change with leaf position and developmental stage, highlighting the dynamic nature of these protective structures. Furthermore, ongoing research into the molecular mechanisms controlling the development of glandular trichomes holds promise for improving plant tolerance to oxidative stress, including $O_3$ exposure. This knowledge could be applied to modify biosynthetic, storage, and secretory pathways in specific plant species and developmental stages to enhance their ability to withstand $O_3$-induced damages.

## 4.4 Impact of VOCs on Air Quality

Emissions of VOCs exert a notable influence on air quality, as they fuel an array of chemical reactions upon entering the atmosphere (Wu et al. 2020). VOCs that are prevalent in the atmosphere are both biologically and anthropogenically derived (Pardo et al. 2022). Biogenic volatile organic compounds (BVOCs) emitted from terrestrial and marine sources collectively account for > 80% of the total global VOC input, despite the fact that volcanic activity and hydrothermal systems only contribute relatively minute amounts (Sahu 2012; Tassi et al. 2013). In urban areas, notably, anthropogenic sources account for the majority of VOC emissions (Coggon et al. 2021). Emissions from vehicles, industries, and biomass burning have been identified as the most significant contributors to ambient anthropogenic volatile organic compound concentrations (AVOCs) (Deng et al. 2018). Although there has been little focus on BVOCs released by urban vegetation, due to their increased chemical reactivity as compared to numerous AVOCs and larger emission proportion of the total VOC at the global scale, BVOCs play a critical role in shaping tropospheric chemistry. These characteristics point to their considerable potential to influence the production of secondary pollutants even in urban environments (Ma et al. 2019).

BVOCs have a substantial influence on biosphere-atmosphere interactions via atmospheric reactivity, aerosol growth processes, cloud formation, and hence radiative balancing (Fig. 4.3). BVOCs can directly affect the regional radiative forcing through aerosol scattering and indirectly by altering cloud dynamics since they are precursors to aerosols. Concomitantly, additional cloud cover results in more diffuse radiation, which escalates net primary production and, in turn, BVOC emissions (Yanez-Serrano et al. 2020). BVOCs, on the other hand, can have further consequences on radiative forcing through their influence on tropospheric $O_3$ formation or by prolonging the lifespan of methane through suppression of its principal sink, the OH radical (Wilson et al. 2019). Moreover, by their disintegration in the atmosphere (with $CO_2$ as an end product) and by perturbation of the lifespan of other greenhouse gases like $O_3$, CO, etc. BVOCs additionally have a considerable impact on radiative forcing (Yanez-Serrano et al. 2020).

The oxidation of VOCs via various oxidants leads to the formation of ground-level $O_3$ and secondary organic aerosols (SOAs), thereby influencing regional air quality as well as climate change (Wu et al. 2020). The regional hydrological cycle heavily relies on BVOCs, as the presence of BVOCs affects associated albedo by changing the amount and size of cloud condensation nuclei (CCN) and cloud droplets, influencing how clouds reflect and absorb light and accordingly the radiation balance and hydrological cycle (Shrivastava et al. 2017). BVOCs and their function in secondary organic aerosol (SOAs), which are the major constituent of fine-mode aerosol mass, may eventually have an impact on the Earth's radiative cooling (Ehn et al. 2014).

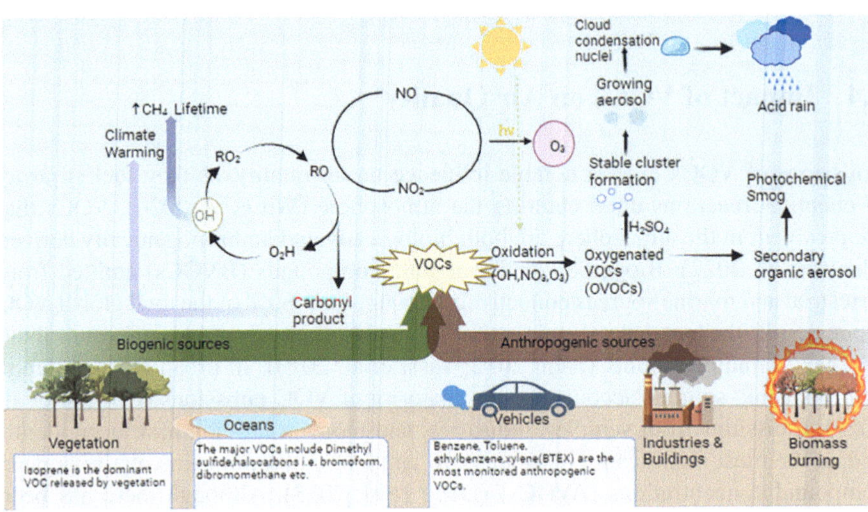


**Fig. 4.3** Influence of BVOCs on biospheric-atmospheric interactions

VOCs comprise a wide spectrum of compounds such as carbonyls, organic acids, alcohols, alkanes, alkenes, aldehydes, esters, ketones, and aromatic hydrocarbons. The physical and chemical characteristics of VOCs, as well as their residence times in the atmosphere (which can range from a few minutes to several months), frequently pose risks to the environment and public health (Carpenter et al. 2007). Because VOCs are potent $O_3$ precursors and provide the necessary OH radicals in ambient air for the generation of ground-level $O_3$, they play critical roles in modulating atmospheric chemistry (Wang et al. 2021). Moreover, VOCs also affect the quality of air by playing crucial roles in the development of photochemical smog, stratospheric $O_3$ depletion, and the production of organic acids that reduce the pH of rainwater and thus, contribute to environmental acidification.

Direct photolysis as well as oxidation of VOCs can have a significant impact on the radical budget and secondary pollutant generation in the atmosphere (Zhang et al. 2021). VOC oxidation and elimination are dependent on its reactivity with hydroxyl and $O_3$ radicals during the day and nitrate radicals during the night (Tan et al. 2019). The contributions of different VOCs to atmospheric chemical reactions that are linked to the formation of secondary pollutants may be visibly and effectively defined by VOC reactivity. In urban and suburban settings having substantially elevated levels of nitrogen oxides ($NO_x = NO + NO_2$), VOCs chemically undergo various reactions leading to the formation of $O_3$ and particulate matter (PM) (Churkina et al. 2017). The dissociation reactions of volatile organic compounds (VOCs) lead to the formation of intermediate radicals known as $RO_2$ and $HO_2$. These radicals, $RO_2$ and $HO_2$, interact with NO, causing the transformation of NO into $NO_2$, which subsequently undergoes photolysis to produce $O_3$ (Lu et al. 2019). Consequently, there exists a strong correlation between the levels of $O_3$ and VOCs at various altitudes due to the pivotal part played by VOCs in fuelling the formation of $O_3$ in the troposphere.

Although plants produce hundreds of distinct VOCs, only a handful of them have a significant impact on air quality. Isoprene, monoterpenes, and sesquiterpenes are the most commonly encountered reactive biogenic VOCs (Ciccioli et al. 2014). While gaseous VOCs exaggerate $O_3$ formation, the translocation of VOCs between gas and particle phases can either impede or accelerate urban $O_3$ formation (Brook et al. 2010). Semi-volatile compounds transfer from the gas phase to the particle phase and are momentarily eliminated from the cycle of $O_3$ formation in the case of high gas-phase concentrations of VOCs with more than five carbon atoms. Isoprene emissions primarily contribute to the formation of ground-level $O_3$, while monoterpene and sesquiterpene emissions may enhance the quantity and mass of particulate matter in the atmosphere (Churkina et al. 2017).

Warming in the twenty-first century is anticipated to increase isoprene emission, which is one of the major volatile organic compounds released by plants and plays a key role in regulating atmospheric composition including concentrations of $O_3$ and particulate matter (PM) (Tai et al. 2013). In densely polluted regions with high nitrogen oxide ($NO_x$) concentrations, isoprene plays a substantial role in the formation of surface $O_3$; yet, in more distant regions, it either directly interacts with $O_3$ or indirectly decreases $O_3$ by sequestering $NO_x$ as isoprene nitrate (Lu et al. 2019). The formation of secondary organic aerosols (SOA) is primarily mediated by isoprene emissions. The oxidation of isoprene is predominantly triggered by hydroxyl radicals (OH), which are produced principally by the photolysis of $O_3$. In the atmosphere, isoprene undergoes a primary reaction with hydroxyl radicals (OH), resulting in the formation of peroxy radicals ($RO_2$) as a result of the addition of oxygen (Wennberg et al. 2018). Under pristine, $HO_2$-dominated (low-NO) conditions, the peroxy radicals derived from isoprene (isoprene $RO_2$) interact with hydroperoxyl radicals ($HO_2$). This interaction leads to the formation of complex organic hydroperoxides known as isoprene hydroxyl hydroperoxides (ISOPOOH, $C_5H_{10}O_3$). Among these hydroperoxides, the most abundant isomers are 1-hydroxyl-2-hydroperoxyl-ISOPOOH and 4-hydroxyl-3-hydroperoxyl-ISOPOOH, referred to as 1,2-ISOPOOH and 4,3-ISOPOOH, respectively. ISOPOOH engages in the oxidation of sulphur dioxide ($SO_2$, aq) within cloud and fog water, resulting in the formation of sulphate $\left(SO_4^{-2}\right)$. $SO_2$, aq combines with water to produce sulphurous acid ($SO_2{\cdot}H_2O$), which breaks down into bisulphite ions ($HSO^{3-}$) when the pH > 2. Under more alkaline conditions (pH > 6), this bisulphite further disintegrates, forming sulphite ions $\left(SO_3^{-2}\right)$. Thus, the oxidation of sulphates by isoprene hydroperoxides plays a significant role in sulphate formation in cloud water, thereby leading to the formation of acid rain (Dovrou et al. 2021). In the gaseous phase, ISOPOOH undergoes oxidation by OH radicals, predominantly yielding isoprene epoxydiols (IEPOX), which undergoes acid-driven multiphase chemistry yielding a wide variety of oxygenated compounds including $C_5$ alkene triols, 2-methyltetrols, hemiacetal dimers, cis- and trans-3-methyl tetra hydro furan-3,4-diols, and organosulphates (Dubois et al. 2021). Concomitantly, these organic aerosol precursor gases transform into more hygroscopic, less volatile, and highly oxidized molecules, leading to the formation of highly oxygenated molecules (HOMs), which in turn have an influential effect on the regional oxidant chemistry (Yan et al. 2020).

VOCs are not only a concern outdoors but also indoors. Many household products, such as paints, solvents, cleaning agents, and air fresheners, release VOCs into the indoor air (Bari et al. 2015). This can lead to "indoor air pollution", which can be particularly problematic because indoor environments are often more poorly ventilated than outdoor spaces (Begum et al. 2009). Chemical reactions between distinct VOCs or between VOC and inorganic gases like $O_3$ or $NO_x$ might occur in indoor air, and these reactions may occasionally be influenced by relative humidity and temperature (Duncan et al. 2018). It has recently been reported that chemical interactions between certain unsaturated VOCs, or SVOCs, and oxidants, such as $O_3$ or $NO_2$, may occur in indoor air under specific conditions, producing new molecules with possible sensory irritation effects. Formaldehyde (HCHO) and volatile organic compounds (VOCs), including Benzene ($C_6H_6$), Toluene ($C_7H_8$), Ethylbenzene ($C_8H_{10}$), Xylene ($C_8H_{10}$), and Styrene ($C_8H_8$), are illustrative examples of indoor air contaminants (Arar and Jung 2021). The influence of volatile organic compounds (VOCs) originating from marine sources on air quality is a significant aspect to consider. These compounds, emitted by marine ecosystems, can have notable effects on the composition and quality of the surrounding air as they play key roles in modulating climate, the global carbon cycle, and atmospheric reactions. Marine-derived VOCs, such as dimethyl sulphide (DMS) and various halocarbons, play a multifaceted role in atmospheric chemistry and the environment (Halsey and Giovannoni 2023). Dimethyl sulphide (DMS), for instance, is released by marine phytoplankton and plays a role in the sulphur cycle. When DMS is emitted into the atmosphere, it undergoes chemical transformations that lead to the formation of sulphate aerosols. These aerosols can act as cloud condensation nuclei, affecting cloud properties and potentially impacting regional climate patterns (Jackson and Gabric 2022). Halocarbons, which include compounds like bromoform and chloroform, are also released from marine environments. These compounds have the potential to contribute to the depletion of $O_3$ in the stratosphere, which in turn can influence the amount of UV radiation reaching the Earth's surface (Keng et al. 2020). In addition to these specific compounds, marine-derived VOCs can interact with other atmospheric components, leading to the formation of secondary pollutants such as $O_3$ and fine particulate matter (Swan et al. 2016). These pollutants have implications for air quality, human health, and even climate. Understanding the role of marine-derived VOCs in affecting air quality is crucial for comprehending the complex interactions between marine ecosystems and the atmosphere. It also highlights the interconnectedness of different environmental systems and the need for a holistic approach to environmental management and research.

To sum up, the impact of volatile organic compounds (VOCs) on air quality is a crucial concern with wide-ranging implications. The quality of the air we breathe is directly influenced by the concentration, dispersion, and effects of VOCs in both indoor and outdoor environments. Poor air quality exacerbates the harmful effects of VOCs, intensifying their contribution to health issues and environmental degradation. Addressing this interplay is essential for creating healthier living spaces and a more sustainable planet. Through vigilant monitoring, effective regulation, and collaborative efforts, we can strive to improve air quality, mitigate VOC-related risks, and pave the way for a cleaner, safer future.

## 4.5 Isoprenoids in Safeguarding Against Oxidative Stress

Several abiotic and biotic stresses including drought, humidity, heavy metal, high light intensity, UV-irradiation, salinity, ground-level $O_3$, mechanical damage, pathogenesis, insecticides, herbicides, and temperature (including freezing) can lead to oxidative stress in plants thereby affecting their normal functioning (Gautam et al. 2022, Vickers et al. 2009). Plants generally respond in a stress-specific manner. Mostly, stresses cause generic responses notably an excess production of reactive oxygen species (ROS). ROS includes singlet oxygen ($_1O^2$), superoxide ($O_2\cdot^-$), hydrogen peroxide ($H_2O_2$), and hydroxyl radicals (·OH) (Madheshiya et al. 2023). Excessive ROS generation leads to lipid peroxidation, inactivation of enzymes, proteins and denaturation of nucleic acids ultimately programming cell death (Jalil et al. 2017; Banerjee and Roychoudhury 2017; Takshak and Agrawal 2018). Several pathways of signalling that stimulate biomolecules that can safeguard against oxidative stresses are developed by plants (Roychoudhury et al. 2013). Although most of the research on plant antioxidants has been focused on non-volatile substances, numerous volatile substances linked to the isoprenoid groups have also played a defensive role against oxidative stress (Mishra et al., 2025a; Takshak and Agrawal 2018, Vickers et al. 2009, Madheshiya et al. 2023, Ansari et al. 2021). Naturally synthesized volatile isoprenoids are a significantly large, lipophilic, and highly diverse group of organic compounds whose chemical skeletons are constituted of C5 building blocks that can be placed together using various ways and have an abundance of modifications (Vickers et al. 2009). Isoprenoids enhance the ability of plants to manage internal oxidative stress (Roychoudhury et al. 2013) and play pivotal roles across all domains of life, from prokaryotes to eukaryotes. They are involved in essential processes such signal transduction and electron transport via quinones, as well as functioning as hormones and antioxidants (Hoshino and Gaucher 2018). It is well documented that isoprene helps to reduce the effects of $O_3$ stress (Loreto and Fineschi 2015), similarly, isoprene emission has also been found to be stimulated by short-term exposure to high $O_3$ concentrations (100–300 ppb) (Calfapietra et al. 2009). Stressed plants can maintain higher isoprenoid emissions, and plants recovering from stress, particularly plants stressed due to drought, would subsequently raise the emission (Brilli et al. 2007). The rate of isoprenoid biosynthesis can be governed by various factors, including gene expression modulation, substrate accessibility, sub-cellular compartmentalization, enzyme activity, and other regulatory influences (Vranová et al. 2012). Two distinct metabolic pathways (Fig. 4.4) are responsible for the generation of the C5 building block of isoprenoids: the MVA (mevalonic acid) pathway in cytoplasm and the MEP (2-C-methyl-Derythritol 4-phosphate) pathway in chloroplast (Lange et al. 2000) also documented in peroxisomes and the membranes of the endoplasmic reticulum (Vranová et al. 2012). Both the MEP and MVA pathways can be interlinked via the translocation of multiple metabolic intermediates across the plastid membrane. (Wu et al. 2006). Sesquiterpene (C15) and triterpene (C30) formation occur in the cytosol, whereas later steps in the formation of hemiterpenes (C5), monoterpenes (C10), diterpenes (C20), and tetraterpenes (C40) are present in the plastids membrane (Vickers et al. 2009, Dudareva et al. 2006). Isoprene synthesis is light-dependent process in higher plants (Sanadze 2004). According to Vranová et al. (2012) light and sugar signalling augments the MEP pathway while attenuate the MVA pathway.

Elevated temperatures damaged the photosynthetic apparatus but presence of isoprene is increasing the thermal tolerance of photosynthetic apparatus. This effect was recently demonstrated in transgenic isoprene-free poplar leaves with isoprene synthase expression genetically knocked down (Peñuelas and Munné-Bosch 2005, Behnke et al. 2007). High temperatures lead to increased membrane permeability and decreased photosynthetic process efficiency. By inhibiting the formation of water channels through the separation of the thylakoid lipid bilayer, isoprene may be capable to restrict membrane leakage at high temperatures (Sharkey et al. 2001). Several studied reported that at elevated temperature, isoprene-emitting leaves cells sustain the same elasticity and fluidity than non-emitting leaves cells which become

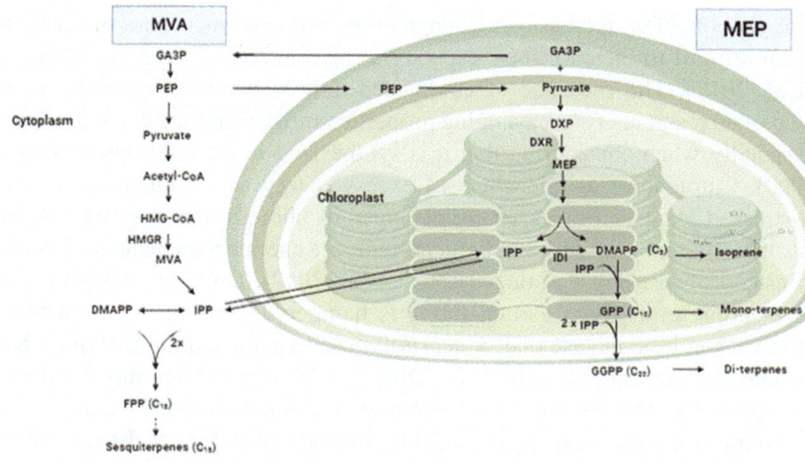


**Fig. 4.4** Two-linked isoprenoid biosynthetic pathway. DXP, 1-deoxy-D-xylulose 5-phosphate; GA3P, glyceraldehyde 3-phosphate; HMG-CoA, 3-hydroxy-3- methylglutaryl-CoA; HMGR, HMG-CoA reductase; PEP, phosphoenolpyruvate; FPP, farnesyl diphosphate; IPP, isopentyl pyrophosphate; DMAPP, dimethylallyl diphosphate; DXR, 1-deoxy-D-xylulose 5-phosphate reductoisomerase; GPP, geranyl diphosphate; GGPP, geranylgeranyl diphosphate

rigid and forfeit fluidity (Pollastri et al. 2019; Velikova et al. 2011). Conjugated double bonds (delocalized p-electrons) facilitated in isoprene transfer easily electron and energy to ROS that mediated stabilization of thylakoid membrane (Velikova et al. 2011). Isoprene stabilizes thylakoid membranes, which indirectly lowers ROS and reactive nitrogen species (RNS) production and improves in electron transport during extreme conditions in stressed leaves (Velikova et al. 2004, 2012). During $O_3$ fumigation, isoprene-emitting plants had significantly lower lipid peroxidation levels by scavenging of $H_2O_2$ (Vickers et al. 2009). Isoprene-emitting plants have chloroplast thylakoid membranes which are more resistant to denaturation, sustaining homeostatic of molecular diffusion and electron transport rate (Pollastri et al. 2021). The primary supporting evidence for this is that, especially in conditions that typically restrict photosynthesis, the non-photochemical quenching of chlorophyll fluorescence (NPQ), which determines the excess light dissipated as heat, increases less in isoprene-emitting leaves than non-isoprene emitting leaves, because isoprene scavenges ROS before they reach the membranes when there is stress, thylakoids are less challenged, which could explain whether NPQ is able to maintain its stability. However, photosynthesis is also at its peak, and NPQ is lower in isoprene-emitting plants than in non-emitting plants (Loreto and Fares 2007; Pollastri et al. 2014). Photosynthesis decreased in both isoprene-emitting and non-isoprene emitting plants during $O_3$ stress (Vickers et al. 2009). Study reported during stress condition isoprene act as hormone like molecules which can play role as a defence priming signal at the gene level, maintain transcripts and proteins of the antioxidant pool of the plant and regulated the growth and development (Zuo et al. 2019). Defence priming, a biological phenomenon, increases the adaptability and responsiveness of plants to stress, permitting it to increase their capacity to resistance and tolerance (Mauch-Mani et al. 2017). The presence of conjugated double bonds in isoprene, as in carotenoids, can result in $O_2\cdot^-$ scavenging properties and appropriate defensive functions, especially when considering that it is synthesized in the chloroplast (Logan et al. 2000). Carotenoids as quenchers of excess energy act as important role in drought-stressed plants suffering severe damaged of photosynthetic system. Additionally, some carotenoids, particularly b-carotene, are chemical quenchers of singlet oxygen instead of just physical ones (Ramel et al. 2012). Drought-stressed plants generate isoprene even when carbon fixing is hampered, leading to a significant increase in carbon loss, in case of *P.* x *acerifolia* used significant amount of carbon into isoprene biosynthesis (Tattini et al. 2014). So, several studies supported that volatile isoprenoids showed similar defence role as compare to others antioxidants under the oxidative stress (Loreto et al. 2004; Peñuelas and Munné-Bosch 2005). The several experimental result suggests that volatile isoprenoids provide additional plant protection that may be utilized in combination with carotenoids and tocopherols or serve as an alternative; however, the precise biophysical and biochemical nature of the process has yet to be identified (Peñuelas and Munné-Bosch 2005). Vickers et al. 2009) proposed a model describing a "unified biochemical process for diverse physiological stressors" which facilitates the synthesis of diverse experimental observations regarding isoprenoids and their role in conferring tolerance to environmental stresses (Fig. 4.5).

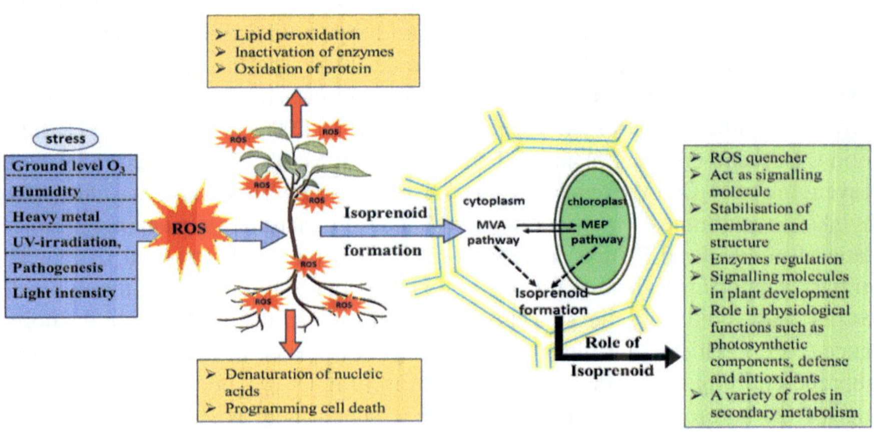


**Fig. 4.5** Role of isoprenoids in plants facing various oxidative stresses

## 4.6 Mode of Action

The mechanism of how volatile isoprenoids function in plants can be challenging to study directly in planta (in the plant) because current techniques often cannot distinguish between cause and effect. However, their activity is believed to be mediated through several potential mechanisms:

1. **Direct Reactions with Oxidizing Species**: Volatile isoprenoids may directly react with oxidizing species, such as reactive oxygen species (ROS). This means that isoprenoids might chemically neutralize or scavenge these harmful oxidizing species, reducing their damaging effects on cellular components.
2. **Indirect Alteration of ROS Signalling**: Isoprenoids could indirectly affect ROS signalling within plant cells. Instead of directly interacting with ROS, they may influence the pathways that regulate ROS production or response. This indirect modulation of ROS signalling can have cascading effects on the plant's ability to manage oxidative stress.
3. **Membrane Stabilization**: Volatile isoprenoids may play a role in stabilizing lipid membranes. By doing so, they can help prevent the peroxidation of lipids, which is a process where lipids are damaged by ROS. Stabilizing membranes can directly impact the oxidative state of plant cells, enhancing their resilience to oxidative stress.

The concept of membrane stabilization is particularly interesting because it suggests that isoprenoids might offer a general, non-specific form of oxidative protection. This protection may not solely rely on direct chemical reactions with ROS but rather on the structural reinforcement of cell membranes, making them less susceptible to damage. To gain a deeper understanding of the antioxidant behaviour of isoprene and other volatile isoprenoids, researchers may explore specific reaction products that result from the oxidation of isoprenoids. Identifying these reaction products could provide insights into the chemical processes by which isoprenoids interact with ROS or other oxidizing species.

In summary, while the precise mechanisms by which volatile isoprenoids protect plants from oxidative stress are complex and challenging to directly observe, they likely involve a combination of direct reactions with oxidizing species, indirect modulation of ROS signalling pathways, and the stabilization of lipid membranes. Further research into the specific chemical interactions and products associated with isoprenoid oxidation can help shed light on these mechanisms and their importance in plant defence against oxidative damage.

### *4.6.1 Membrane Stabilization*

It was initially postulated that the effect of isoprene occurs through stabilizing membranes. Since it is synthesized in the chloroplast and exhibits lipophilic characteristics, isoprene is expected to partition into the lipid phases of thylakoid membranes. Heat stress causes membranes to become more permeable, which in turn reduces the effectiveness of membrane-associated activities like photosynthesis. Therefore, it was hypothesized that the protective effect operates through the physical stabilization of hydrophobic contacts (between lipids, proteins, and/or proteins and lipids) when temperatures rise. Isoprene does, in fact, partition towards the centre of phospholipid membranes, as predicted by molecular dynamics simulations (Siwko et al. 2007). As a result, membrane order is improved without substantially altering the membrane's dynamic qualities. The presence of isoprene aids photosynthetic activities under heat stress, and decreased rates of electron transfer are found in photosynthetic membranes following suppression by fosmidomycin (Velikova et al. 2005a, b). The hydrophobicity of most volatile isoprenoids suggests that this process may be applicable to a wide variety of isoprenoids. It is challenging, however, to verify this mechanistic idea directly in an in vivo system.

Despite this circumstantial evidence in favour of a role for membrane stabilization in the isoprene effect, this explanation still faces certain physicochemical hurdles. Isoprene, being a hydrocarbon, is very hydrophobic; it is almost insoluble in pure water, and the Henry's Law constant has been calculated at KH B 3.1 (Milne et al. 1995), making it almost insoluble in saltwater as well. As the location of isoprene formation is the chloroplast, it seems reasonable to suppose that, even at modest emission rates, lipid membranes (especially chloroplast membranes) must be saturated with isoprene by the time isoprene is detected at the leaf surface. Once membranes are saturated at a certain temperature, every further increase in isoprene synthesis must increase the rate at which isoprene travels through the membrane. Since it has been observed that isoprene fumigant does not provide any further protection, it follows that membranes are always isoprene-saturated in emitting species (Logan et al. 1999). Interestingly, separated membranes do not benefit from exogenously provided isoprene under heat stress, perhaps because they do not produce their own isoprene (Logan et al. 1999). Contrarily, isoprene fumigation may partly counteract the effects of fosmidomycin-induced MEP pathway suppression on photosynthesis when temperatures are high (Sharkey et al. 2008a, b; Velikova et al. 2006). These seemingly conflicting findings may be explained, at least in part, by the fact that isoprene seems to have different effects on isolated membranes than it does on whole plant membranes. Other metabolites from the MEP route (suggested by partial complementation) may also aid in protecting photosynthesis from heat stress. The isoprene effect should be largely species-independent if it were just a physical stabilization of membranes. In species that do not naturally release isoprene, however, fumigation with isoprene has been shown to provide protection in some situations but not in others (Logan et al. 1999; Velikova et al. 2005a, b; Sasaki et al. 2007; Sharkey et al. 2008a, b). This variation may result from species-specific changes in membrane characteristics, although these differences have yet to be uncovered.

### 4.6.2 *Antioxidant Hypothesis*

Isoprene plays a role in mitigating the effects of ROS by engaging in processes that utilize its conjugated double-bond structure (Loreto et al. 2001; Loreto et al. 2001; Affek and Yakir 2002; Velikova et al. 2004). This implies an alternative mechanistic explanation for its function. Isoprenoids with a high degree of reduction in the environment can react with ROS like $O_3$, singlet oxygen, nitrogen oxides ($NO_x$), and hydroxyl radicals (OH). These reactions can occur within the cells, particularly on membranes, or in the moist surroundings outside the cells, including the gaps between cells and the leaf surface. Reactions involving the lipid phase of a leaf are also possible.

When plants undergo stress, they generate various ROS, which are known as oxidizing agents. The growth of plants can be influenced by their antioxidant defence mechanisms, which protect cells from excessive ROS. These defences can include both lipophilic (fat-soluble) and hydrophilic (water-soluble) antioxidant compounds. Examples of antioxidants include tocopherols, carotenoids, and enzymes like glutathione reductase and dehydroascorbate reductase. These enzymes help convert highly reactive ROS into less harmful substances or water, such as SOD, peroxidase, and catalase which are found in chloroplasts and contribute to the regulation of ROS levels in lipid and water phases. Hydrogen peroxide is produced during the detoxification of more harmful ROS and plays a role in stress responses.

Plant cells maintain a steady supply of ROS even under normal conditions, and ROS serve important functions in various cellular responses. The production of ROS in the photosynthetic process and its signalling in membranes are essential for photosynthesis. Proper regulation of ROS production is crucial in normal circumstances, as imbalances in ROS levels can disrupt cellular processes, requiring a significant amount of antioxidants and antioxidant enzymes to scavenge excess ROS and protect cells from damage. Tocopherols, zeaxanthin, and lutein are examples of non-volatile isoprenoids that play a role in this non-enzymatic antioxidant defence system (Munne-Bosch and Alegre 2002). Additionally, carnosic acid may indirectly scavenge ROS through hydroxyl-dependent processes. Volatile isoprenoids typically contain reactive double bonds, either terminal or conjugated, and they are suggested to be part of the body's antioxidant defence system.

## 4.7 Cooperation of Isoprene and NO in Reducing Oxidative Stress

Nitric oxide (NO) and isoprene are two volatile substances generated in the leaves. It has been suggested that both substances play a significant protective role against oxidative stress (Velikova et al. 2008). NO has been recognized as an important bioactive molecule and generated by plants through the process of nitrogen metabolism, non-enzymatic processes and although it is frequently regulated by the enzyme nitrate reductase. NO participates in numerous physiological processes, including serving as a plant growth regulator, modulating reproductive functions, influencing respiratory metabolism, regulating senescence, and mediating stress responses. Initially characterized in mammals, NO also functions as a signalling molecule (Hari et al. 2003; Schmidt and Walter 1994; Hiscock et al. 2007; Wendehenne et al. 2004). Several studied reported the interactive role of isoprene and NO in plant protection by alteration of physiological and biochemical parameters. The RB (4,5,6,7-tetrachloro-2′,4′,5′,7′-tetraiodofluorescein) treatment caused to emit much more NO in poplar leaves. In isoprene-emitting leaves treated with RB, the rate of NO emission was approximately six times higher. In fosmidomycin-fed, isoprene-inhibited leaves under RB treatment, produced maximum NO. Thus, NO and isoprene appear to work simultaneously to reduce the adverse impacts of oxidative stress on the photosynthetic machinery. $O_3$ stress induced the production of free radical which increased the level of MDA, which was more significant in isoprene-inhibited leaves. However, with NO reduced free radical and MDA level in both poplar leaves and reed leaves (Velikova et al. 2008; Velikova et al. 2005a, b). The interaction between isoprene and NO influences the advancement of signalling pathways, such as those involving MAPK (mitogen-activated protein kinase) signalling pathway and the stress-related hormone signalling pathway (Jalil and Ansari 2020). Loreto et al. (2001) observed that isoprene secretion reduced leaf necrosis under $O_3$ stress. A unique signalling mechanism drives hypersensitivity and cellular death in response to several oxidative stresses. This involvement producing NO and interacting with rising ROS levels. Interestingly, cellular death does not occur at NO and $H_2O_2$. In fact, a minor rise in NO can activate genes signalling and produce metabolites for plant defence system (Mishra et al, 2025b; Pasqualini et al. 2003; Delledonne et al. 2001) in presence of NO, isoprene induced the $O_3$ formation. Velikova et al. (2005a, b) reported their findings that under $O_3$ stress isoprene-emitting leaves experience less $H_2O_2$ formation, antioxidant enzyme activation, and membrane peroxidation than non-isoprene-emitting leaves and also provided the idea about isoprene mediates NO to improve leaf resistance to $O_3$. Isoprene reduces oxidative stress at the primary metabolism level, which may be an indirect cause of the decreased NO production. The efficient use of photochemical energy for photosynthesis results in less ROS-producing photoreduction of oxygen in isoprene-emitting leaves compared to isoprene-inhibited leaves. However, isoprene could also have effectively regulated NO levels in the leaf, probably reacting with NO radicals. By preventing the build-up of NO and of its reactive products to toxic concentration, isoprene might modulate the signalling pathway activated by NO-ROS interactions and ultimately contribute to cellular death (Velikova et al. 2005a, b). Combining isoprene and NO prevents plants against oxidative stress more effectively than releasing simply one component (Velikova et al. 2008).

## 4.8 The Evolution of Stress Tolerance Function: From Bryophytes to Higher Plants

The plant kingdom has undergone remarkable adaptations throughout its evolutionary history to cope with various environmental stresses (Negin and Moshelion 2016). From the early emergence of bryophytes to the complexity of higher plants, the evolution of stress tolerance mechanisms has played a crucial role in enabling these organisms to thrive in diverse habitats (Rensing 2018). This short note explores the evolutionary trajectory of stress tolerance functions, highlighting key adaptations and molecular mechanisms that have emerged over time.

1. *Bryophytes—Pioneers of Stress Tolerance*

   Bryophytes, including mosses, liverworts, and hornworts, represent some of the earliest land plants (Shaw et al. 2011). As pioneers in colonizing terrestrial habitats, they faced numerous challenges, particularly dehydration stress due to their lack of well-developed vascular systems. Bryophytes evolved strategies such as desiccation tolerance, the ability to withstand complete water loss, and anhydrobiosis, a state of suspended animation under extreme drought conditions. These adaptations involved the accumulation of protective solutes and the expression of unique stress-responsive genes.
2. *Vascular Plant Evolution—Adaptations in Abundance*

   The evolution of vascular plants marked a significant step in the plant kingdom. With the development of vascular tissues, xylem, and phloem, higher plants gained the ability to transport water, nutrients, and signalling molecules more efficiently (Lucas et al. 2013). This transition also paved the way for more complex stress tolerance mechanisms. Cuticle formation, stomatal regulation, and the synthesis of osmoprotectants became essential strategies to combat water scarcity and excessive transpiration. Additionally, the evolution of root systems allowed for better water and nutrient uptake, enhancing overall stress resilience.
3. *Hormonal Signalling and Stress Responses*

   As plants continued to diversify, the emergence of stress-responsive hormonal signalling pathways became pivotal. Abscisic acid (ABA), a key stress hormone, evolved as a central regulator of various stress responses. ABA-mediated stomatal closure, triggered by water deficit, helps plants conserve water while maintaining gas exchange (Hsu et al. 2021). Similarly, the role of jasmonic acid (JA) and salicylic acid (SA) in defence against herbivores and pathogens, respectively, became essential elements of stress tolerance mechanisms in advanced plants (Luo et al. 2019).
4. *Genetic and Molecular Adaptations*

   The evolution of stress tolerance in plants involves intricate genetic and molecular changes (Gupta and Huang 2014). The expansion of gene families, such as those encoding heat shock proteins (HSPs) and late embryogenesis abundant (LEA) proteins, facilitated enhanced stress protein production and protein stabilization under adverse conditions (Priya et al. 2019). The evolution of transcription factors, such as DREB and WRKY, enabled the fine-tuning of stress-responsive gene expression. These genetic adaptations contributed to the versatility of stress tolerance mechanisms across different plant species.

5. *Evolution of Photosynthesis Under Stress*

   Photosynthesis, a fundamental process in plants, has also evolved to cope with stress. The emergence of C4 and CAM photosynthesis in response to water and heat stress allowed certain plant species to optimize carbon fixation while minimizing water loss (Shah et al. 2022). These pathways exemplify how plants have diversified their metabolic strategies to adapt to specific stressors.

6. *Modern Agricultural and Biotechnological Implications*

   Understanding the evolutionary journey of stress tolerance mechanisms in plants has significant implications for modern agriculture and biotechnology. Genetic engineering techniques allow for the incorporation of stress-responsive genes into crop plants, enhancing their ability to withstand environmental stresses such as salinity, drought, and temperature extremes (Munaweera et al. 2022). By tapping into the genetic reservoir of stress adaptations accumulated over millions of years, scientists can develop resilient and productive crop varieties to ensure food security in a changing climate.

The evolution of stress tolerance functions in plants, from the simple bryophytes to the intricate mechanisms of higher plants, showcases the remarkable capacity of these organisms to adapt to diverse and often harsh environments. Through a combination of physiological, genetic, and molecular adaptations, plants have developed an array of strategies to withstand stressors. This journey of evolution continues to influence modern agricultural practices and holds the promise of sustainable food production in an increasingly challenging world.

## 4.9 Conclusion

The intricate interplay between emissions of volatile organic carbon (VOC) and the heightened stress of tropospheric $O_3$ presents a multifaceted and pressing challenge for the cultivation of crops. As the global population continues its expansion, the imperative of ensuring food security becomes increasingly pronounced. However, the adverse effects of elevated $O_3$ levels on crop productivity cannot be underestimated. $O_3$-induced stress translates into reduced crop yields, diminished crop quality, and financial losses for farmers. It extends its repercussions to global food security and the sustainability of agricultural practices.

It is paramount to grasp the intricate relationship between VOC emissions and $O_3$ stress to formulate effective strategies that mitigate these deleterious consequences. VOCs, which encompass compounds such as isoprene and other biogenic substances released by plants, are far from passive substances merely released into the atmosphere. They assume dynamic roles in shaping plant physiology and environmental interactions. On one hand, VOCs have demonstrated their protective capabilities, aiding plants in managing the oxidative stress induced by $O_3$ exposure. Conversely, they can also contribute to the creation of ground-level $O_3$, thereby compounding the problem.

Isoprene, as one of the major biogenic volatile organic compounds (BVOCs) emitted by plants, plays a pivotal role in shaping atmospheric chemistry. Its selective emission by "emitter" species suggests that it may be a key player in regulating plant responses to various stresses, both biotic and abiotic. Moreover, isoprene's potential as a natural antioxidant highlights its capacity to stabilize plant cell membranes and protect against oxidative stress induced by factors like $O_3$ and high temperatures. This dual role as both a protective mechanism and a plant regulator underscores the intricate web of relationships within ecosystems and the complex nature of plant physiology. Addressing this intricate predicament necessitates a multidisciplinary approach. Collaborative efforts among agronomists, atmospheric scientists, ecologists, and policymakers are indispensable for the development of strategies to alleviate $O_3$-induced stress on crops. These strategies may encompass the development of crop varieties that tolerate $O_3$, enhanced monitoring systems with early warning capabilities, and the reduction of human-generated emissions contributing to $O_3$ formation. The intersection of VOC emissions and the heightened stress of tropospheric $O_3$ in crop production is a pivotal area of study with extensive ramifications. By acquiring a more profound comprehension of these interactions and implementing efficacious mitigation measures, we can progress towards the safeguarding of crop yields, the assurance of food security, and the advancement of sustainable agricultural practices in an ever-changing world.

## References

Affek HP, Yakir D (2002) Protection by isoprene against singlet oxygen in leaves. Plant Physiol 129(1):269–277

Ansari N, Agrawal M, Agrawal SB (2021) An assessment of growth, floral morphology, and metabolites of a medicinal plant *Sida cordifolia* L. under the influence of elevated $O_3$. Environ Sci Pollut Res 28:832–845

Arar M, Jung C (2021) Improving the indoor air quality in nursery buildings in United Arab Emirates. Int J Environ Res Public Health 18(22):12091

Asensi-Fabado MA, Amtmann A, Perrella G (2017) Plant responses to abiotic stress: the chromatin context of transcriptional regulation. Biochimica et Biophysica Acta (BBA)-Gene Regulatory Mechanisms 1860(1):106–122

Banerjee A, Roychoudhury A (2017) Abiotic stress, generation of reactive oxygen species, and their consequences: an overview. In: Reactive oxygen species in plants: boon or bane-revisiting the role of ROS, pp 23–50

Bari MA, Kindzierski WB, Wheeler AJ, Héroux MÈ, Wallace LA (2015) Source apportionment of indoor and outdoor volatile organic compounds at homes in Edmonton, Canada. Build Environ 90:114–124

Begum BA, Paul SK, Hossain MD, Biswas SK, Hopke PK (2009) Indoor air pollution from particulate matter emissions in different households in rural areas of Bangladesh. Build Environ 44(5):898–903

Behnke K, Ehlting B, Teuber M, Bauerfeind M, Louis S, Hänsch R, Polle A, Bohlmann J, Schnitzler JP (2007) Transgenic, non-isoprene emitting poplars don't like it hot. Plant J 51(3):485–499

Brilli F, Barta C, Fortunati A, Lerdau M, Loreto F, Centritto M (2007) Response of isoprene emission and carbon metabolism to drought in white poplar (*Populus alba*) saplings. New Phytol 175(2):244–254

Brilli F, Gioli B, Zona D, Pallozzi E, Zenone T, Fratini G, Calfapietra C, Loreto F, Janssens IA, Ceulemans R (2014) Simultaneous leaf-and ecosystem-level fluxes of volatile organic compounds from a poplarbased SRC plantation. Agric for Meteorol 187:22–35

Brook RD, Rajagopalan S, Pope III CA, Brook JR, Bhatnagar A, Diez-Roux AV, Holguin F, Hong Y, Luepker RV, Mittleman MA, Peters A (2010) Particulate matter air pollution and cardiovascular disease: an update to the scientific statement from the American Heart Association. Circulation 121(21):2331–2378

Butler T, Lupascu A, Nalam A (2020) Attribution of ground-level $O_3$ to anthropogenic and natural sources of nitrogen oxides and reactive carbon in a global chemical transport model. Atmos Chem Phys 20(17):10707–10731

Calfapietra C, Fares S, Loreto F (2009) Volatile organic compounds from Italian vegetation and their interaction with $O_3$. Environ Pollut 157(5):1478–1486

Cappellin L, Loreto F, Biasioli F, Pastore P, McKinney K (2019) A mechanism for biogenic production and emission of MEK from MVK decoupled from isoprene biosynthesis. Atmos Chem Phys 19(5):3125–3135

Carpenter PC, Ciccioli P, Goldstein A, Hamilton JF, Hoffmann T, Lewis AC, Mannozzi M, O'Doherty S, Reimann S, Roberts J, Rudolph J (2007) Volatile organic compounds in the atmosphere. Blackwell, Oxford

Carter WP (1994) Development of $O_3$ reactivity scales for volatile organic compounds. Air Waste 44(7):881–899

Chameides WL (1989) The chemistry of $O_3$ deposition to plant leaves: role of ascorbic acid. Environ Sci Technol 23(5):595–600

Churkina G, Kuik F, Bonn B, Lauer A, Grote R, Tomiak K, Butler TM (2017) Effect of VOC emissions from vegetation on air quality in Berlin during a heatwave. Environ Sci Technol 51(11):6120–6130

Ciccioli P, Centritto M, Loreto F (2014) Biogenic volatile organic compound emissions from vegetation fires. Plant, Cell Environ 37(8):1810–1825

Coggon MM, Gkatzelis GI, McDonald BC, Gilman JB, Schwantes RH, Abuhassan N, Aikin KC, Arend MF, Berkoff TA, Brown SS, Campos TL (2021) Volatile chemical product emissions enhance $O_3$ and modulate urban chemistry. Proc Natl Acad Sci 118(32):e2026653118

Delledonne M, Zeier J, Marocco A, Lamb C (2001) Signal interactions between nitric oxide and reactive oxygen intermediates in the plant hypersensitive disease resistance response. Proc Natl Acad Sci 98(23):13454–13459

Deng C, Jin Y, Zhang M, Liu X, Yu Z (2018) Emission characteristics of VOCs from on-road vehicles in an urban tunnel in eastern China and predictions for 2017–2026. Aerosol Air Qual Res 18(12):3025–3034

Dovrou E, Bates KH, Rivera-Rios JC, Cox JL, Shutter JD, Keutsch FN (2021) Towards a chemical mechanism of the oxidation of aqueous sulfur dioxide via isoprene hydroxyl hydroperoxides (ISOPOOH). Atmos Chem Phys 21(11):8999–9008

Dubois C, Cholleton D, Gemayel R, Chen Y, Surratt JD, George C, Rairoux P, Miffre A, Riva M (2021) Decrease in sulfate aerosol light backscattering by reactive uptake of isoprene epoxydiols. Phys Chem Chem Phys 23(10):5927–5935

Dudareva N, Negre F, Nagegowda DA, Orlova I (2006) Plant volatiles: recent advances and future perspectives. Crit Rev Plant Sci 25(5):417–440

Duncan SM, Sexton KG, Turpin BJ (2018) Oxygenated VOC s, aqueous chemistry, and potential impacts on residential indoor air composition. Indoor Air 28(1):198–212

Dunker AM, Koo B, Yarwood G (2016) $O_3$ sensitivity to isoprene chemistry and emissions and anthropogenic emissions in central California. Atmos Environ 145:326–337

Ehn M, Thornton JA, Kleist E, Sipilä M, Junninen H, Pullinen I, Springer M, Rubach F, Tillmann R, Lee B, Lopez-Hilfiker F (2014) A large source of low-volatility secondary organic aerosol. Nature 506(7489):476–479

Fittschen C, Assaf E, Vereecken L (2017) Experimental and theoretical investigation of the reaction NO+ OH+ O2 → HO2+ NO2. J Phys Chem A 121(24):4652–4657

Fuchs H, Hofzumahaus A, Rohrer F, Bohn B, Brauers T, Dorn HP, Häseler R, Holland F, Kaminski M, Li X, Lu K (2013) Experimental evidence for efficient hydroxyl radical regeneration in isoprene oxidation. Nat Geosci 6(12):1023–1026

Fuentes JD, Lerdau M, Atkinson R, Baldocchi D, Bottenheim JW, Ciccioli P, Lamb B, Geron C, Gu L, Guenther A, Sharkey TD (2000) Biogenic hydrocarbons in the atmospheric boundary layer: a review. Bull Am Meteor Soc 81(7):1537–1575

Gautam AK, Gupta GS, Madheshiya P (2022) Role of biochemical and biophysical characteristics in management of o 3 injury in plants through soil nutrient amendments. Plant Arch 22(1):09725210

Gilma JB, Kuster WC, Goldan PD, Herndon SC, Zahniser MS, Tucker SC, Brewer WA, Lerner BM, Williams EJ, Harley RA, Fehsenfeld FC (2009) Measurements of volatile organic compounds during the 2006 TexAQS/GoMACCS campaign: industrial influences, regional characteristics, and diurnal dependencies of the OH reactivity. J Geophys Res Atmos 114(D7).

Goodger JQ, Senaratne SL, Nicolle D, Woodrow IE (2018) Differential metabolic specialization of foliar oil glands in Eucalyptus brevistylis Brooker (Myrtaceae). Tree Physiol 38(10):1451–1460

Greenberg JP, Asensio D, Turnipseed A, Guenther AB, Karl T, Gochis D (2012) Contribution of leaf and needle litter to whole ecosystem BVOC fluxes. Atmos Environ 59:302–311

Guenther AB, Jiang X, Heald CL, Sakulyanontvittaya T, Duhl TA, Emmons LK, Wang X (2012) The model of emissions of gases and aerosols from nature version 2.1 (MEGAN2. 1): an extended and updated framework for modeling biogenic emissions. Geosci Model Dev 5(6):1471–1492

Guenther A (2013) Biological and chemical diversity of biogenic volatile organic emissions into the atmosphere. International Scholarly Research Notices

Gupta B, Huang B (2014) Mechanism of salinity tolerance in plants: physiological, biochemical, and molecular characterization. Int J Genom

Halsey KH, Giovannoni SJ (2023) Biological controls on marine volatile organic compound emissions: a balancing act at the sea-air interface. Earth Sci Rev 240:104360

Hari P, Raivonen M, Vesala T, Munger JW, Pilegaard K, Kulmala M (2003) Ultraviolet light and leaf emission of $NO_x$. Nature 422(6928):134–134

Hiscock S, Bright J, McInnis SM, Desikan R, Hancock JT (2007) Signaling on the stigma: potential new roles for ROS and NO in plant cell signaling. Plant Signal Behav 2(1):23–24

Hoshino Y, Gaucher EA (2018) On the origin of isoprenoid biosynthesis. Mol Biol Evol 35(9):2185–2197

Hsu PK, Dubeaux G, Takahashi Y, Schroeder JI (2021) Signaling mechanisms in abscisic acid-mediated stomatal closure. Plant J 105(2):307–321

Jackson R, Gabric A (2022) Climate change impacts on the marine cycling of biogenic sulfur: a review. Microorganisms 10(8):1581

Jacob DJ, Heikes EG, Fan SM, Logan JA, Mauzerall DL, Bradshaw JD, Singh HB, Gregory GL, Talbot RW, Blake DR, Sachse GW (1996) Origin of $O_3$ and $NO_x$ in the tropical troposphere: a photochemical analysis of aircraft observations over the South Atlantic basin. J Geophys Res Atmos 101(D19):24235–24250

Jacobs MI, Darer AI, Elrod MJ (2013) Rate constants and products of the OH reaction with isoprene-derived epoxides. Environ Sci Technol 47(22):12868–12876

Jalil SU, Ahmad I, Ansari MI (2017) Functional loss of GABA transaminase (GABA-T) expressed early leaf senescence under various stress conditions in *Arabidopsis thaliana.* Curr Plant Biol 9:11–22

Jalil SU, Ansari MI (2020) Isoprenoids in plant protection against abiotic stress. In: Protective chemical agents in the amelioration of plant abiotic stress: biochemical and molecular perspectives, pp 424–436

Jardine KJ, Monson RK, Abrell L, Saleska SR, Arneth A, Jardine A, Ishida FY, Serrano AMY, Artaxo P, Karl T, Fares S (2012) Within-plant isoprene oxidation confirmed by direct emissions of oxidation products methyl vinyl ketone and methacrolein. Glob Change Biol 18(3):973–984

Jud W, Fischer L, Canaval E, Wohlfahrt G, Tissier A, Hansel A (2016) Plant surface reactions: an opportunistic $O_3$ defence mechanism impacting atmospheric chemistry. Atmos Chem Phys 16(1):277–292

Kanakidou M, Sfakianaki M, Probst A (2022) Impact of air pollution on terrestrial ecosystems. Atmospheric chemistry in the mediterranean region: volume 2-from air pollutant sources to impacts. Springer, Cham, pp 511–542

Keng FSL, Phang SM, Abd Rahman N, Leedham Elvidge EC, Malin G, Sturges WT (2020) The emission of volatile halocarbons by seaweeds and their response towards environmental changes. J Appl Phycol 32:1377–1394

Lange BM, Rujan T, Martin W, Croteau R (2000) Isoprenoid biosynthesis: the evolution of two ancient and distinct pathways across genomes. Proc Natl Acad Sci 97(24):13172–13177

Li S, Tosens T, Harley PC, Jiang Y, Kanagendran A, Grosberg M, Jaamets K, Niinemets Ü (2018) Glandular trichomes as a barrier against atmospheric oxidative stress: relationships with $O_3$ uptake, leaf damage, and emission of LOX products across a diverse set of species. Plant, Cell Environ 41(6):1263–1277

Ling ZH, Guo H, Lam SHM, Saunders SM, Wang TJJoGRA (2014) Atmospheric photochemical reactivity and $O_3$ production at two sites in Hong Kong: application of a master chemical mechanism–photochemical box model. J Geophys Res Atmos 119(17):10567–10582

Liu Y, Monod A, Tritscher T, Praplan AP, DeCarlo PF, Temime-Roussel B, Quivet E, Marchand N, Dommen J, Baltensperger U (2012) Aqueous phase processing of secondary organic aerosol from isoprene photooxidation. Atmos Chem Phys 12(13):5879–5895

Logan BA, Anchordoquy TJ, Monson RK, Pan RS (1999) The effect of isoprene on the properties of spinach thylakoids and phosphatidyicholine liposomes. Plant Biol 1(6):602–606

Logan BA, Monson RK, Potosnak MJ (2000) Biochemistry and physiology of foliar isoprene production. Trends Plant Sci 5(11):477–481

Loreto F, Fares S (2007) Is $O_3$ flux inside leaves only a damage indicator? Clues from volatile isoprenoid studies. Plant Physiol 143(3):1096–1100

Loreto F, Fineschi S (2015) Reconciling functions and evolution of isoprene emission in higher plants. New Phytol 206(2):578–582

Loreto F, Schnitzler JP (2010) Abiotic stresses and induced BVOCs. Trends Plant Sci 15(3):154–166

Loreto F, Mannozzi M, Maris C, Nascetti P, Ferranti F, Pasqualini S (2001) $O_3$ quenching properties of isoprene and its antioxidant role in leaves. Plant Physiol 126(3):993–1000

Loreto F, Pinelli P, Manes F, Kollist H (2004) Impact of O3 on monoterpene emissions and evidence for an isoprene-like antioxidant action of monoterpenes emitted by *Quercus ilex* leaves. Tree Physiol 24(4):361–367

Lu X, Zhang L, Shen L (2019) Meteorology and climate influences on tropospheric $O_3$: a review of natural sources, chemistry, and transport patterns. Curr Pollut Rep 5:238–260

Lucas WJ, Groover A, Lichtenberger R, Furuta K, Yadav SR, Helariutta Y, He XQ, Fukuda H, Kang J, Brady SM, Patrick JW (2013) The plant vascular system: evolution, development and functions f. J Integr Plant Biol 55(4):294–388

Luo J, Xia W, Cao P, Xiao ZA, Zhang Y, Liu M, Zhan C, Wang N (2019) Integrated transcriptome analysis reveals plant hormones jasmonic acid and salicylic acid coordinate growth and defense responses upon fungal infection in poplar. Biomolecules 9(1):12

Ma M, Gao Y, Wang Y, Zhang S, Leung LR, Liu C, Wang S, Zhao B, Chang X, Su H, Zhang T (2019) Substantial $O_3$ enhancement over the North China Plain from increased biogenic emissions due to heat waves and land cover in summer 2017. Atmos Chem Phys 19(19):12195–12207

Madheshiya P, Gupta GS, Sahoo A, Tiwari S (2023) Role of elevated $O_3$ on development and metabolite contents of lemongrass [*Cymbopogon flexuosus* (Steud.) (Wats.)] Metabolites 13(5):597

Manisalidis I, Stavropoulou E, Stavropoulos A, Bezirtzoglou E (2020) Environmental and health impacts of air pollution: a review. Front Public Health 8:14

Mauch-Mani B, Baccelli I, Luna E, Flors V (2017) Defense priming: an adaptive part of induced resistance. Annu Rev Plant Biol 68:485–512

Milne PJ, Riemer DD, Zika RG, Brand LE (1995) Measurement of vertical distribution of isoprene in surface seawater, its chemical fate, and its emission from several phytoplankton monocultures. Mar Chem 48(3–4):237–244

Mishra AK, Agrawal SB, Tiwari S (2025a) Unveiling oxidative stress: surface ozone triggers phenylpropanoid pathway shifts and metabolite rewiring in PBW-550 wheat. Plant Cell Rep 44(8):185

Mishra AK, Gupta S, Tiwari S (2025b) Differential antioxidant and metabolic responses to ozone stress in ozone-sensitive and tolerant wheat cultivars treated with nitric oxide donor. Sci Total Environ 1002:180588

Mishra AK, Gupta GS, Singh AA, Agrawal SB, Tiwari S (2024a) Can fertilization of $CO_2$ heal the ozone-injured agroecosystems? Atmos Pollut Res 102046

Mishra AK, Gupta GS, Agrawal SB, Tiwari S (2024b) Divergent responses of ascorbate and glutathione pools in ozone-sensitive and ozone-tolerant wheat cultivars under elevated ozone and carbon dioxide interaction. J Hazard Mater 472:134453

Munaweera TIK, Jayawardana NU, Rajaratnam R, Dissanayake N (2022) Modern plant biotechnology as a strategy in addressing climate change and attaining food security. Agric Food Secur 11(1):1–28

Munné-Bosch S, Alegre L (2002) The function of tocopherols and tocotrienols in plants. Crit Rev Plant Sci 21(1):31–57

Negin B, Moshelion M (2016) The evolution of the role of ABA in the regulation of water-use efficiency: from biochemical mechanisms to stomatal conductance. Plant Sci 251:82–89

Pardo M, Offer S, Hartner E, Di Bucchianico S, Bisig C, Bauer S, Pantzke J, Zimmermann EJ, Cao X, Binder S, Kuhn E (2022) Exposure to naphthalene and β-pinene-derived secondary organic aerosol induced divergent changes in transcript levels of BEAS-2B cells. Environ Int 166:107366

Pasqualini S, Piccioni C, Reale L, Ederli L, Della Torre G, Ferranti F (2003) $O_3$-induced cell death in tobacco cultivar Bel W3 plants. The role of programmed cell death in lesion formation. Plant Physiol 133(3):1122–1134

Paulot F, Henze DK, Wennberg PO (2012) Impact of the isoprene photochemical cascade on tropical $O_3$. Atmos Chem Phys 12(3):1307–1325

Peeters J, Müller JF, Stavrakou T, Nguyen VS (2014) Hydroxyl radical recycling in isoprene oxidation driven by hydrogen bonding and hydrogen tunneling: the upgraded LIM1 mechanism. J Phys Chem A 118(38):8625–8643

Peñuelas J, Munné-Bosch S (2005) Isoprenoids: an evolutionary pool for photoprotection. Trends Plant Sci 10(4):166–169

Pollastri S, Tsonev T, Loreto F (2014) Isoprene improves photochemical efficiency and enhances heat dissipation in plants at physiological temperatures. J Exp Bot 65(6):1565–1570

Pollastri S, Jorba I, Hawkins TJ, Llusià J, Michelozzi M, Navajas D, Peñuelas J, Hussey PJ, Knight MR, Loreto F (2019) Leaves of isoprene-emitting tobacco plants maintain PSII stability at high temperatures. New Phytol 223(3):1307–1318

Pollastri S, Baccelli I, Loreto F (2021) Isoprene: An antioxidant itself or a molecule with multiple regulatory functions in plants? Antioxidants 10(5):684

Priya M, Dhanker OP, Siddique KH, HanumanthaRao B, Nair RM, Pandey S, Singh S, Varshney RK, Prasad PV, Nayyar H (2019) Drought and heat stress-related proteins: an update about their functional relevance in imparting stress tolerance in agricultural crops. Theor Appl Genet 132:1607–1638

Ramel F, Birtic S, Cuiné S, Triantaphylidès C, Ravanat JL, Havaux M (2012) Chemical quenching of singlet oxygen by carotenoids in plants. Plant Physiol 158(3):1267–1278

Rensing SA (2018) Great moments in evolution: the conquest of land by plants. Curr Opin Plant Biol 42:49–54

Rivera-Rios JC, Nguyen TB, Crounse JD, Jud W, St. Clair JM, Mikoviny T, Gilman JB, Lerner BM, Kaiser JB, De Gouw J, Wisthaler A (2014) Conversion of hydroperoxides to carbonyls in field and laboratory instrumentation: Observational bias in diagnosing pristine versus anthropogenically controlled atmospheric chemistry. Geophys Res Lett 41(23):8645–8651

Roychoudhury A, Paul S, Basu S (2013) Cross-talk between abscisic acid-dependent and abscisic acid-independent pathways during abiotic stress. Plant Cell Rep 32:985–1006

Sahoo A (2023) Chapter fourteen isoprene emission in plants: a tool to handle $O_3$ stress ansuman sahoo and supriya tiwari

Sahoo A, Tiwari S (2021) Chapter Fourteen: Isoprene emission in plants: a tool to handle ozone stress

Sahu LK (2012) Volatile organic compounds and their measurements in the troposphere. Curr Sci 1645–1649

Salter L, Hewitt CN (1992) $O_3$-hydrocarbon interactions in plants. Phytochemistry 31(12):4045–4050

Sanadze GA (2004) Biogenic isoprene (a review). Russ J Plant Physiol 51:729–741

Sasaki K, Saito T, Lämsä M, Oksman-Caldentey KM, Suzuki M, Ohyama K, Muranaka T, Ohara K, Yazaki K (2007) Plants utilize isoprene emission as a thermotolerance mechanism. Plant Cell Physiol 48(9):1254–1262

Sauer F, Schäfer C, Neeb P, Horie O, Moortgat GK (1999) Formation of hydrogen peroxide in the ozonolysis of isoprene and simple alkenes under humid conditions. Atmos Environ 33(2):229–241

Schmidt HH, Walter U (1994) NO at work. Cell 78(6):919–925

Shah WH, Saleem S, Mushtaq NU, Rasool A, Tahir I, Rehman RU (2022) C4 and CAM plants with better resilience to environmental stresses. In: Photosynthesis and respiratory cycles during environmental stress response in plants. Apple Academic Press, pp 163–191

Sharkey TD, Yeh S (2001) Isoprene emission from plants. Annu Rev Plant Biol 52(1):407–436

Sharkey TD, Chen X, Yeh S (2001) Isoprene increases thermotolerance of fosmidomycin-fed leaves. Plant Physiol 125(4):2001–2006

Sharkey TD, Wiberley AE, Donohue AR (2008a) Isoprene emission from plants: why and how. Ann Bot 100:1–14

Sharkey TD, Wiberley AE, Donohue AR (2008b) Isoprene emission from plants: why and how. Ann Bot 101(1):5–18

Shaw AJ, Szövényi P, Shaw B (2011) Bryophyte diversity and evolution: windows into the early evolution of land plants. Am J Bot 98(3):352–369

Shrivastava M, Cappa CD, Fan J, Goldstein AH, Guenther AB, Jimenez JL, Kuang C, Laskin A, Martin ST, Ng NL, Petaja T (2017) Recent advances in understanding secondary organic aerosol: Implications for global climate forcing. Rev Geophys 55(2):509–559

Simon H, Reff A, Wells B, Xing J, Frank N (2015) $O_3$ trends across the United States over a period of decreasing $NO_x$ and VOC emissions. Environ Sci Technol 49(1):186–195

Siwko ME, Marrink SJ, de Vries AH, Kozubek A, Uiterkamp AJS, Mark AE (2007) Does isoprene protect plant membranes from thermal shock? A molecular dynamics study. Biochimica et Biophysica Acta (BBA)-Biomembranes 1768(2):198–206

Swan HB, Crough RW, Vaattovaara P, Jones GB, Deschaseaux ES, Eyre BD, Miljevic B, Ristovski ZD (2016) Dimethyl sulfide and other biogenic volatile organic compound emissions from branching coral and reef seawater: potential sources of secondary aerosol over the Great Barrier Reef. J Atmos Chem 73:303–328

Tai AP, Mickley LJ, Heald CL, Wu S (2013) Effect of CO2 inhibition on biogenic isoprene emission: Implications for air quality under 2000 to 2050 changes in climate, vegetation, and land use. Geophys Res Lett 40(13):3479–3483

Takshak S, Agrawal SB (2018) Interactive effects of supplemental ultraviolet-B radiation and indole-3-acetic acid on Coleus forskohlii Briq.: alterations in morphological-, physiological-, and biochemical characteristics and essential oil content. Ecotoxicol Environ Saf 147:313–326

Tan Z, Lu K, Jiang M, Su R, Wang H, Lou S, Fu Q, Zhai C, Tan Q, Yue D, Chen D (2019) Daytime atmospheric oxidation capacity in four Chinese megacities during the photochemically polluted season: a case study based on box model simulation. Atmos Chem Phys 19(6):3493–3513

Tassi F, Capecchiacci F, Giannini L, Vougioukalakis GE, Vaselli O (2013) Volatile organic compounds (VOCs) in air from Nisyros Island (Dodecanese Archipelago, Greece): natural versus anthropogenic sources. Environ Pollut 180:111–121

Tattini M, Velikova V, Vickers C, Brunetti C, Di Ferdinando M, Trivellini A, Fineschi S, Agati G, Ferrini F, Loreto F (2014) Isoprene production in transgenic tobacco alters isoprenoid, non-structural carbohydrate and phenylpropanoid metabolism, and protects photosynthesis from drought stress. Plant Cell Environ 37(8):1950–1964

Tissier A, Morgan JA, Dudareva N (2017) Plant volatiles: going 'in'but not 'out'of trichome cavities. Trends Plant Sci 22(11):930–938

Velikova V, Edreva A, Loreto F (2004) Endogenous isoprene protects *Phragmites australis* leaves against singlet oxygen. Physiol Plant 122(2):219–225

Velikova V, Pinelli P, Loreto F (2005a) Consequences of inhibition of isoprene synthesis in *Phragmites australis* leaves exposed to elevated temperatures. Agr Ecosyst Environ 106(2–3):209–217

Velikova V, Pinelli P, Pasqualini S, Reale L, Ferranti F, Loreto F (2005b) Isoprene decreases the concentration of nitric oxide in leaves exposed to elevated $O_3$. New Phytol 166(2):419–426

Velikova V, Loreto F, Tsonev T, Brilli F, Edreva A (2006) Isoprene prevents the negative consequences of high temperature stress in *Platanus orientalis* leaves. Funct Plant Biol 33(10):931–940

Velikova V, Fares S, Loreto F (2008) Isoprene and nitric oxide reduce damages in leaves exposed to oxidative stress. Plant Cell Environ 31(12):1882–1894

Velikova V, Várkonyi Z, Szabó M, Maslenkova L, Nogues I, Kovács L, Peeva V, Busheva M, Garab G, Sharkey TD, Loreto F (2011) Increased thermostability of thylakoid membranes in isoprene-emitting leaves probed with three biophysical techniques. Plant Physiol 157(2):905–916

Velikova V, Sharkey TD, Loreto F (2012) Stabilization of thylakoid membranes in isoprene-emitting plants reduces formation of reactive oxygen species. Plant Signal Behav 7(1):139–141

Vickers CE, Gershenzon J, Lerdau MT, Loreto F (2009) A unified mechanism of action for volatile isoprenoids in plant abiotic stress. Nat Chem Biol 5(5):283–291

Vranová E, Coman D, Gruissem W (2012) Structure and dynamics of the isoprenoid pathway network. Mol Plant 5(2):318–333

Wagner GJ (1991) Secreting glandular trichomes: more than just hairs. Plant Physiol 96(3):675–679

Wang W, van der AR, Ding J, van Weele M, Cheng T (2021) Spatial and temporal changes of the $O_3$ sensitivity in China based on satellite and ground-based observations. Atmos Chem Phys 21(9):7253–7269

Wedow JM, Ainsworth EA, Li S (2021) Plant biochemistry influences tropospheric $O_3$ formation, destruction, deposition, and response. Trends Biochem Sci 46(12):992–1002

Wendehenne D, Durner J, Klessig DF (2004) Nitric oxide: a new player in plant signalling and defence responses. Curr Opin Plant Biol 7(4):449–455

Wennberg PO, Bates KH, Crounse JD, Dodson LG, McVay RC, Mertens LA, Nguyen TB, Praske E, Schwantes RH, Smarte MD, St Clair JM (2018) Gas-phase reactions of isoprene and its major oxidation products. Chem Rev 118(7):3337–3390

Wilson SR, Madronich S, Longstreth JD, Solomon KR (2019) Interactive effects of changing stratospheric $O_3$ and climate on tropospheric composition and air quality, and the consequences for human and ecosystem health. Photochem Photobiol Sci 18(3):775–803

Wu S, Schalk M, Clark A, Miles RB, Coates R, Chappell J (2006) Redirection of cytosolic or plastidic isoprenoid precursors elevates terpene production in plants. Nat Biotechnol 24(11):1441–1447

Wu C, Wang C, Wang S, Wang W, Yuan B, Qi J, Wang B, Wang H, Wang C, Song W, Wang X (2020) Measurement report: Important contributions of oxygenated compounds to emissions and chemistry of volatile organic compounds in urban air. Atmos Chem Phys 20(23):14769–14785

Yan C, Nie W, Vogel AL, Dada L, Lehtipalo K, Stolzenburg D, Wagner R, Rissanen MP, Xiao M, Ahonen L, Fischer L (2020) Size-dependent influence of $NO_x$ on the growth rates of organic aerosol particles. Sci Adv 6(22):eaay4945

Yáñez-Serrano AM, Bourtsoukidis E, Alves EG, Bauwens M, Stavrakou T, Llusià J, Filella I, Guenther A, Williams J, Artaxo P, Sindelarova K (2020) Amazonian biogenic volatile organic compounds under global change. Glob Change Biol 26(9):4722–4751

Ye L, Wang X, Fan S, Chen W, Chang M, Zhou S, Wu Z, Fan Q (2016) Photochemical indicators of $O_3$ sensitivity: application in the Pearl River Delta, China. Front Environ Sci Eng 10:1–14

Zenone T, Hendriks C, Brilli F, Fransen E, Gioli B, Portillo-Estrada M, Schaap M, Ceulemans R (2016) Interaction between isoprene and $O_3$ fluxes in a poplar plantation and its impact on air quality at the European level. Sci Rep 6(1):32676

Zhang K, Huang L, Li Q, Huo J, Duan Y, Wang Y, Yaluk E, Wang Y, Fu Q, Li L (2021) Explicit modeling of isoprene chemical processing in polluted air masses in suburban areas of the Yangtze River Delta region: radical cycling and formation of $O_3$ and formaldehyde. Atmos Chem Phys 21(8):5905–5917

Zuo Z, Weraduwage SM, Lantz AT, Sanchez LM, Weise SE, Wang J, Childs KL, Sharkey TD (2019) Isoprene acts as a signaling molecule in gene networks important for stress responses and plant growth. Plant Physiol 180(1):124–152

# Chapter 5
# Combined Effect of UV Radiation and Tropospheric Ozone on Crop Physiology, Yield and Quality

**Aditya Abha Singh, Deepanshi Jaiswal, Kshama Rai, and S. B. Agrawal**

**Abstract** For the past few decades, ozone ($O_3$) in stratosphere as well as in troposphere has been identified as a global environmental concern. Stratospheric $O_3$ depletion by human-induced emissions of CFCs (chlorofluorocarbons), HCFCs, and halons leads to an enhancement in ultraviolet-B (UV-B) radiation on Earth's surface. These two abiotic stresses (UV-B and $O_3$) negatively impact the biochemistry and physiology of the plants. Besides individual, the collective UV-B and $O_3$ exposure induces oxidative stress in crops. The harmful UV radiation promotes the overproduction of ROS causing oxidative stress, degradation of photosynthetic pigments, and simultaneously affecting the working of photosynthetic efficiency and photosynthetic machinery. Researches have revealed the reduction in biomass and yield under $O_3$ stress in plants. Moreover, similar responses have been observed under $O_3$ stress too. In plants, the UV-B and $O_{3,}$ in combination, results in impaired plant growth, decline in biomass, reduction in gas exchange parameters, photosynthetic efficiency and stomatal conductance as well as reductions in yield and also alterations in nutritional quality of the produce. This chapter addresses the individual as well as the combined effects of UV-B and $O_3$ on biomass, physiology, yield, and seed quality chiefly on crop plants. Both stresses can individually exert negative impacts on plants and are generally milder compared to elevated levels of $O_{3,}$ and their combined effect is synergistic that can further exacerbate these negative impacts. Nevertheless, the mode of interaction of UV-B and $O_3$ can also be less than additive and the response can be more damaging under individual exposure of $O_{3.}$

**Keywords** Biomass · Physiology · Nutritional quality · Yield · Ozone · UV-B

A. A. Singh
Department of Botany, University of Lucknow, Lucknow, Uttar Pradesh, India

D. Jaiswal
Department of Botany, Dr. Harisingh Gour Vishwavidyalaya (A Central University), Sagar, M.P, India

K. Rai · S. B. Agrawal (✉)
Department of Botany, Institute of Science, Banaras Hindu University, Varanasi, Uttar Pradesh, India
e-mail: sbagrawal56@gmail.com

 
A. K. Pandey and E. Oksanen (eds.), *Tropospheric Ozone and Food Security*,
https://doi.org/10.1007/978-3-032-03608-7_5

## 5.1 Introduction

Depletion of $O_3$ layer have led to enhancement in the magnitude of biologically active UV-B radiations striking the Earth's surface (Laube et al. 2014). As abiotic stress, UV-B induces oxidative stress in plants, having a dual impact on their life by targeting various cellular components, including DNA, photosynthetic machinery, membranes, antioxidants, phytohormones, and secondary metabolites leading to both beneficial and harmful effects (Jansen et al. 1998; Takshak and Agrawal 2019). UV-B radiations (280–315 nm) are biologically active, with low doses found to be more effective in inducing positive changes in gene expression, physiology and metabolite accumulation (Heijde and Ulm 2012; Schreiner et al. 2012). While high UV doses have found to be detrimental for humans and plants. UV-B exposure affects plants system through two distinct mechanisms: either disrupting the physiological processes that impact growth and yield, or triggering molecular alterations resulting in to heritable modifications (Zlatev et al. 2012). Plant responses to UV-B radiation exhibit variability based on the UV treatment protocol, the level of the adaptation and acclimation to UV-B and interactions with other environmental factors.

Since industrial revolution, anthropogenic influences have resulted in an increase in tropospheric ozone ($O_3$), which is a secondary air pollutant created by the reactions between primary pollutants like volatile organic compounds, nitrogen dioxides, methane, and non-methane compounds under bright sunshine (Sampedro et al. 2020; Singh et al. 2023; Yuan et al. 2025). Being a greenhouse gas, through atmospheric chemical reactions, it has long-term influences on global climate change (Moore and MacCracken 2009). Worldwide ambient $O_3$ has enhanced twice the concentration since industrial revolution (Monks et al. 2015), and more specifically in the tropical regions of the world (Singh and Agrawal 2017). Ozone is a concern as it can severely affect the humans as well as plants (Agathokleous et al. 2020). Moreover, $O_3$ is equally hazardous to cultivated crops, trees, and semi-natural communities including grasslands (Agathokleous et al. 2015). The effects of elevated $O_3$ is multidimensional having effects on physiological, biochemical and molecular processes which influences the plant development, growth and yield (Frei et al. 2012; Malik et al. 2024). In vegetation, $O_3$ gain entry through the stomata during the gaseous exchange and creates enormous amount of reactive oxygen species (ROS) due to the chain reactions that deleteriously affect several biochemical, molecular, and physiological processes in plants (Foyer and Noctor 2005; Yadav et al. 2019; Zhang et al. 2019). Furthermore, it induces the anti-oxidative defense mechanism to counteract the oxidative stress caused due to $O_3$ resulting in biochemical tunings and fine metabolic adjustments, transcription of genes related to enzymatic or non-enzymatic antioxidants, secondary metabolites and heat shock proteins (Mittler 2002; Natali et al. 2018). The plants defense mechanism has enzymatic as well as non-enzymatic modules that help in overpowering the ROS created in apoplast and the symplast (Tiwari and Agrawal 2018). The low levels of ROS generated act as early messengers in signaling cascade pathways, producing a relay of downstream signaling and also inducing the defense responses (Vainonen and Kangasjärvi 2015), via the enzymatic antioxidants, including superoxide dismutase (SOD), ascorbate peroxidase (APX), catalase (CAT), glutathione reductase (GR), etc. (Blokhina et al. 2003) and non-enzymatic antioxidants such as thiols (protein and non-protein), ascorbic acid (AA), phenolics, flavonoids, carotenoids, polyamines, vitamin E ($\alpha$-tocopherol), etc. (Caregnato et al. 2013; Fatima et al. 2018). Ozone also interrupts the photosynthetic light reactions, chlorophyll fluorescence kinetics and the dark reactions of photosynthesis as well (Tiwari and Agrawal 2018). Furthermore, $O_3$ results in visible injury, and the ROS produced predominantly alters numerous physiological processes and metabolic pathways and simultaneously elicits the defense system, which put off the photosynthates away from growth and production causing variations in carbon allocation eventually declining the yield and quality (Ashmore 2005; Singh et al. 2015a, b; Ainsworth 2017). The combined effect of $O_3$ and ultraviolet-B (UV-B) radiation on crops is a critical concern for agricultural productivity, as these stressors can synergistically intensify damage to plant growth, photosynthetic efficiency, and overall yield (Mao et al. 2017; Tripathi et al. 2019). Ozone, a highly reactive oxidant, and UV-B radiation, which can damage DNA and disrupt cellular functions, together increase stress on plants beyond their individual impacts (Castagna and Ranieri 2009; Su et al. 2022). Research suggests that this combination can significantly reduce chlorophyll content, biomass, and seed nutritional quality, posing a threat to crop productivity and food security (Runeckles and Krupa 1994; Rao et al. 1996; Tripathi et al. 2019). Understanding these interactions is essential for developing strategies to mitigate their effects on crops. This chapter highlights the associated physiological, yield, and quality responses under $O_3$ and UV stress and together in their combination.

## 5.2 Crop Responses Towards UV Radiation

Due to their immobile nature, plants are inherently exposed to incident UV exposure. Plant expresses alterations in morphological, physiological, and biochemical processes and display genetic level modifications under different doses of UV exposures (Kumari et al. 2009a, b; Takshak and Agrawal 2019). Several studies have revealed that the increased UV radiation poses a significant threat to plant survival, considered as one of the greatest ecological concerns (Jovanić et al. 2022). Furthermore, these harmful UV radiation can trigger metabolic activities that promote the excess generation of reactive oxygen species (ROS) leading to oxidative stress, thereby unbalancing the normal functioning of the metabolic pathways that may ultimately lead to cell death (Anjum et al. 2015). These alterations affect the fundamental metabolic functioning such as photosynthetic machinery which is necessary for the food production and the multidimensional growth of the plants.

Studies suggest that small amount of UV radiation alters the developmental progression of photosynthetic ability in plants (Teramura and Caldwell 1981). Macromolecules (DNA and proteins) and lipid membranes are the preferred sensitive targets of UV-B in plant cells (Jansen et al. 1998). UV-B radiation triggers the activation of

UV RESISTANCE LOCUS 8 (UVR8) (UV-B-specific signalling component), a key regulator that controls the expression of multiple genes responsible for protecting plants against UV-B damages. Under the defense adaptation, plants stimulate the accumulation of flavonoids, phenols, alkaloids, and essential oil production (Kumari et al. 2009a). In crop plants, yield either in terms of biomass or number of economically important part is an important parameter to evaluate the growth under UV exposure. So, here in this section we discuss about the photosynthetic adaptations, growth in terms of biomass and yield and other qualitative parameters in crop plants under the UV-B radiations.

### *5.2.1 Physiological Responses*

Physiological processes serve as the key mechanisms that support and regulate plant growth. UV radiations are inherent and unavoidable component of sunlight so it's impossible to eradicate through secondary management. But excess UV-B stimulates oxidative stress and hinders the normal physiological processes such as photochemical efficiency of chlorophyll, transpiration, photosynthetic rate and stomatal conductance, etc. which ultimately impairs the growth and metabolic function. Besides, the disturbances in biochemical processes such as availability of low level of metabolites for the pigment formation, disruption of photosynthetic pigments and membrane proteins (D1/D2) also targets the physiological machinery in plants under UV exposure. Previous reports have suggested some targeted processes of photosynthesis, which are (1) impairment and uncoupling between PS-I (photo system) and PS-II, (2) impairment of electron pool between $Q_A$ (quinon) to $Q_B$. (3) reduced C-fixation and oxygen evolution, (4) ATP synthesis, (5) decline in RuBisCO activity, and (6) photosynthetic storage.

Previous studies have reported that photosynthetic processes are vulnerable to the harmful effects of UV-B radiation, as several components of photosynthetic apparatus were found to be affected, indicating that nuclear encoded genes showed more sensitivity than chloroplast encoded genes (Zlatev et al. 2012). Jovanić et al. (2022) found variable response of five different corn hybrids indicating differential response of photosynthetic machinery due the variation in relative composition of photosynthetic pigments, therefore two of them showed resistance against UV-B radiation. Ri et al. (2024) also observed reduction in photosynthetic activity of Schisandra chinensis after 30 days of UV exposure. A study by Yu et al. (2013) on super-high-yield hybrid rice (Liangyoupeijiu) indicated improved photosynthetic performance, characterized by elevated chlorophyll levels and enhanced electron transport rate (ETR) in photosystem II (PSII), but reduced $Ca^{2+}$-ATPase activity and phosphorylation. In this study, ultrastructure organization of chloroplast showed more compact arrangements of stromal thylakoids in UV-B treated plants, which might get involved in protecting chlorophyll and therefore PS II. Besides, reduced photosynthetic efficiency was found which was associated with the reduced activity of ribulose-1,5-bisphosphate carboxylase/oxygenase (Rubisco). It was found that prolonged exposure of UV-B led to decline in mRNA levels of Rubisco subunits, hence under the oxidative modifications large subunit (LSU, 54 KDa) could be changed into 66 kDa protein. Reduction in photosynthetic machinery was observed in both long and short term UV-B exposures in *Vitis vinifera* at 1 and 10 days after fruit set and also at the maturity of fruits, however, recovery was only found at veraison (onset of grape's ripening) (Martínez-Lüscher et al. 2013). This transient recovery at veraison possibly is associated with increase in UV shielding compounds and flavonols which serves as ROS scavengers. UV-B induced flavonols and other antioxidant formation is through activation of flavonol synthase and upregulation of genes of different antioxidants by multiple responses through UV-B photoreception. Similar reduction was observed in both the male and female plants of *Morus alba*, without disturbing the stomatal conductance (Chen et al. 2016).

Ambasht and Agrawal (1998) observed a decline in photosynthesis that is probably linked to the reduction in stomatal conductance and concentrations of photosynthetic pigments in rice plants. Wang et al. (2021) reported significant reductions in gas exchange thus reducing intercellular $CO_2$ concentration ($C_i$) that affects $CO_2$ assimilation efficiency resulting in reduced photosynthesis under UV-B exposure in mango leaves. Similarly, a study conducted on two varieties of *Olea europaea* (Giarraffa and Olivastra seggeianese) showed differential response and displayed variability in photosynthetic stability between these varieties under UV-B (Piccini et al. 2020). This study suggests that distinct profiles of antioxidant responses and photosynthetic efficiencies involved in activating the defense response already after the first week of treatment, therefore Giarraffa performed better than Olivastra seggeianese.

A comprehensive study by Kataria et al. (2019) on UV-B and wheat suggests that UV-B imposes detrimental effects on various photosynthetic parameters including chlorophyll fluorescence, Hill reaction, electron transport rate (ETR), and enzymes of the dark reaction like carbonic anhydrase (CA), Rubisco, phosphoenolpyruvate carboxylase (PEPC), malic dehydrogenase (MDH), and chlorophyllase. Yao et al. (2005) reported that quantum yield of PS-II was declined in *Cucumis sativus* under UV-B associated with an increase in initial fluorescence ($F_0$). Contrary to this, Ali et al. (2024) reported improved leaf fluorescence performance index under two different doses of UV-B. Most studies have shown that UV-B radiation adversely impacts the photosynthetic machinery of plants, depending on the treatments and growth stages.

### 5.2.2 Biomass and Yield

Any morphological or physiological disturbances in plants can ultimately lead to the alteration in biomass and consequently the yield. Usually, in most of the plants, biomass was reduced in presence of UV-B because of the adverse consequences of UV-B on photosynthetic processes. A reduction in biomass of *Lactuca sativa* was reported under the UV-B treatment which was correlated with inhibition of photosynthesis, leading to reduction in biomass. Plants treated with UV-B, redirect their energy allocation, prioritizing other physiological processes (defense mechanisms) over biomass production (Rajabbeigi et al. 2013). Similarly, the reduction in biomass of root, stems, leaves and also in the total biomass was reported in *Morus alba* under UV-B treatment and the reduction was more prominent in female plant than male plants which was associated with more allocation of biomass towards leaves in female which could help in increasing carbohydrate necessary for the reproduction, while in males it was more allocated towards roots for better mineral and water absorption and thus strengthening adaptability (Chen et al. 2016). Adverse effects of UV-B was found in the form of biomass reduction in *Lolium perrene* therefore affecting the plant yield (Comont et al. 2013). UV-B induced reduction in biomass of wheat plants has been reported by Ambasht and Agrawal (2003). Similar reduction of biomass in linseed has been reported by Tripathi and Agrawal (2013). Further, reduction in biomass of two varieties of sunflower (DRSF 108 and Sungold) was associated with accumulation of phytohormones (abscisic acid and zeatin riboside) involved in reducing the concentration of indole acetic acid and gibberellic acid, hence decreasing the plant height and ultimately the biomass under UV-B (Tripathi et al. 2019).

Liu et al. (2013) found reductions in seed yield of soybean cultivars under UV-B exposure associated with reduction in pod number. Besides, exposures during the early flowering stage alters the availability of resources to developing reproductive structures, impacting flower and pod abscission, and ultimately affecting the final pod count at harvest. Reduced growth and yield was reported in *Thymus vulgaris* under UV-B treatment. Santin et al. (2022) reported reduced biomass and stem length of microgreens (vegetable greens harvested with cotyledon leaves and one true leaf set, but it is different from sprouts) in linseed under UV-B radiation treatment. Kataria et al. (2019) has reported the harmful effect of UV-B on biomass and yield in wheat, correlated with reduced number and weight of grains. However, Usano-Alemany and Panjai (2015) detected an increased plant growth, biomass and secondary metabolism in Lavandin as it is well adapted in high UV-B conditions. So, majority of the studies have consistently shown that UV radiation has detrimental effects on plants growth, biomass production and yield of the plant (Table 5.1).

### *5.2.3 Qualitative Parameters of Seed*

Seed quality is an important attribute which determines the performance of the seed when sown or stored (Hampton 2002). Further, it should be maintained in high qualitative characters for the better germination and overall growth of the plant. Liu et al. (2013) found the reduction of 12% in seed size under the UV-B treatment in soybean cultivars, which was predominantly correlated with a decrease in cotyledon

**Table 5.1** Responses of plants against UV-B stress

*Physiological responses*

| Plant/species/cultivar | Experimental setup | Parameter | Change recorded | References |
|---|---|---|---|---|
| *Trifolium repens L* | 13.3 kJ $m^{-2}d^{-1}$ | Leaf water potential | Increased | Hofmann et al. (2003) |
| | | Quantum yield | Not change | |
| *Arabidopsis thaliana* | 9.96 kJ $m^{-2}d^{-1}$ | Photosynthetic rate | No change | Lake et al. (2009) |
| | | Transpiration | Increased | |
| | | Stomatal conductance | Increased | |
| | | Quantum yield | Decreased | Schultze and Bilger (2019) |
| *Cucumis sativus* | 3.5 kJ $m^{-2}$ | Photosynthetic efficiency | Decreased | Yao et al. (2005) |
| *Cymbopogon citratus* | 1.8 kJ $m^{-2}d^{-1}$<br>3.6 kJ $m^{-2}d^{-1}$ | Photosynthetic rate | No change | Kumari and Agrawal (2010) |
| | | Stomatal conductance | Decreased | |
| | | Transpiration rate | Decreased | |
| | | Water use efficiency | Decreased | |
| | | Photosynthetic efficiency | Decreased | |
| *Glycine max* | 7.2 kJ $m^{-2}d^{-1}$ | Photosynthetic rate | Decreased | Choudhary and Agrawal (2016) |
| | | Stomatal conductance | Decreased | |
| | | Transpiration rate | Decreased | |
| | | Water use efficiency | Decreased | |
| | | Photosynthetic efficiency | Decreased | |
| *Vigna radiata* | 7.2 kJ $m^{-2}d^{-1}$ | Photosynthetic rate | Decreased | Choudhary and Agrawal (2014) |
| | | Stomatal conductance | Decreased | |
| | | Water use efficiency | Decreased | |

(continued)

**Table 5.1** (continued)

| | | | | |
|---|---|---|---|---|
| *Physiological responses* | | | | |
| *Coleus forskohlii* | 3.6 kJ $m^{-2}d^{-1}$ | Photosynthetic rate | Decreased | Takshak and Agrawal (2015) |
| | | Stomatal conductance | Decreased | |
| | | Water use efficiency | Decreased | |
| | | Photosynthetic efficiency | Decreased | |
| *Oryza sativa* L var. *Liangyoupeijiu* | | Photosynthetic rate | Decreased | Yu et al. (2013) |
| | | Stomatal conductance | Decreased | |
| | | Ribulose-1,5-bisphosphate carboxylase/oxygenase (Rubisco) activity | Decreased | |
| *Vitis vinifera* (cv. Tempranillo) | 5.98 and 9.66 kJ $m^{-2}$ $d^{-1}$ | Photosynthesis | Decreased | Martínez-Lüscher et al. (2013) |
| | | Stomatal conductance | Decreased | |
| | | Sub-stomatal $CO_2$ concentration | Decreased | |
| | | Photosystem II (PSII) efficiency | Decreased | |
| *Mangifera indica* | 24 and 96 kJ $m^{-2}d^{-1}$ | Photosynthetic rate | Decreased | Wang et al. (2021) |
| | | Stomatal conductance | Decreased | |
| | | Transpiration rate | Decreased | |
| | | Intercellular $CO_2$ concentration | Decreased | |
| | | Chlorophyll a/b | Decreased | |
| ***Biomass and yield*** | | | | |
| *Lactuca sativa* | 8.2 Ws $m^{-2}$ | Total Biomass | Decreased | Rajabbeigi et al. (2013) |
| *Morus alba* | 3.78 kJ $m^{-2}$ $d^{-1}$ | Root Biomass | Decreased | Chen et al. (2016) |
| | | Stem Biomass | Decreased | |

(continued)

**Table 5.1** (continued)

| *Physiological responses* | | | | |
|---|---|---|---|---|
| | | Leaf Biomass | Decreased | |
| *Triticum aestivum* | 7.1 kJ $m^{-2}$ | Total Biomass | Decreased | Ambasht and Agrawal (2003) |
| *Lolium perrene* | 2.3, 3.2, 4.1, 5.0, and 5.7 kJ $m^{-2}$ $day^{-1}$ | Biomass and Yield | Decreased | Comont et al. (2013) |
| *Glycine max* | 13 kJ $m^{-2}d^{-1}$ | Seed yield | Decreased | Liu et al. (2013) |
| *Vigna radiata* | 12.2 kJ $m^{-2}d^{-1}$ | Potential yield | Decreased | Rajendiran and Ramanujam (2004) |
| *Brassica rapa* | 5.1 kJ $m^{-2}$<br>7.4 kJ $m^{-2}$ | Seed yield | No change | Demchik and Day (1996) |
| *Triticum durum* | 1.34 J $cm^{-2}$ to 6.33 J $cm^{-2}$ | Grain mass | Increased | Al-oudat et al. (1998) |
| *Fagopyrum tataricum* | Spring 5.3 kJ<br>8.5 kJ<br>Autumn 5.3 kJ<br>8.5 kJ | Seed yield | Decreased | Yao et al. (2006) |
| *Phaseolus mungo* | 0.5 $Wm^{-2}$ | Pod number/plant<br>Seed number/pod | Decreased | Jayakumar et al. (2003) |
| *Zea mays* | 3.16 kJ $m^{-2}$ $day^{-1}$ | Grain yield | Decreased | Correia et al. (2005) |
| *Raphanus sativus* | 7.2 kJ $m^{-2}$ $day^{-1}$ | Yield | Decreased | Singh et al. (2011) |
| *Vigna radiata* | 7.2 kJ $m^{-2}$ $day^{-1}$ | Yield | Decreased | Choudhary and Agrawal (2014) |

cell number. Besides, the decline in seed ABA concentration, resulting in lower cotyledon cell number in UV-exposed plants. A study by Al-Oudat et al. (1998) denoted that drought resistant wheat variety (*Triticum durum* var. Horani) showed resistance against UV-B treatment through increased grain mass and number of spikes, whereas drought sensitive broad bean (*Vicia faba*) did not show any variations under exposed and non-exposed UV-B conditions.

Seed quality analysis by Teramura et al. (1990) showed that seed protein content was reduced in two varieties of Soybean (Essex and Williams); however, seed lipid content showed variable results having an increase as well as decrease in the Williams variety in different years of experimentation. A study conducted on wild legume and a crop legume (*Abrus precatorius* L. and *Vigna mungo* (L.) Hepper) revealed that total free phenols and tannins were increased under the UV-B exposure and it serves as an external stimulus for the accumulation of these bioactive compounds (Doss et al. 2022).

Demchik and Day (1996) indicated that seed abortion rates were stimulated in *Brassica rapa* pollinated with pollen grains obtained from UV-treated stock. Reddy et al. (2016) stated reductions in some seed quality parameters (protein, palmitic and oleic acids), whereas oil content, linolenic and linoleic acid were increased in UV-B-treated plants of soybean. Similarly, Singh and Choudhary (2025) reported reduced grain quality (total soluble sugars, reducing sugars, total free amino acids, starch, and soluble protein) in pearl millet after supplemental UV-B exposure, indicated a declined quality response index. Contradictory to this, UV-B priming experiments showed positive results in rice seeds indicating reduced panicle sterility and an increased harvest index, and also these imprints were maintained throughout the life cycle of rice plants (Sen and Puthur 2020). So, the previous studies suggest variable response of UV-B on qualitative parameters of seed.

## 5.3 Crop Responses Towards Ozone Stress

### *5.3.1 Physiological Responses*

The $O_3$ entry in a plant's apoplastic space is coordinated by the stomatal pores and possessing an oxidative nature, it generates excessive reactive oxygen species (ROS). Ozone's oxidizing activities are chiefly due to stomatal uptake, and damage is proportional to the absorbed dose. Since $O_3$ causes injury only when there is stomatal uptake and the flux-based approaches interprets the stomatal regulation of $O_3$ uptake (Fiscus et al. 2005). While, non-stomatal factors also play role. Feng et al. (2016) analyzed five modern wheat cultivars (Jiaxing 002, Yangmai 15, Yangmai16, Yannong 19 and Yangfumai 2) to govern the influences of elevated $O_3$ concentration and found that lower photosynthetic rates across the cultivars were chiefly due to non-stomatal factors like reduced electron transport rates and maximum carboxylation capacity.

Long-term exposure to $O_3$ elevation unfavorably affects the Ribulose 1,5 bisphosphate carboxylase/oxygenase (RuBisCO) carboxylation efficiency, reduces the photosynthesis ($P_s$), and stomatal conductance ($g_s$), enhanced transpiration rate and stomatal sluggishness (Paoletti and Grulke 2010; Ainsworth et al. 2012; Emberson et al. 2018; Xu et al. 2022). Impact of ozone exposure on stomatal conductance and photosynthetic performance in different crop species has been provided in Table 2. Photosynthetic decline in $O_3$ treated plants is associated with structural impairment of thylakoids, reduction in excitation energy efficiency, capture, and impairment of electron transport in PSI and PSII (Calatayud and Barreno 2001; Fiscus et al. 2005). Feng et al. (2007) have observed that elevated $O_3$ concentration of 105 ppb for 8 h reduced the photosynthetic rate by 42.4% in wheat. The light reactions that deal with the photosynthetic electron transport are also negatively afflicted by $O_3$ (Fiscus et al. 2005) like reductions in electron transport rate and $F_v/F_m$ ratio has been observed under the $O_3$ exposure. Photosynthetic efficiency or the ratio of ($F_v/F_m$) quantifies the maximum efficiency of PS II, and its decline advocates a reduction in linear electron flow through PS II (Rai et al. 2016). While, it is the carbon reduction reactions of photosynthesis that are usually known to have greater sensitivity to $O_3$ generated ROS, leading to drops in carbon assimilation capacity and carboxylation efficiency (Fiscus et al. 2005; Ainsworth et al. 2012).

Meta-analysis studies with rice (Ainsworth 2008) and wheat (Feng et al. 2008), respectively, manifested variable grades of deleterious response of $P_s$ under $O_3$ treatment. Photosynthetic rate showed linear reductions with enhancing $O_3$ levels under variable $O_3$ concentrations ranging from 37–116 ppb in two soybean cultivars IA3010 and Dwight (Sun et al. 2014). Noontime photosynthetic rate measurements taken four times over the growth period for consecutive two years manifested a seasonal average linear drop in photosynthetic rate of approximately 0.06–0.21 μmol $m^{-2}$ $s^{-1}$ per ppb rise in $O_3$ concentration (Betzelberger 2012). Between the cultivars, Saurabh 950 and NDR 97, sensitive cultivar of rice, i.e., NDR 97 showed higher $P_s$ while reduction in $g_s$ as compared to the tolerant cultivar Saurabh 950 (Rai and Agrawal 2008).

The depression in $g_s$ can be due to the regulatory practices originated by the rise of internal $CO_2$ concentration ($C_i$) as a result of lower $CO_2$ fixation rate in $O_3$-exposed leaves (Vavasseur and Raghavendra 2005), but direct $O_3$ influences on guard cells could also be involved (Vaultier and Jolivet 2015). Plants exposed to $O_3$ display a decline in $g_s$ as a usual response, which is rather a defensive measure, diminishing the $O_3$ flux through stomatal pores (Hoshika et al. 2013) and the higher sensitivity of crop species towards $O_3$ have directly been linked to the stomatal conductance (Biswas et al. 2008; Osborne et al. 2016). Wheat cultivars Soissons and Cap Horn at 120 ppb $O_3$ exposure manifested up to 60% decline in $g_s$ in both the cultivars (Bagard et al. 2015). Also, the closure of stomata due to $O_3$ is a secondary response directed by the enhanced internal $CO_2$ concentration ($C_i$) in the sub-stomatal cavity (Paoletti et al. 2006). The stomatal closure usually results in an improved water use efficiency (WUE) (Bailey et al. 2019).

It has also been found that RuBisCO enzyme of the Calvin cycle involved in $CO_2$ carboxylation reaction is highly susceptible to $O_3$ stress. Emberson et al. (2018) have documented that $O_3$ predominantly affects RuBisCO content rather than its activation. Approximately across the 20 experimental studies, Galmés et al. (2013) reviewed that $O_3$ leads to a reduction in the activation state of RuBisCO rather than its concentration. Considering the past studies, Goumenaki et al. (2010) found that though the transcript expression of both larger subunit (LSU) as well as smaller subunit (SSU) of RuBisCO were adversely affected by $O_3$ treatment, the gene expression and the stability of rbcS mRNA were comparatively affected more.

Some documentations indicates that photosynthetic responses may depend on the leaf age, senescence or year of cultivar release. Betzelberger et al. (2010) observed that under elevated $O_3$ exposure, older and mature leaves had comparatively reduced photosynthetic rate across the six soybean cultivars indicating that physiological effects of $O_3$ can be dependent on age. Similarly, Morgan et al. (2006) detected that the photosynthetic impairment triggered due to $O_3$ was more prominent on the mature plants and/or older leaves. Stomatal conductance of leaves of a C4 plant maize and C3 plant wheat declined with time owing to the joint effects of foliar ageing and the progressive shading of the measured leaves by emerging/developing foliage (Bagard et al. 2015). Fiscus et al. (2005) have suggested that photosynthesis is rather affected more due to the accelerated leaf senescence than the depression in photosynthetic capacity by the elevated $O_3$. When the leaves are approximately 50% senescenced, flag leaves of wheat no longer produce the photosynthates as suggested by Araus and Labrana (1991). The cultivars response of a plant may also depend on its year of release. Singh et al. (2018) have reported that modern cultivars were more sensitive towards $O_3$ stress due to higher reductions in $P_s$ rate in them compared to the older cultivars. Moreover, Biswas et al. (2008) detected that the modern cultivars manifested higher relative reduction in $P_s$ than the older wheat cultivars due to enhanced foliar $O_3$ influx due to higher $g_s$ in modern cultivars.

### *5.3.2 Biomass and Yield*

Plants under stress usually have more assimilate requirements towards the sink organs for protection and repair which alters the regular source and sink balance in plants (Anderson 2003). A general decline in plant biomass has mostly been observed due to $O_3$, because of the adverse effect of $O_3$ on physiological processes. The responses to $O_3$ exposure involve considerable changes in plant development and growth leading to differential biomass allocation in plants (Morgan et al. 2006; Ahmad et al. 2013). Saitanis and Karandinos (2002) detected 86, 59 and 89% reduction in leaf, stem and root biomass upon exposure to 180 ppb of $O_3$ in controlled chamber environment. A reduction of 39% in wheat cultivars biomass have been recorded by Biswas et al. (2008) under the $O_3$ exposure of 82 ppb. Respective decline of 11.6, 36.1%, and 13, 40.2% were detected for the maize cultivars HQPM1 and DHM117 in biomass accumulation at ambient and ambient +30 ppb $O_3$ dose. At 40 and 80 DAG, respective decline of 22.2% and 28.8% in wheat biomass has been reported in elevated $O_3$ environment for timely sown HD2967 cultivar as compared to ambient $O_3$ (Ghosh et al. 2020).

**Table 5.2** Effect of $O_3$ treatment on stomatal conductance and photosynthetic rate of various crop plants

| S. No | Crops/ cultivars | $O_3$ concentration | Stomatal conductance ($g_s$) | Photosynthetic rate ($P_s$)/$CO_2$ assimilation | References |
|---|---|---|---|---|---|
| 1 | *Triticum aestivum* | 150% of ambient | 68 (↓) | 61(↓) | Ji-Ling et al. (2009) |
| 2 | *Triticum aestivum* Twenty cultivars | 82 ppb for 7 h | 8 (↓) | 24 (↓) | Biswas et al. (2008) |
| 3 | *Triticum aestivum* PBW 343 M 533 | 50.2 ppb (8 h) | 33.2 (↓) 43.6 (↓) | 19 (↓) 18.4 (↓) | Rai and Agrawal (2014) |
| 4 | *Triticum aestivum* Fourteen cultivars | Seasonal mean 47.2 ppb | 10.3–31.5 (↓) | 8.0–32.8 (↓) | Singh et al. (2018) |
| 5 | *Triticum aestivum* | Ambient + 20 ppb | 26.2 (↓) | 33.9 (↓) | Ghosh et al. (2020) |
| 6 | *Triticum aestivum* HUW234 HD3118 | Ambient + 20 ppb | 23.1 (↓) 19.2 (↓) | 20.8 (↓) 14.8 (↓) | Yadav et al. (2020b) |
| 7 | *Oryza sativa* WYJ3 SY 63 | 1.5 × ambient (13.8–74.2) for 7 h | NS (↓) 33% (↓) | 14.8 (↓) 27.1 (↓) | Pang et al. (2009) |
| 8 | *Zea mays* Chambord | 80 ppb | 60 (↓) | 40 (↓) | Bagard et al. (2015) |
| 9 | *Glycine max* | 1.2 × current concentration | 1 (↓) | NS | Bernacchi et al. (2006) |
| 10 | *Glycine max* 10 cultivars | 2 × ambient $O_3$ (46 ppb) | 15 (↓) | 11 (↓) | Betzelberger et al. (2010) |
| 11 | *Glycine max* L Nine cultivars | E-$O_3$, ambient air + 40 ppb | NS | 45 (↓) across all the cultivars | Zhang et al. (2014) |
| 12 | *Glycine max* Mandarin (Ottawa) Fiskeby III | 70 ppb | 77(↓) 30(↓) | 38(↓) 5(↓) | Bailey et al. (2019) |
| 13 | *Phaseolus vulgaris* | 150 ppb | 59(↓) | 64(↓) | Guidi et al. (2002) |

(continued)

**Table 5.2** (continued)

| S. No | Crops/ cultivars | $O_3$ concentration | Stomatal conductance ($g_s$) | Photosynthetic rate ($P_s$)/$CO_2$ assimilation | References |
|---|---|---|---|---|---|
| 14 | *Phaseolus vulgaris* S156 | 60 ppb (12 h) | 52.6 (↓) | 38 (↓) | Flowers et al. (2007) |
| 15 | *Phaseolus vulgaris* Camellino | 165 ppb for 3 h | 26 (↓) | 36 (↓) | Guidi et al. (2009) |
| 16 | *Vigna radiata* cv HUM-2 HUM-6 | + 10 ppb elevated $O_3$ (63.4 ppb) | 38.8 (↓) 24.8 (↓) | 28.3 (↓) 19 (↓) | Mishra and Agrawal (2015) |
| 17 | *Brassica campestris* Vardan Sanjukta | + 10 ppb elevated $O_3$ (49.4 ppb) | 66 (↓) 66.6 (↓) | 46.6 (↓) 49.3 (↓) | Tripathi and Agrawal (2012) |
| 18 | *Brassica napus* | 50, 100 ppb | 15.2(↓) | 21.3(↓) | Feng et al. (2006) |
| 19 | *Arachis hypogaea* NC-VII | 48 ppb for 12 h | NS | 21 (↓) | Booker et al. (2007) |
| 20 | *Gossypium hirsutum* | 100 ppb | 69(↓) | 33(↓) | Zouzoulas et al. (2009) |

Ghosh et al. (2020) documented that $O_3$ causes substantial losses in crop production globally. Seeds are among the key plant component for carbon loading and a plant's yield relies on the amount up to which the C requirement can be accomplished under the $O_3$ stress (Emberson et al. 2018). Wheat cultivar (HUW-234) under ambient $O_3$ in OTCs suffered depression in yield (Rai et al. 2007). Mishra et al. (2013) under ambient and elevated $O_3$ concentrations found the similar results with cultivars K-9107 and HUW-37. In an experimentation with rice, Rai and Agrawal (2008) reported that the fast-growing rice cultivar facing higher yield losses displayed less reduction in photosynthesis with respect to the slow growing cultivar having lower yield losses. In a meta-analysis, average ambient $O_3$ concentrations of 40 ppb reduced the productivity of important food crops (soybean, wheat, rice, and potato) significantly by nearly 10% with respect to $O_3$-filtered air (Feng and Kobayashi 2009). The probable rise in background $O_3$ concentrations is projected to influence the soybean yield losses up to 9–19% by 2030 (Avnery et al. 2011). Similarly, Feng et al. (2019) found relative yield loss of 6% and 8%, for wheat and rice, respectively. An $O_3$ risk assessment showed that the international productivity of maize, soybean, rice, and wheat was reduced annually by 6.1%, 12.4%, 4.4%, and 7.1%, respectively (an average of 2010–2012) (Mills et al. 2018).

The $O_3$ response to yield components is usually dependent on the species but may also be linked to the shortening of the maturing time (Emberson et al. 2018). According to Fiscus et al. (2005), the loss of economic yield as a result of $O_3$ exposure is a direct result of the loss of photosynthetic activity, chiefly during the reproductive stage, due to the reduced leaf area growth. It is also noteworthy that the effect of $O_3$ on wheat production was predominantly by a drop in grain mass, since the influence on grain number is comparatively lesser (Broberg et al. 2015). Ozone-induced yield loss in maize cultivar Zhengdan 958 exposed to a range of chronic $O_3$ was chiefly due to a decline in kernel weight rather than the number of kernels (Peng et al. 2020). It has been suggested by Oksanen et al. (2013) that in future, more unexplored species like maize and pea, and some local varieties of staple crops should be researched for their $O_3$ responses (Table 5.3).

### *5.3.3 Qualitative Parameters of Seed*

Apart from the effects on yield component of the plants, $O_3$ exposure also causes alterations in the nutritional quality of the seeds. A general reduction in starch content in seeds have been observed in many of the crop plants. Distortions in visual appearance due to chalkiness of grains (Wang et al. 2014) and reductions in grain size (Frei et al. 2012), have been documented. Nutritional parameters of grains showed reduced starch accumulation and an increased reducing sugars and total soluble sugars in rice (Sarkar et al. 2015). Similarly, under ambient + 20 ppb $O_3$ treatment (NF+), starch amount declined significantly by 4.6 and 6.1% in both the tested modern and old cultivars and increased total soluble sugar and reducing sugar as reported by Yadav et al. (2020a, b). According to Zheng et al. (2013) concentration of amylose reduced due to the $O_3$ exposure. Elevated $O_3$ led to the reduction in starch by approximately 5.6–12% in C-306 and 11.7–11.8% in HD-2967 as compared to ambient $O_3$ concentration (Yadav et al. 2019). Pleijel et al. (2018) investigated the effects of ambient surface $O_3$ versus pre-industrial $O_3$ concentrations on wheat using results from thirty-three experimental results, and it was concluded that with respect to pre-industrial levels, prevailing $O_3$ concentrations negatively affected the starch content by 3.0% and starch yield by 10.9%.

Routinely it has been observed that the concentration of minerals and nitrogen in crops under $O_3$ treatment are positively affected (Emberson et al. 2018). There has been found a significant positive effect of $O_3$ on the seed protein concentration in crops like rice and wheat (Broberg et al. 2020). The consequences of elevated $O_3$ on nutrient quality of rice and winter wheat grain were explored for the two consecutive years, 2007 and 2008 in Yangtze River Delta, China. The protein content and lysine increased in rice and winter wheat. Moreover, the N content enhanced under elevated $O_3$ treatment in both the years while C–N ratios declined significantly (Zheng et al. 2013). Seed protein and nitrogen contents in the rice cultivars Shivani and Malviya dhan36 were increased due to $O_3$ treatment (Sarkar et al. 2015). In meta-analysis investigations, also an enhanced protein content was detected, whereas protein yield

**Table 5.3** Responses of yield parameters of various agricultural crop plants to different concentration of $O_3$

| S. No | Species/cultivar | Experimental setup | $O_3$ conc | Parameters | Percent Yield loss | References |
|---|---|---|---|---|---|---|
| 1 | *Triticum aestivum* Giza 68 | Open Field | 77–166 ppb | Seed yield $Pot^{-1}$ (g) | 9–46 | Ali et al. (2008) |
| 2 | *Triticum aestivum*<br>Sufi<br>Bijoy | Greenhouse | 60 ppb<br>100 ppb | Yield $plant^{-1}$ | 11.5<br>44.5<br>33.2<br>45.6 | Akhtar et al. (2010) |
| 3 | *Triticum aestivum*<br>HUW-37<br>K-9107 | OTCs | Ambient $O_3$ (48.5–52.7 ppb) + 10 ppb (4 h) | Yield (g $plant^{-1}$) | Average loss in two years =<br>25.7<br>27.4 | Mishra et al. (2013) |
| 4 | *Triticum aestivum*<br>PBW 343<br>M 533 | OTCs | First year-50.2 (M 12)<br>Second year-53.2 (M 12) | Weight of grains (g $plant^{-1}$) | Average loss in two years =<br>17.6<br>16.4 | Rai and Agrawal (2014) |
| 5 | *Triticum aestivum*<br>LOK-1<br>HUW 510 | OTCs | 58.2 (M 12) | Weight of grains (g $plant^{-1}$) | 7.3<br>16.2 | Singh et al. (2015b) |
| 6 | *Triticum aestivum*<br>Fourteen cultivars | OTCs | Ambient + 30 ppb | Weight of grains $plant^{-1}$ | 10.0–31.3 | Singh et al. (2018) |
| 7 | *Triticum aestivum*<br>HD3086<br>HUW468 | OTCs | Ambient<br>+ 20 ppb | Weight of grains $plant^{-1}$ | 33.3<br>36.2 | Yadav et al. (2020a) |
| 8 | *Triticum aestivum*<br>HD 2967 | OTCs | Ambient<br>+ 20 ppb | 1000 seeds weight | 4.5 | Ghosh et al. (2020) |

(continued)

**Table 5.3** (continued)

| S. No | Species/cultivar | Experimental setup | $O_3$ conc | Parameters | Percent Yield loss | References |
|---|---|---|---|---|---|---|
| 9 | *Oryza sativa*<br>MR 84<br>MR 185 | OTCs | 32.5 ppb (8 h mean) | Yield plant$^{-1}$ (g) | 3.4<br>6.3 | Ishii et al. (2004) |
| 10 | *Oryza sativa* | OTCs | 62 ppb | Seed yield (g) | 24 | Ainsworth (2008) |
| 11 | *Oryza sativa* Wujing 15 (WJ15), yangdao 6 (YD6), Shanyou<br>63 (SY63), Liangyoupeijiu (LYP) | $O_3$-FACE system | 7 h mean, 56 ppb for SY6, 54 ppb for YD6, 56 ppb for WJ15 and 59 ppb for LYP | Yield plant $^{-1}$ (g) | 17.5 in SY6 and 15.0 in LYP | Shi et al. (2009) |
| 12 | *Oryza sativa*<br>Shivani<br>Malviya dhan 36 | OTCs | 62.2 ppb<br>72.6 ppb | Number of grains plant$^{-1}$ | 17.3<br>27.2<br>19.8<br>28.8 | Sarkar et al. (2015) |
| 13 | *Oryza sativa*<br>Kasalath<br>Koshihikari | Greenhouse | 71 ppb | Filled kernel | 5.2<br>4.2 | Sawada et al. (2016) |
| 14 | *Oryza sativa*<br>Three indica cultivars<br>Three japonica cultivars | Green house type gas fumigation system | 100 ppb | Grain yield | Average yield loss across six cultivars<br>25.2–57.4<br>Indica cultivars (42.6)<br>Japonica cultivars (34.4) | Shao et al. (2020) |
| 15 | *Oryza sativa*<br>Fifteen cultivars | OTCs | 50 ppb | Thousand grain weight | 0.62 | Ramya et al. (2021) |

(continued)

**Table 5.3** (continued)

| S. No | Species/cultivar | Experimental setup | $O_3$ conc | Parameters | Percent Yield loss | References |
|---|---|---|---|---|---|---|
| 16 | *Zea mays*<br>HQPM1<br>DHM117 | OTCs | (55.6 ppb)<br>Ambient $O_3$ + 15 ppb (5 h)<br>Ambient + 30 ppb (5 h) | Weight of kernels plant$^{-1}$ | 10.1<br>5.5<br>9.5<br>13.8 | Singh et al. (2014) |
| 17 | *Zea mays*<br>Zhengdan 958 | OTCs | Non-Filtered + 60 ppb $O_3$ | Hundred kernels weight (g plant$^{-1}$) | 3.1 | Peng et al. (2021) |
| 18 | *Zea mays*<br>Zhengdan 958 | OTCs | 85.7<br>98.8 | Weight of kernels plant$^{-1}$ | 16.6<br>32.1 | Peng et al. (2021) |
| 19 | *Brassica campestris*<br>Kranti | OTCs | 42–54 ppb | Yield (g plant$^{-1}$) | 16.4 | Singh et al. (2009) |
| 20 | *Brassica campestris*<br>Vardan<br>Aashirwad | OTCs | 28–59 ppb (12 h) | Weight of seeds (g plant$^{-1}$) | 7<br>19.3 | Singh et al. (2012) |
| 21 | *Brassica campestris*<br>Sanjukta<br>Vardan | OTCs | Ambient + 10 ppb (3 h) | Test weight | 12.5<br>47 | Tripathi and Agrawal (2012) |
| 22 | *Brassica napus*<br>Click<br>Phoenix | Glasshouse | 30–110 ppb | Thousand seed weight | 20<br>40 | Roberts et al. (2022) |
| 23 | *Brassica rapa* | OTCs | 150 ppb<br>200 ppb | Tuber weight | 69.1<br>85.9 | Kovilpillai et al. (2023) |
| 24 | *Glycine max* Forrest<br>Essex | OTCs | 62.4 ppb (7 h) day-15 days a week | Seed yield (g m$^{-2}$) | 32<br>10 | Chernikova et al. (2000) |

(continued)

**Table 5.3** (continued)

| S. No | Species/cultivar | Experimental setup | $O_3$ conc | Parameters | Percent Yield loss | References |
|---|---|---|---|---|---|---|
| 25 | *Glycine max* | OTCs | First year-75 ppb<br>Second year-67 ppb | Total seed weight | 23.9 (field)<br>38.9 (field) | Booker and Fiscus (2005) |
| 26 | *Glycine max* | OTCs | 60–75 ppb | Grain yield (kg $m^{-2}$) | 47 | Bou Jaoude et al. (2008) |
| 27 | *Glycine max*<br>PK472<br>Bragg | OTCs | 100 ppb (4 h) | Yield $plant^{-1}$ (g) | 33.6<br>30 | Singh et al. (2010) |
| 28 | *Glycine max*<br>Seven soybean genotypes | Fully open air agricultural field | 38–120 ppb | Seed yield kg $ha^{-1}$ | 37–39 | Betzelberger et al. (2012) |
| 29 | *Glycine max*<br>Nine cultivars | OTCs | E-$O_3$, ambient air + 40 ppb | Seed yield (g $plant^{-1}$) | 32–46 | Zhang et al. (2014) |
| 30 | *Phaseolus vulgaris*<br>Borlotto Nano<br>Lingua di Fuoco | OTCs | 4.6 ppm h (cumulative exposure) | Seed weight $plant^{-1}$ (g) | 40.6 | Gerosa et al. (2009) |
| 31 | *Phaseolus vulgaris* Giza 3 | Open Field | 77–166 ppb | Seed yield $Pot^{-1}$ (g) | 20–45 | Ali et al. (2008) |
| 32 | *Vigna radiata*<br>Six cultivars | OTCs | Ambient $O_3$ (64 ppb) + 10 ppb elevated $O_3$ | Weight of seeds (g $m^{-2}$) | 10.2–15.4 | Chaudhary and Agrawal (2015) |
| 33 | *Solanum tuberosum* Hela | Closed exposure chamber | 65 ppb<br>(24 h mean) 110 ppb (4 h $day^{-1}$) | Fresh weight of tuber | 24<br>11 | Kollner and Krause (2000) |

(continued)

**Table 5.3** (continued)

| S. No | Species/cultivar | Experimental setup | $O_3$ conc | Parameters | Percent Yield loss | References |
|---|---|---|---|---|---|---|
| 34 | *Solanum tuberosum*<br>Bintje | OTCs | 43 ppb (12 h mean) | Tuber Fresh weight (g) | 11.4 | Persson et al. (2003) |
| 35 | *Solanum tuberosum*<br>Bintje<br>Kardal | OTCs | 31 ppb<br>57 ppb | Number of tubers | 30<br>20.4 (Kardal)<br>19.2 (Bintje)<br>11.4 (Kardal) | Piikki et al. (2004) |

OTCs = Open Top Chambers

was reduced in wheat (Feng et al. 2008; Feng and Kobayashi 2009). Pleijel et al. (2006) found that older cultivar of wheat 'Lantvete' when exposed to elevated $O_3$ displayed increased protein and amino acid content. Protein content in wheat grains declined significantly by 15.3 and 18.8% in EO (NFC + 10 ppb $O_3$) in cultivars K-9107 and HUW-37, respectively, with respect to NFCs (Mishra et al. 2013). While considering the seed quality of an old (HUW234) and a modern (HD3118) cultivar of wheat treated with elevated $O_3$, Yadav et al. (2020a, b) found that the reduction in protein content was comparatively lower in old (6%) than the modern (7.8%) cultivar in NF + treatment. The augmentation in foliar protein content was more prominent in cultivar L81 than the cultivar Nipponbare and the tendency of increased grain protein was reverse to that of the nitrogen levels, which were affected negatively by the $O_3$ (Jing et al. 2016). Protein values for the wheat grain was 15% in the NF + post ($O_3$ treatment twenty-six days post onset of anthesis), and it was substantially higher than the concentration in the NF environment (13.7%) (Pleijel et al. 1998).

## 5.4 Combined Effect of $O_3$ and UV-B Radiation on Crops

The joint effects of $O_3$ and UV-B radiation on plants can lead to notable changes in growth patterns, development, leaf morphology, photosynthesis, and overall productivity (Schnitzler et al. 1999; Wang et al. 2015). Both stressors can individually exert negative impacts on plants, but the consequences of increased UV-B radiation on plants are generally milder compared to elevated levels of $O_3$ and their combined effect is synergistic that can further exacerbate these negative impacts (Mao et al. 2017; Tripathi et al. 2019). The greater impact of $O_3$ on plants is due to its highly reactive nature as an oxidant, which makes it very toxic to vegetation even at lower concentrations. Additionally, $O_3$ readily reacts with water and chemicals present in the sub-stomatal apoplast (Schraudner et al. 1998; Yadav et al. 2023). Conversely, plants possess a range of protective and repair mechanisms that can effectively counteract the damage inflicted by elevated levels of UV-B radiation (Rai and Agrawal 2022). We insist here that the available literature on the combined effect of $O_3$ and UV-B radiation on crop yields and plant growth is limited (Agrawal et al. 1991; Booker and Miller 1992; Ambasht and Agrawal 2003; Tripathi and Agrawal 2013; Mao et al. 2017; Tripathi et al. 2011, 2019). There are several important factors to consider when examining the mutual effects of UV-B and $O_3$ on crops and we will focus here on physiological, biomass, yield and nutritional quality traits.

### *5.4.1 Physiological Responses*

Ozone and UV-B radiation can affect various physiological processes in crops, including photosynthesis, stomatal behavior, antioxidant defense mechanisms, gene expression, and secondary metabolite production. Both stressors can influence the stomatal behavior in plants, affecting gas exchange and water loss (Schnitzler et al. 1999). The combined effect may result in altered stomatal responses, influencing the overall water-use efficiency and nutrient uptake of crops. Ambasht and Agrawal (2003) reported that combined effect of elevated $O_3$ (0.07 $\mu$mol mol$^{-1}$) and enhanced UV-B radiation (7.1 kJ m$^{-2}$) intensifies the negative impacts, resulting in greater reductions in chlorophyll, carotenoid content, photosynthesis, ascorbic acid, and carbon assimilation in wheat plants. Though the observed combined effects were less than from the additive responses.

When plants are exposed to $O_3$ or UV-B radiation, ROS levels rise, often surpassing the plant's ability to manage them with its natural defence mechanisms. Mao et al. (2017) observed that the combined effect of $O_3$ (110 ppb) and UV-B radiation (9.50 kJ d$^{-1}$ m$^{-2}$) resulted in higher concentrations of malondialdehyde (MDA) and increased relative electrical conductivity in soybean roots. According to that report, the activity of SOD, CAT, and POD under the combined stress of elevated $O_3$ and increased UV-B radiation was merely additive, reflecting the individual effects of each stressor. In an investigation carried out by Tripathi et al. (2011), it was observed that the combined exposure of $O_3$ (+ 10 ppb) and UV-B radiation (+7.2 kJ m$^{-2}$ d$^{-1}$) led to a further induction in the activity of three antioxidant enzymes viz. SOD, APX, and CAT in the leaves of *Linum usitatissimum* L.

The effects of combined $O_3$ and UV-B radiation on the concentrations of secondary metabolites such as flavonoids, phenolic acids, terpenoids, and UV-absorbing properties in crops are still largely unexplored. The interaction between $O_3$ and UV-B radiation may increase the production of certain metabolites in crops, potentially impacting their nutritional quality, flavour, and resistance (Ambasht and Agrawal 2003). The combined stress of $O_3$ and UV-B radiation often resulted in greater oxidative damage than each stressor individually. This increased damage led to alterations in secondary metabolites, particularly flavonoids, as well as changes in endogenous hormone levels and enzyme activities in soybean roots (Mao et al. 2017).

Ozone and UV-B radiation up-regulate the expression of genes involved in stress responses, such as those related to ROS detoxification, DNA repair, and the synthesis of protective compounds (Castagna and Ranieri 2009; Su et al. 2022). The combined effect may modulate gene expression patterns, influencing the activation of stress-related pathways and adaptive responses in crops.

Some studies suggested that the repair of $O_3$-induced damage during elevated UV-B radiation exposure only takes place if $O_3$ and UV-B have distinct action sites in the plant (Rao et al. 1996; Van De Staaij et al. 1997). While $O_3$ and UV-B radiation have different ways of penetrating into plants, both primarily inflict damage on the mesophyll of leaves. Ozone enters leaf mesophyll cells via the stomata, whereas UV-B radiation must penetrate through the epidermal layer to affect the plant tissues. However, as UV-B radiation traverse the epidermis, most of it is largely absorbed by pigments, resulting in only a small fraction reaching the mesophyll cells (Van De Staaij et al. 1997). Research suggests that the damage caused by UV-B radiation primarily affects macromolecules like DNA directly, whereas $O_3$ primarily affects the permeability of cell membrane. However, in both cases, a portion of the damage is attributed to the generation of oxygen free radicals within the plant cells, which contributes to the harmful consequences of both UV-B radiation and $O_3$ on plants (Rao et al. 1996).

### 5.4.2 *Biomass and Yield*

The combined effects of $O_3$ and UV-B radiation significantly impact crop productivity, leading to greater growth inhibition, reduced biomass accumulation, and increased yield loss compared to the effects of each stressor individually (Miller et al. 1994; Hao et al. 2000; Tripathi and Agrawal 2013). Similarly, pre-exposure of $O_3$ was found to enhance the negative impact of UV-B radiation on biomass production of *A. thaliana* (Rao and Ormrod 1995). Some studies have reported that the combined effect of elevated $O_3$ and increased UV-B radiation significantly worsens the reduction in photosynthetic capacity, thereby restricting the crop's ability to produce and accumulate biomass, which ultimately impacts the overall productivity of soybean (Booker and Miller 1992; Miller et al. 1994; Mao et al. 2017).

Ozone and UV-B radiation can negatively impact reproductive development and ultimately reduce crop yield. Ozone interferes with floral development, pollen viability, and fertilization, leading to reduced fruit set and seed production (Gupta et al. 2022). UV-B radiation affects flower morphology, pollen tube growth, and pollen germination, impairing successful pollination, and fertilization (Cetinbas et al. 2022). The cumulative effect of $O_3$ and UV- B radiation synergistically impairs reproductive processes, resulting in decreased yield and lower crop productivity (Feder and Shrier 1990; Runeckles and Krupa 1994; Rao et al. 1996; Tripathi et al. 2019). It's important to note that the severity of the combined effects may also depend on the specific environmental conditions, exposure levels, and duration of stress and be influenced by additional components for instance nutrient availability, humidity, and temperature, which may further modulate the combined impacts on crop biomass and yield.

### 5.4.3 *Qualitative Parameters of Seed*

Ozone exposure affects seed nutrient content, impairing nutritional quality. UV-B radiation impacts seed coat permeability, thickness, and composition, potentially affecting seed protection and germination. The combined effect can alter these seed quality traits, influencing the nutritional value and overall quality of the harvested seeds. For instance, the combined effect of UV-B radiation and elevated $O_3$ has been found to exacerbate the rancidification of oil in two cultivars of sunflower, rendering it undesirable for human consumption (Tripathi et al. 2019). Similarly, Tripathi and Agrawal (2013) found that $O_3$ alone had a more severe impact on plant health than the combined exposure to both UV-B radiation and $O_3$. Their study indicated that while simultaneous exposure to UV-B and $O_3$ did cause damage, the individual

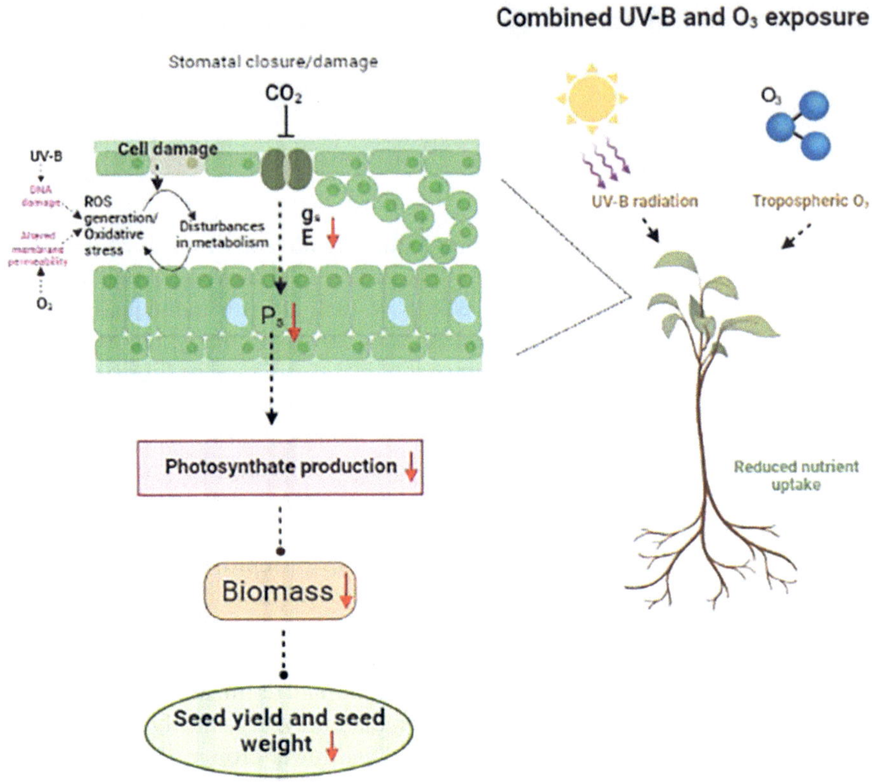


Fig. 5.1 The combined effects of UV-B and $O_3$ on plant

stress caused by $O_3$ was more detrimental overall. However, the combined effect of UV-B radiation and $O_3$ treatment exacerbated the levels of mineral nutrients in the oil of two cultivars of linseed (*Linum usitatissimum* L.). Figure 5.1 illustrates the combined effect of UV-B and ozone exposure on plants. UV-B and ozone lead to reactive oxygen species generation and cause metabolic disturbances. This results in reduced photosynthetic activity, biomass production, and yield.

## 5.5 Mechanism of Action of Combined UV-B Radiation and $O_3$ Treatment

The mechanisms underlying these interactive effects are complex and can involve multiple physiological processes. For example, UV-B radiation can enhance $O_3$ uptake by increasing stomatal conductance and facilitating $O_3$ diffusion into leaf tissues. This can exacerbate $O_3$-induced oxidative stress and damage to cellular components. Conversely, $O_3$ can increase plant sensitivity to UV-B radiation by damaging the UV-absorbing compounds and antioxidants that plants produce in response to UV-B exposure. Consequently, the combined stress of UV-B radiation and $O_3$ can lead to more severe reductions in photosynthetic efficiency and growth compared to the effects of each stressor alone (Van De Staaij et al. 1997).

However, it is crucial to recognize that the combined effects of $O_3$ and UV-B radiation on crops can differ based on crop species, cultivars, and their specific sensitivities. Different crops and varieties may exhibit varying levels of sensitivity or tolerance to these stressors. Some crop varieties may possess genetic traits that confer better resilience to the combined stress, resulting in milder reductions in growth, yield, and productivity compared to more sensitive varieties (Mukherjee et al. 2020).

## 5.6 Management and Mitigation

To mitigate the combined influences of $O_3$ and UV-B radiation on crops, various strategies can be employed:

**Crop Breeding**: Developing crop cultivars with increased tolerance or resistance to both $O_3$ and UV-B radiation can help minimize the negative impacts.

**Crop Management Practices**: Implementing good agricultural practices, such as optimizing irrigation, nutrient management, and pest control, can help enhance crop resilience and minimize stress.

**Protective Measures**: Implementing physical barriers, such as screens or shading materials, can reduce UV-B radiation levels reaching the crops. Similarly, implementing $O_3$-reducing measures, such as controlling industrial and vehicular emissions, can help minimize $O_3$ exposure.

**Enhanced UV-B Radiation Screening**: Implementing greenhouse or polytunnel systems with UV-B-blocking films or coatings can reduce UV-B radiation exposure while allowing optimal light transmission for photosynthesis.

**Monitoring and Forecasting**: Monitoring $O_3$ and UV-B radiation levels and integrating this information into crop management decision-making can help farmers anticipate and respond to potential stress events.

## 5.7 Future Perspective

I. Different crops may exhibit varying sensitivities to $O_3$ and UV-B radiation. Future research can focus on understanding crop-specific responses to these stressors and identifying cultivars or genetic traits that confer tolerance to the combined effect.
II. Advances in genetic engineering techniques can be employed to create crop varieties that are more resistant to the combined effects of $O_3$ and UV-B radiation. Genetic modifications may include boosting the activity of antioxidant enzymes or increasing the production of UV-B-absorbing compounds to reduce damage from these stressors.
III. $O_3$ levels can vary spatially, and accurately monitoring $O_3$ concentrations in agricultural areas can help farmers implement appropriate mitigation strategies. Future perspectives may involve the integration of $O_3$ monitoring technologies with precision agriculture techniques to provide real-time information on $O_3$ levels and enable timely interventions.
IV. Developing and implementing protective technologies can help mitigate the negative effects of $O_3$ and UV-B radiation on crops. These may include the use of shading materials, UV-absorbing films or coatings, and $O_3$-resistant structures like greenhouses or protective enclosures.
V. As climate change continues to alter environmental conditions, including $O_3$ levels and UV-B radiation intensity, future perspectives may involve the development of adaptation strategies for crop production. This may include modifications in planting schedules, altered crop management practices, or the utilization of novel agricultural techniques that can buffer against the mutual impacts of $O_3$ and UV-B radiation.
VI. Integrated models and forecasting systems that incorporate atmospheric chemistry, climate data, and crop growth models can provide valuable insights into the combined effects of $O_3$ and UV-B radiation on crop yields. These tools can assist policymakers, researchers, and farmers in making well-informed decisions about crop management strategies, land-use planning, and the development of effective agricultural policies. By providing detailed insights into how crops respond to various environmental stresses, these tools enable more precise recommendations for improving crop resilience, optimizing resource use, and addressing potential challenges in agriculture. This, in turn, can lead to more sustainable farming practices and enhanced food security.

## 5.8 Conclusion

UV-B and $O_3$, both acts as abiotic stress factor in plants having detrimental effects on crop physiology, biomass, yield as well as quality. Understanding the combined effects of $O_3$ and UV-B radiation on crops is crucial for assessing the potential risks on crop productivity, quality, and overall agricultural sustainability, and implementing appropriate management strategies.

This knowledge can guide breeding programs, agronomic practices, and environmental management strategies to boost crop resilience and maintain agricultural productivity amid shifting environmental conditions. By integrating insights into how crops respond to stressors such as $O_3$ and UV-B radiation, breeding programs can develop more robust crop varieties with enhanced tolerance to these stresses. Agronomic practices can be tailored to optimize growing conditions and mitigate adverse effects, while environmental management strategies can focus on protecting crops from excessive stress. Ultimately, these efforts help ensure that agricultural systems remain productive and sustainable, even as environmental challenges evolve.

**Acknowledgment** SBA is thankful to National Academy of Sciences for providing "NASI-Senior Scientist" fellowship.

## References

Agathokleous E, Saitanis CJ, Koike T (2015) Tropospheric $O_3$, the nightmare of wild plants: a review study. J Agric Meteorol 71:142–152

Agathokleous E, Saitanis CJ, Feng Z, De Marco A, Araminiene V, Domingos M, Sicard P, Paoletti E (2020) Ozone biomonitoring: a versatile tool for science, education and regulation. Curr Opin Environ Sci Health 18:7–13

Agrawal M, Agrawal SB, Krizek DT, Kramer GF, Lee EH, Mirecki RM, Rowland RA (1991) Physiological and morphological responses of snapbean plants to ozone stress as influenced by pretreatment with UV-B radiation. In: Wattal PN, Ort DR, Gnanam A, Teramura AH (eds) In Abrol, YP, Govindjee. Impact of global climatic changes on photosynthesis and plant productivity. Proceedings of the indo-US workshop, New Delhi, pp 8–12

Ahmad MN, Büker P, Khalid S, Van Den Berg L, Shah HU, Wahid A, Emberson L, Power SA, Ashmore M (2013) Effects of ozone on crops in north-west Pakistan. Environ Pollut Mar 174:244–249

Ainsworth EA (2008) Rice production in a changing climate: a meta-analysis of responses to elevated carbon dioxide and elevated ozone concentration. Glob Change Biol 14:1642–1650

Ainsworth EA (2017) Understanding and improving global crop response to ozone pollution. Plant J 90:886–897

Ainsworth EA, Yendrek CR, Sitch S, Collins WJ, Emberson LD (2012) The effects of tropospheric ozone on net primary productivity and implications for climate change. Annu Rev Plant Biol 63:637–661

Akhtar N, Yamaguchi M, Inada H, Hoshino D, Kondo T, Izuta T (2010) Effects of ozone on growth, yield and leaf gas exchange rates of two Bangladeshi cultivars of wheat (Triticum aestivum L.). Environ Pollut 158:1763–1767

Ali A, Alfarhan A, Robinson E, Bokhari N, Al-Rasheid K, Al-Quraishy S (2008) Tropospheric ozone effects on the productivity of some crops in central Saudi Arabia. Am J Environ Sci 4:631–7

Ali A, Santoro P, Mori J, Ferrante A, Cocetta G (2024) Effect of UV-B elicitation on spearmint's (Mentha spicata L.) morpho-physiological traits and secondary metabolites production. Plant Growth Regul 104:63–76

Al-Oudat M, Baydoun SA, Mohammad A (1998) Effects of enhanced UV-B on growth and yield of two Syrian crops wheat (*Triticum durum* var. Horani) and broad beans (*Vicia faba)* under field conditions. Environ Exp Bot 40:11–16

Ambasht NK, Agrawal M (1998) Influence of supplemental UV-B radiation on photosynthetic characteristics of rice plants. Photosynthetica 34:401–408

Ambasht NK, Agrawal M (2003) Effects of enhanced UV-B radiation and tropospheric ozone on physiological and biochemical characteristics of field grown wheat. Biol Plant 47:625–628

Andersen CP (2003) Source-sink balance and carbon allocation below ground in plants exposed to ozone. New Phytol 157:213–228

Anjum NA, Sofo A, Scopa A, Roychoudhury A, Gill SS, Iqbal M, Lukatkin AS, Pereira E, Duarte AC, Ahmad I (2015) Lipids and proteins—major targets of oxidative modifications in abiotic stressed plants. Environ Sci Pollut Res 22:4099–4121

Araus JL, Labrana X (1991) Leaf photosynthesis and chloroplast senescence patterns in wheat flag leaves during grain filling. Photosynthetica 25:33–37

Ashmore MR (2005) Assessing the future global impacts of ozone on vegetation. Plant Cell Environ 28:949–964

Avnery S, Mauzerall DL, Liu J, Horowitz LW (2011) Global crop yield reductions due to surface ozone exposure: 1. Year 2000 crop production losses and economic damage. Atmos Environ 45:2284–2296

Bagard M, Jolivet Y, Hasenfratz-Sauder MP, Gérard J, Dizengremel P, Le Thiec D (2015) Ozone exposure and flux-based response functions for photosynthetic traits in wheat, maize and poplar. Environ Pollut 206:411–420

Bailey A, Burkey K, Taggart M, Rufty T (2019) Leaf traits that contribute to differential ozone response in ozone-tolerant and sensitive soybean genotypes. Plants 8:235

Bernacchi CJ, Leakey AD, Heady LE, Morgan PB, Dohleman FG, McGrath JM, Ort DR (2006) Hourly and seasonal variation in photosynthesis and stomatal conductance of soybean grown at future CO2 and ozone concentrations for 3 years under fully open air field conditions. Plant Cell Environ 29:2077–2090

Betzelberger AM, Gillespie KM, Mcgrath JM, Koester RP, Nelson RL, Ainsworth EA (2010) Effects of chronic elevated ozone concentration on antioxidant capacity, photosynthesis and seed yield of 10 soybean cultivars. Plant Cell Environ 33:1569–1581

Betzelberger A (2012) Current and future consequences of tropospheric ozone on soybean biochemistry, physiology and yield. University of Illinois at Urbana-Champaign

Biswas DK, Xu H, Li YG, Sun JZ, Wang XZ, Han XG, Jiang GM (2008) Genotypic differences in leaf biochemical, physiological and growth responses to ozone in 20 winter wheat cultivars released over the past 60 years. Glob Change Biol 14:46–59

Blokhina O, Virolainen E, Fagerstedt KV (2003) Antioxidants, oxidative damage and oxygen deprivation stress: a review. Ann Bot 91:179–194

Booker FL, Miller JE (1992) Effect of ozone and UV-B radiation on pigments, biomass and peroxidase activity in Soybean. In: tropospheric ozone and the environment: II. Effects, modeling and control air and waste management association. Pittsburgh, PA, pp 489–503

Booker FL, Burkey KO, Pursley WA, Heagle AS (2007) Elevated carbon dioxide and ozone effects on peanut: I. Gas-exchange, biomass, and leaf chemistry. Crop Sci 47:1475–1487

Broberg MC, Feng Z, Xin Y, Pleijel H (2015) Ozone effects on wheat grain quality—a summary. Environ Pollut 197:203–213

Broberg MC, Daun S, Pleijel H (2020) Ozone induced loss of seed protein accumulation is larger in soybean than in wheat and rice. Agronomy 10:357

Calatayud A, Barreno E (2001) Chlorophyll a fluorescence, antioxidant enzymes and lipid peroxidation in tomato in response to ozone and benomyl. Environ Pollut 115:283–289

Caregnato FF, Bortolin RC, Junior AM, Moreira JC (2013) Exposure to elevated ozone levels differentially affects the antioxidant capacity and the redox homeostasis of two subtropical *Phaseolus vulgaris* L. varieties. Chemosphere 93:320–330

Castagna A, Ranieri A (2009) Detoxification and repair process of ozone injury: from $O_3$ uptake to gene expression adjustment. Environ Pollut 157:1461–1469

Cetinbas A, Toksoz O, Piccini C (2022) Effects of UV-B radiation on the performance, antioxidant response and protective compounds of hazelnut pollen. Plants 11:2574

Chaudhary N, Agrawal SB (2015) The role of elevated ozone on growth, yield and seed quality amongst six cultivars of mung bean. Ecotoxicol Environ Safe 111:286–294

Chen M, Huang Y, Liu G, Qin F, Yang S, Xu X (2016) Effects of enhanced UV-B radiation on morphology, physiology, biomass, leaf anatomy and ultrastructure in male and female mulberry (*Morus alba*) saplings. Environ Exp Bot 129:85–93

Chernikova T, Robinson JM, Lee EH, Mulchi CL (2000) Ozone tolerance and antioxidant enzyme activity in soybean cultivars. Photosyn Res 64:15–26

Choudhary KK, Agrawal SB (2014) Cultivar specificity of tropical mung bean (Vigna radiata L.) to elevated ultraviolet-B: changes in antioxidative defense system, nitrogen metabolism and accumulation of jasmonic and salicylic acids. Environ Exp Bot 99:122–132

Choudhary KK, Agrawal SB (2016) Assessment of fatty acid profile and seed mineral nutrients of two soybean (*Glycine max L.*) cultivars under elevated ultraviolet-B: role of ROS, pigments and antioxidants. Photochem Photobiol 92:134–143

Comont D, Winters A, Gomez LD, Mcqueen-Mason SJ, Gwynn-Jones D (2013) Latitudinal variation in ambient UV-B radiation is an important determinant of *Lolium perenne* forage production, quality, and digestibility. J Exp Bot 64:2193–2204

Correia CM, Pereira JMM, Coutinho JF, Björn LO, Torres-Pereira JM (2005) Ultraviolet-B radiation and nitrogen affect the photosynthesis of maize: a Mediterranean field study. Eur J Agron 22:337–347

Demchik SM, Day TA (1996) Effect of enhanced UV-B radiation on pollen quantity, quality, and seed yield in *Brassica rapa* (Brassicaceae). Am J Bot 83:573–579

Doss A, Esther A, Rajalakshmi R (2022) Influence of UV-B treatment on the accumulation of free phenols and tannins in the legumes of *Abrus precatorius* L. and *Vigna mungo* (L.) Hepper. Phytomed Plus 2:100189

Emberson LD, Pleijel H, Ainsworth EA, Van den Berg M, Ren W, Osborne S, Mills G, Pandey D, Dentener F, Büker P, Ewert F (2018) Ozone effects on crops and consideration in crop models. Eur J Agron 100:19–34

Fatima A, Singh AA, Mukherjee A, Agrawal M, Agrawal SB (2018) Ascorbic acid and thiols as potential biomarkers of ozone tolerance in tropical wheat cultivars. Ecotoxicol Environ Safe 171:701–708

Feder WA, Shrier R (1990) Combination of UV-B and ozone reduces pollen tube growth more than either stress alone. Environ Exp Bot 30:451–454

Feng Z, Kobayashi K (2009) Assessing the impacts of current and future concentrations of surface ozone on crop yield with meta-analysis. Atmos Environ 43:1510–1519

Feng ZZ, Yao FF, Chen Z, Wang XK, Zheng QW, Feng ZW (2007) Response of gas exchange and yield components of field-grown *Triticum aestivum* L. to elevated ozone in China. Photosynthetica 45:441–446

Feng Z, Kobayashi K, Ainsworth EA (2008) Impact of elevated ozone concentration on growth, physiology, and yield of wheat (*Triticum aestivum* L.): a meta-analysis. Glob Change Biol 14:2696–2708

Feng Z, Wang X, Zheng Q, Feng Z, Xie J, Chen Z (2006) Response of gas exchange of rape to ozone concentration and exposure regime. Acta Ecol Sin 26:823–829

Feng Z, Wang L, Pleijel H, Zhu J, Kobayashi K (2016) Differential effects of ozone on photosynthesis of winter wheat among cultivars depend on antioxidative enzymes rather than stomatal conductance. Sci Tot Environ 572:404–411

Feng Z, De Marco A, Anav A, Gualtieri M, Sicard P, Tian H, Fornasier F, Tao F, Guo A, Paoletti E (2019) Economic losses due to ozone impacts on human health, forest productivity and crop yield across China. Environ Int 131:104966

Fiscus EL, Booker FL, Burkey KO (2005) Crop responses to ozone: uptake, modes of action, carbon assimilation and partitioning. Plant Cell Environ 28:997–1011

Flowers MD, Fiscus EL, Burkey KO, Booker FL, Dubois JJ (2007) Photosynthesis, chlorophyll fluorescence, and yield of snap bean (Phaseolus vulgaris L.) genotypes differing in sensitivity to ozone. Environ Exp Bot 61:190–198

Foyer CH, Noctor G (2005) Oxidant and antioxidant signalling in plants: a re-evaluation of the concept of oxidative stress in a physiological context. Plant Cell Environ 28:1056–1071

Frei M, Kohno Y, Tietze S, Jekle M, Hussein MA, Becker T, Becker K (2012) The response of rice grain quality to ozone exposure during growth depends on ozone level and genotype. Environ Pollut 163:199–206

Galmés J, Aranjuelo I, Medrano H, Flexas J (2013) Variation in Rubisco content and activity under variable climatic factors. Photosynth Res 117:73–90

Gerosa G, Marzuoli R, Rossini M, Panigada C, Meroni M, Colombo R, Faoro F, Iriti M (2009) A flux-based assessment of the effects of ozone on foliar injury, photosynthesis, and yield of bean (Phaseolus vulgaris L. cv. Borlotto Nano Lingua di Fuoco) in open-top chambers. Environ Pollut 157:1727–1736

Ghosh A, Pandey AK, Agrawal M, Agrawal SB (2020) Assessment of growth, physiological, and yield attributes of wheat cultivar HD 2967 under elevated ozone exposure adopting timely and delayed sowing conditions. Environ Sci Pollut Res 27:17205–17220

Goumenaki E, Taybi T, Borland A, Barnes J (2010) Mechanisms underlying the impacts of ozone on photosynthetic performance. Environ Exp Bot 69:259–266

Guidi L, Degl'Innocenti E, Soldatini GF (2002) Assimilation of CO2, enzyme activation and photosynthetic electron transport in bean leaves, as affected by high light and ozone. New Phytol 156:377–388

Guidi L, Degl'Innocenti E, Martinelli F, Piras M (2009) Ozone effects on carbon metabolism in sensitive and insensitive Phaseolus cultivars. Environ Exp Bot 66:117–125

Gupta A, Yadav DS, Agrawal SB, Agrawal M (2022) Individual effects of high temperature and tropospheric ozone on tomato: a review. J. Plant Growth Regul 1–23

Hampton JG (2002) What is seed quality? Seed Sci Technol 30:1–10

Hao X, Hale BA, Ormrod DP, Papadopoulos AP (2000) Effects of pre-exposure to ultraviolet-B radiation on responses of tomato (*Lycopersicon esculentum* cv. New Yorker) to ozone in ambient and elevated carbon dioxide. Environ Pollut 110:217–224

Heijde M, Ulm R (2012) UV-B photoreceptor-mediated signalling in plants. Trends Plant Sci 17(4): 230–237

Hofmann RW, Campbell BD, Bloor SJ, Swinny EE, Markham KR, Ryan KG, Fountain DW (2003) Responses to UV-B radiation in *Trifolium repens* L. physiological links to plant productivity and water availability. Plant Cell Environ 26:603–612

Hoshika Y, Watanabe M, Inada N, Koike T (2013) Model-based analysis of avoidance of ozone stress by stomatal closure in Siebold's beech (*Fagus crenata*). Ann Bot 112:1149–1158

Ishii S, Marshall FM, Bell JN, Abdullah AM (2004) Impact of ambient air pollution on locally grown rice cultivars (Oryza sativa L.) in Malaysia. Water Air Soil Pollut 154:187–201

Jansen MAK, Gaba V, Greenberg BM (1998) Higher plants and UV-B radiation: balancing damage, repair and acclimation. Trends Plant Sci 3:131–135

Jaoude MB, Katerji N, Mastrorilli M, Rana G (2008) Analysis of the ozone effect on soybean in the Mediterranean region: II. The consequences on growth, yield and water use efficiency. Eur J Agron 28:519–525

Jayakumar M, Amudha P, Kulandaivelu G (2003) Changes in growth and yield of *Phaseolus mungo* L. induced by UV-A and UV-B enhanced radiation. J Plant Biol 46:59–61

Jing L, Dombinov V, Shen S, Wu Y, Yang L, Wang Y, Frei M (2016) Physiological and genotype-specific factors associated with grain quality changes in rice exposed to high ozone. Environ Pollut 210:397–408

Ji-Ling CA, Liang WA, Qing ZE, Liang J, Hao-Ye TA, Zu-Bin XI, Gang LI, Jian-Guo ZH, Kobayashi K (2009) Characteristics of photosynthesis in wheat cultivars with different sensitivities to ozone under O3-free air concentration enrichment conditions. Acta Agronomica Sinica. 35:1500–1507

Jovanić BR, Radenković B, Despotović-Zrakić M, Bogdanović Z, Barać D (2022) Effect of UV-B radiation on chlorophyll fluorescence, photosynthetic activity and relative chlorophyll content of five different corn hybrids. J Photochem Photobiol 10:100115

Kataria S, Jain M, Kanungo M, Sharma S (2019) Wheat responses and tolerance to UV-B radiation: an overview. Wheat Prod Chang Environ 175–196

Köllner B, Krause GH (2000) Changes in carbohydrates, leaf pigments and yield in potatoes induced by different ozone exposure regimes. Agric Ecosyst Environ 78:149–158

Kovilpillai B, Nedumaran S, Mani S, Raja Mani J, Natarajan S, Ramasamy J (2023) Impacts of elevated ozone and ozone protectants on plant growth, nutrients, biochemical and yield properties of turnip (Brassica Rapa L.). Ozone: Sci Eng 45:475–487

Kumari R, Agrawal SB, Singh S, Dubey NK (2009a) Supplemental ultraviolet-B induced changes in essential oil composition and total phenolics of *Acorus calamus* L. (sweet flag). Ecotoxicol Environ Saf 72:2013–2019

Kumari R, Singh S, Agrawal SB (2009b) Effects of supplemental ultraviolet-b radiation on growth and physiology of *Acorus calamus* L. (sweet flag). Acta Biol Cracoviensia Ser Bot 51:19–27

Kumari R, Agrawal SB (2010) Supplemental UV-B induced changes in leaf morphology, physiology and secondary metabolites of an Indian aromatic plant *Cymbopogon citratus* (DC) Staph under natural field conditions. Int J Environ Stud 67:655–675

Lake JA, Field KJ, Davey MP, Beerling DJ, Lomax BH (2009) Metabolomic and physiological responses reveal multi-phasic acclimation of Arabidopsis thaliana to chronic UV radiation. Plant Cell Environ 32:1377–1389

Laube JC, Newland MJ, Hogan C, Brenninkmeijer CA, Fraser PJ, Martinerie P, Oram D, Reeves C, Röckmann T, Schwander J, Witrant E, Sturges WT (2014) Newly detected ozone-depleting substances in the atmosphere. Nat Geosci 7(4): 266–269

Liu B, Liu X, Li Y-S, Herbert SJ (2013) Effects of enhanced UV-B radiation on seed growth characteristics and yield components in soybean. F Crop Res 154:158–163

Malik P, Kaur K, Kaur P, Seni S, Kaur H (2024) Breeding climate-smart cultivars: can transcriptomics enable strategic cereal breeding?. In Omics and System Biology Approaches for Delivering Better Cereals, pp 201–231. CRC Press

Mao B, Wang Y, Zhao TH, Tian RR, Wang W, Ye JS (2017) Combined effects of elevated $O_3$ concentrations and enhanced UV-B radiation of the biometric and biochemical properties of soybean roots. Front Plant Sci 8:1–14

Martínez-Lüscher J, Morales F, Delrot S, Sánchez-Díaz M, Gomés E, Aguirreolea J, Pascual I (2013) Short- and long-term physiological responses of grapevine leaves to UV-B radiation. Plant Sci 213:114–122

Miller JE, Booker FL, Fiscus EL, Heagle AS, Pursley WA, Vozzo SF, Heck WW (1994) Ultraviolet-B radiation and ozone effects on growth, yield, and photosynthesis of Soybean. J Environ Qual 23:83–91

Mills G, Sharps K, Simpson D, Pleijel H, Frei M, Burkey K, Emberson L, Uddling J, Broberg M, Feng Z, Kobayashi K (2018) Closing the global ozone yield gap: quantification and cobenefits for multistress tolerance. Glob Change Biol 24:4869–4893

Mills G, Harmens H, Wagg S, Sharps K, Hayes F, Fowler D, Sutton M, Davies B. Ozone impacts on vegetation in a nitrogen enriched and changing climate. Environ Pollut 208:898–908

Mishra AK, Agrawal SB (2015) Biochemical and physiological characteristics of tropical mung bean (Vigna radiata L.) cultivars against chronic ozone stress: an insight to cultivar-specific response. Protoplasma 252:797–811

Mishra AK, Rai R, Agrawal SB (2013) Differential response of dwarf and tall tropical wheat cultivars to elevated ozone with and without carbon dioxide enrichment: growth, yield and grain quality. Field Crops Res 145:21–32

Mittler R (2002) Oxidative stress, antioxidants and stress tolerance. Trends Plant Sci 7:405–410

Monks PS, Archibald AT, Colette A, Cooper O, Coyle M, Derwent R, Fowler D, Granier C, Law KS, Mills GE, Stevenson DS, Tarasova O, Thouret V, von Schneidemesser E, Sommariva R, Wild O, Williams ML (2015) Tropospheric ozone and its precursors from the urban to the global scale from air quality to short-lived climate forcer. Atmos Chem Phys 15:8889–8973

Moore FC, MacCracken MC (2009) Lifetime-leveraging: an approach to achieving international agreement and effective climate protection using mitigation of short-lived greenhouse gases. Int J Clim Chang 1:42–62

Morgan PB, Mies TA, Bollero GA, Nelson RL, Long SP (2006) Season-long elevation of ozone concentration to projected 2050 levels under fully open-air conditions substantially decreases the growth and production of soybean. New Phytol 170:333–343

Mukherjee A, Singh Yadav D, Agrawal SB, Agrawal M (2020) Ozone a persistent challenge to food security in India: Current status and policy implications. Curr Opin Environ Sci Heal 19:100220

Natali L, Vangelisti A, Guidi L, Remorini D, Cotrozzi L, Lorenzini G, Nali C, Pellegrini E, Trivellini A, Vernieri P, Landi M (2018) How *Quercus ilex* L. saplings face combined salt and ozone stress: a transcriptome analysis. BMC Genom 19:1–8

Oksanen E, Pandey V, Pandey AK, Keski-Saari S, Kontunen-Soppela S, Sharma C (2013) Impacts of increasing ozone on Indian plants. Environ Pollut 177:189–200

Osborne SA, Mills G, Hayes F, Ainsworth EA, Büker P, Emberson L (2016) Has the sensitivity of soybean cultivars to ozone pollution increased with time? An analysis of published dose–response data. Glob Change Biol 22:3097–3111

Pang J, Kobayashi K, Zhu J (2009) Yield and photosynthetic characteristics of flag leaves in Chinese rice (Oryza sativa L.) varieties subjected to free-air release of ozone. Agric Ecosyst Environ 132:203–211

Paoletti E (2006) Impact of ozone on Mediterranean forests: a review. Environ Pollut 144:463–474

Paoletti E, Grulke NE (2010) Ozone exposure and stomatal sluggishness in different plant physiognomic classes. Environ Pollut 158:2664–2671

Peng J, Shang B, Xu Y, Feng Z, Calatayud V (2020) Effects of ozone on maize (*Zea mays* L.) photosynthetic physiology, biomass and yield components based on exposure-and flux-response relationships. Environ Pollut 256:113466

Peng J, Xu Y, Shang B, Agathokleous E, Feng Z (2021) Effects of elevated ozone on maize under varying soil nitrogen levels: Biomass, nitrogen and carbon, and their allocation to kernel. Sci Total Environ 765:144332

Persson K, Danielsson H, Sellden G, Pleijel H (2003) The effects of tropospheric ozone and elevated carbon dioxide on potato (Solanum tuberosum L. cv. Bintje) growth and yield. Sci Total Environ 310:191–201

Piccini C, Cai G, Dias MC, Romi M, Longo R, Cantini C (2020) UV-B radiation affects photosynthesis-related processes of two Italian *Olea europaea* (L.) varieties differently. Plants 9:1712

Piikki K, Sellden G, Pleijel H (2004) The impact of tropospheric O3 on leaf number duration and tuber yield of the potato (Solanum tuberosum L.) cultivars Bintje and Kardal. Agric Ecosyst Environ 104:483–492

Pleijel H, Eriksen AB, Danielsson H, Bondesson N, Selldén G (2006) Differential ozone sensitivity in an old and a modern Swedish wheat cultivar—grain yield and quality, leaf chlorophyll and stomatal conductance. Environ Exp Bot 56:63–71

Pleijel H, Broberg MC, Uddling J, Mills G (2018) Current surface ozone concentrations significantly decrease wheat growth, yield and quality. Agric Ecosyst Environ 104(483–92613):687–692

Pleijel H, Danielsson H, Gelang J, Sild E, Selldén G (1998) Growth stage dependence of the grain yield response to ozone in spring wheat (*Triticum aestivum* L.) Agric Ecosyst Environ 70:61–68.

Rai R, Agrawal M (2008) Evaluation of physiological and biochemical responses of two rice (*Oryza sativa* L.) cultivars to ambient air pollution using open top chambers at a rural site in India. Sci Total Environ 407:679–691

Rai R, Agrawal M (2014) Assessment of competitive ability of two Indian wheat cultivars under ambient O3 at different developmental stages. Environ Sci Pollut Res 21:1039–1053

Rai K, Agrawal SB (2022) Effects of elevated ultraviolet-B on the floral and leaf characteristics of a medicinal plant Wedelia chinensis (Osbeck) Merr. along with essential oil contents. Trop Ecol 64:307–323

Rai R, Agrawal M, Agrawal SB (2007) Assessment of yield losses in tropical wheat using open top chambers. Atmos Environ 41:9543–9554

Rai R, Singh AA, Agrawal SB, Agrawal M (2016) Tropospheric $O_3$: a cause of concern for terrestrial plants. In: Plant responses to air pollution, pp 165–195

Rajabbeigi E, Eichholz I, Beesk N, Ulrichs C, Kroh LW, Rohn S, Huyskens-Keil S (2013) Interaction of drought stress and UV-B radiation-impact on biomass production and flavonoid metabolism in lettuce (*Lactuca sativa* L.). J Appl Bot Food Qual 86:190–197

Rajendiran K, Ramanujam MP (2004) Improvement of biomass partitioning, flowering and yield by triadimefon in UV-B stressed *Vigna radiata* (L.) Wilczek. Biol Plant 48:145–148

Ramya A, Dhevagi P, Priyatharshini S, Saraswathi R, Avudainayagam S, Venkataramani S (2021) Response of rice (Oryza sativa L.) cultivars to elevated ozone stress. Environ Monit Assess 193:808

Rao MV, Ormrod DP (1995) Impact of UV-B and $O_3$ on the oxygen free radical scavenging system in Arabidopsis thaliana genotypes differing in flavonoid biosynthesis. Photochem Photobiol 62:719–726

Rao MV, Paliyath G, Ormrod DP (1996) Ultraviolet-B and ozone-induced biochemical changes in antioxidant enzymes of Arabidopsis thaliana. Plant Physiol 110:125–136

Reddy KR, Patro H, Lokhande S, Bellaloui N, Gao W (2016) Ultraviolet-B radiation alters soybean growth and seed quality. Food Nutr Sci 07:55–66

Ri I, Pak S, Pak U, Yun C, Tang Z (2024) How does UV-B radiation influence the photosynthesis and secondary metabolism of Schisandra chinensis leaves? Ind Crop Prod 208:117832

Roberts HR, Dodd IC, Hayes F, Ashworth K (2022) Chronic tropospheric ozone exposure reduces seed yield and quality in spring and winter oilseed rape. Agric For Meteorol 316:108859

Runeckles VC, Krupa SV (1994) The impact of UV-B radiation and ozone on terrestrial vegetation. Environ Pollut 83:191–213

Saitanis CJ, Karandinos MG (2002) Effects of ozone on tobacco (*Nicotiana tabacum* L.) varieties. J Agron Crop Sci 188:51–58

Sampedro J, Waldhoff ST, Van de Ven DJ, Pardo G, Van Dingenen R, Arto I, del Prado A, Sanz MJ (2020) Future impacts of ozone driven damages on agricultural systems. Atmos Environ 231:117538

Santin M, Sciampagna MC, Mannucci A, Puccinelli M, Angelini LG, Tavarini S, Accorsi M, Incrocci L, Ranieri A, Castagna A (2022) Supplemental UV-B exposure influences the biomass and the content of bioactive compounds in *Linum usitatissimum* L sprouts and microgreens. Hortic 8:213

Sarkar A, Singh AA, Agrawal SB, Ahmad A, Rai SP (2015) Cultivar specific variations in antioxidative defense system, genome and proteome of two tropical rice cultivars against ambient and elevated ozone. Ecotoxicol Environ Saf 115:101–111

Sawada H, Tsukahara K, Kohno Y, Suzuki K, Nagasawa N, Tamaoki M (2016) Elevated ozone deteriorates grain quality of Japonica Rice cv. Koshihikari, even if it does not cause yield reduction. Rice 9:7

Schnitzler JP, Langebartels C, Heller W, Liu J, Lippert M, Dohring T, Bahnweg G, Sandermann H (1999) Ameliorating effect of UV-B radiation on the response of Norway spruce and Scots pine to ambient ozone concentrations. Glob Chang Biol 5:83–94

Schraudner M, Moeder W, Wiese C, Van Camp W, Inzé D, Langebartels C, Sandermann H (1998) Ozone-induced oxidative burst in the ozone biomonitor plant, tobacco Bel W3. Plant J 16:235–245

Schreiner M, Mewis I, Huyskens-Keil S, Jansen MAK, Zrenner R, Winkler JB, O'Brien N, Krumbein A (2012) UV-Binduced secondary plant metabolites-potential benefits for plant and human health. Crit Rev Plant Sci 31(3): 229–240

Schultze M, Bilger W (2019) Acclimation of Arabidopsis thaliana to low temperature protects against damage of photosystem II caused by exposure to UV-B radiation at 9 °C. Plant Physiol Biochem 134:73–80

Sen A, Puthur JT (2020) Influence of different seed priming techniques on oxidative and antioxidative responses during the germination of Oryza sativa varieties. Physiol Mol Biol Plants 26:551–565

Shao Z, Zhang Y, Hu S, Jing L, Wang Y, Wang Y, Yang L (2023) Identification of key responsive leaf traits for ozone tolerance in six modern indica and japonica rice cultivars (Oryza sativa L.) over two years. Agric Ecosyst Environ 349:108451

Shi G, Yang L, Wang Y, Kobayashi K, Zhu J, Tang H, Pan S, Chen T, Liu G, Wang Y (2009) Impact of elevated ozone concentration on yield of four Chinese rice cultivars under fully open-air field conditions. Agric Ecosyst Environ 131:178–184

Singh AA, Agrawal SB (2017) Tropospheric ozone pollution in India: effects on crop yield and product quality. Environ Sci Pollut Res 24:4367–4382

Singh AA, Agrawal SB, Shahi JP, Agrawal M (2014) Assessment of growth and yield losses in two Zea mays L. cultivars (quality protein maize and nonquality protein maize) under projected levels of ozone. Environ Sci Pollut Res 21:2628–2641

Singh AA, Singh S, Agrawal M, Agrawal SB (2015a) Assessment of ethylene diurea-induced protection in plants against ozone phytotoxicity. Rev Environ Contam Toxicol 233:129–184

Singh E, Tiwari S, Agrawal M (2010) Variability in antioxidant and metabolite levels, growth and yield of two soybean varieties: an assessment of anticipated yield losses under projected elevation of ozone. Agric Ecosyst Environ 135:168–177

Singh P, Agrawal M, Agrawal SB (2009) Evaluation of physiological, growth and yield responses of a tropical oil crop (Brassica campestris L. var. Kranti) under ambient ozone pollution at varying NPK levels. Environ Pollut 157:871–880

Singh P, Singh S, Agrawal SB, Agrawal M (2012) Assessment of the interactive effects of ambient O3 and NPK levels on two tropical mustard varieties (Brassica campestris L.) using open top chambers. Environ Monit Assess 184:5863–5874

Singh P, Agrawal M, Agrawal SB, Singh S, Singh A (2015b) Genotypic differences in utilization of nutrients in wheat under ambient ozone concentrations: growth, biomass and yield. Agric Ecosyst Environ 199:26–33

Singh P, Choudhary KK (2025) UV-B orchestration of growth, yield and grain quality traits highlights modifications of source-to-sink relationship in pearl millet cultivars. Physiol Plant 177:e70141

Singh AA, Fatima A, Mishra AK, Chaudhary N, Mukherjee A, Agrawal M, Agrawal SB (2018) Assessment of ozone toxicity among 14 Indian wheat cultivars under field conditions: growth and productivity. Environ Monit Assess 190:1–4

Singh AA, Ghosh A, Agrawal M, Agrawal SB (2023) Secondary metabolites responses of plants exposed to ozone: an update. Environ Sci Pollut Res 13:1–32

Singh S, Kumari R, Agrawal M, Agrawal SB (2011) Modification in growth, biomass and yield of radish under supplemental UV-B at different NPK levels. Ecotoxicol Environ Saf 74:897–903

Su M, Wang S, Li C, Zhang Z, Wang N, Li B, Chen X (2022) Ultraviolet-B-induced MdWRKY71-L expression regulates anthocyanin synthesis in apple. Environ Exp Bot 201:105000

Sun J, Feng Z, Ort DR (2014) Impacts of rising tropospheric ozone on photosynthesis and metabolite levels on field grown soybean. Plant Sci 226:147–161

Takshak S, Agrawal SB (2015) Defence strategies adopted by the medicinal plant *Coleus forskohlii* against supplemental ultraviolet-B radiation: augmentation of secondary metabolites and antioxidants. Plant Physiol Biochem 97:124–138

Takshak S, Agrawal SB (2019) Defense potential of secondary metabolites in medicinal plants under UV-B stress. J Photochem Photobiol B Biol 193:51–88

Teramura AH, Caldwell MM (1981) Effects Of ultraviolet-B irradiances on soybean. Iv. leaf ontogeny as a factor in evaluating ultraviolet-B irradiance effects on net photosynthesis. Am J Bot 68:934–941

Teramura AH, Sullivan JH, Lydon J (1990) Effects of UV-B radiation on soybean yield and seed quality: a 6-year field study. Physiol Plant 80:5–11

Tiwari S, Agrawal M (2018) Effect of ozone on physiological and biochemical processes of plants. In: Tropospheric ozone and its impacts on crop plants: a threat to future global food security, pp 65–113

Tripathi R, Agrawal SB (2012) Effects of ambient and elevated level of ozone on Brassica campestris L. with special reference to yield and oil quality parameters. Ecotoxicol Environ Saf 85:1–12

Tripathi R, Agrawal SB (2013) Interactive effect of supplemental ultraviolet B and elevated ozone on seed yield and oil quality of two cultivars of linseed (*Linum usitatissimum* L.) carried out in open top chambers. J Sci Food Agric 93:1016–1025

Tripathi R, Sarkar A, Pandey Rai S, Agrawal SB (2011) Supplemental ultraviolet-B and ozone: impact on antioxidants, proteome and genome of linseed (*Linum usitatissimum* L. cv. Padmini). Plant Biol 13:93–104

Tripathi R, Rai K, Singh S, Agrawal M, Agrawal SB (2019) Role of supplemental UV-B in changing the level of ozone toxicity in two cultivars of sunflower: growth, seed yield and oil quality. Ecotoxicology 28:277–293

Usano-Alemany J, Panjai L (2015) Effects of increasing doses of UV-B on main phenolic acids content, antioxidant activity and estimated biomass in Lavandin (*Lavandula* x *intermedia*)

Vainonen JP, Kangasjärvi J (2015) Plant signalling in acute ozone exposure. Plant Cell Environ 38:240–252

Van De Staaij JWM, Tonneijck AEG, Rozema J (1997) The effect of reciprocal treatments with ozone and ultraviolet-B radiation on photosynthesis and growth of perennial grass *Elymus athericus*. Environ Pollut 97:281–286

Vaultier MN, Jolivet Y (2015) Ozone sensing and early signaling in plants: an outline from the cloud. Environ Exp Bot 114:144–152

Vavasseur A, Raghavendra AS (2005) Guard cell metabolism and $CO_2$ sensing. New Phytol 165:665–682

Wang Y, Song Q, Frei M, Shao Z, Yang L (2014) Effects of elevated ozone, carbon dioxide, and the combination of both on the grain quality of Chinese hybrid rice. Environ Pollut 189:9–17

Wang H, Ma X, Zhang L, Siemann E, Zou J (2015) UV-B has larger negative impacts on invasive populations of *Triadica sebifera* but ozone impacts do not vary. J Plant Ecol 9:61–68

Wang H, Guo Y, Zhu J, Yue K, Zhou K, Li J, Huang X (2021) Characteristics of mango leaf photosynthetic inhibition by enhanced UV-B radiation. Hortic 7:557

Xu Y, Feng Z, Peng J, Tarvainen L (2022) Elevated ozone decreases the activity of Rubisco in poplar but not its activation under fluctuating light. Tree Physiol 42:1762–1775

Yadav DS, Mishra AK, Rai R, Chaudhary N, Mukherjee A, Agrawal SB, Agrawal M (2020a) Responses of an old and a modern Indian wheat cultivar to future $O_3$ level: physiological, yield and grain quality parameters. Environ Pollut 259:113939

Yadav DS, Rai R, Mishra AK, Chaudhary N, Mukherjee A, Agrawal SB, Agrawal M (2019) ROS production and its detoxification in early and late sown cultivars of wheat under future O3 concentration. Sci Total Environ 659: 200-210

Yadav DS, Rai R, Mishra AK, Chaudhary N, Mukherjee A, Agrawal SB, Agrawal M (2020b) ROS production and its detoxification in early and late sown cultivars of wheat under future $O_3$ concentration. Sci Total Environ 659:200–210

Yadav DS, Rai K, Gupta A, Agrawal SB, Agrawal M (2023) Cross-talk between ROS and phytohormones signaling determines crop sensitivity against ozone. In: Ozone pollution and plant health; series: Advances in botanical research 108. Academic Press, pp 47–81

Yao Y, Li Y, Yang Y, Li C (2005) Effect of seed pretreatment by magnetic field on the sensitivity of cucumber (Cucumis sativus) seedlings to ultraviolet-B radiation. Environ Exp Bot 54(3): 286–294

Yao Y, Xuan Z, Li Y, He Y, Korpelainen H, Li C (2006) Effects of ultraviolet-B radiation on crop growth, development, yield and leaf pigment concentration of tartary buckwheat (Fagopyrum tataricum) under field conditions. Eur J Agron 25:215–222

Yu GH, Li W, Yuan ZY, Cui HY, Lv CG, Gao ZP, Han B, Gong YZ, Chen GX (2013) The effects of enhanced UV-B radiation on photosynthetic and biochemical activities in super-high-yield hybrid rice Liangyoupeijiu at the reproductive stage. Photosynthetica 51:33–44

Yuan X, Hu X, Gun S, Qu L, Yan Y, Agathokleous E (2025) Ozone and nitrogen deposition effects on plant physiology and growth vary among tree species, not plant functional types. Ecosyst Health Sustain 11:0348

Zhang J, Gao F, Jia H, Hu J, Feng Z (2019) Molecular response of poplar to single and combined ozone and drought. Sci Total Environ 655:1364–1375

Zheng F, Wang X, Zhang W, Hou P, Lu F, Du K, Sun Z (2013) Effects of elevated $O_3$ exposure on nutrient elements and quality of winter wheat and rice grain in Yangtze River Delta, China. Environ Pollut 179:19–26

Zhang W, Wang G, Liu X, Feng Z (2014) Effects of elevated O3 exposure on seed yield, N concentration and photosynthesis of nine soybean cultivars (Glycine max (L.) Merr.) in Northeast China. Plant Sci 226:172–181

Zlatev ZS, Lidon FJC, Kaimakanova M (2012) Plant physiological responses to UV-B radiation. Emirates J Food Agric 24:481–501

Zouzoulas D, Koutroubas SD, Vassiliou G, Vardavakis E (2009) Effects of ozone fumigation on cotton (Gossypium hirsutum L.) morphology, anatomy, physiology, yield and qualitative characteristics of fibers. Environ Exp Bot 67:293–303

# Chapter 6
# Plant Responses Under Combined Effect of Atmospheric Hydrogen Sulfide and Tropospheric Ozone

**Hauwa Muhammad Lawan and Ashutosh K. Pandey**

**Abstract** Ground-level ozone, also known as tropospheric ozone, is a significant air contaminant which poses serious threats to agricultural productivity and ecosystem health. This chapter examines the detrimental effects of ozone on crop yields, emphasizing its role in reducing photosynthesis rates, biomass accumulation, and fruit quality. The concentration of ground-level ozone has more than doubled since pre-industrial times, with its impacts varying based on concentration levels, specific crop species, and exposure duration. The document highlights the increasing emissions of ozone precursors in Asia, contrasting with the reductions observed in Europe and North America. Additionally, it discusses the interaction of ozone with other pollutants, such as hydrogen sulfide ($H_2S$), and the complex signaling mechanisms in plants that mediate their responses to these stressors. Understanding these dynamics is crucial for developing strategies to breed ozone-tolerant crops and mitigate the opposing effects of air pollution on agriculture.

**Keywords** Crops · Hydrogen sulfide · Ozone · Antioxidant defenses

## 6.1 Introduction

By the end of the twentieth century, environmental issues became a major concern for human survival and well-being. The startling rate at which air pollutant concentrations are rising is among the biggest environmental obstacles fronting the modern universe (Nguyen et al. 2022). Over the last few decades, the issue of tropospheric $O_3$ pollution in the atmosphere has grown significantly and become a major world

H. M. Lawan
Food Science and Technology, Department of Life Sciences, Sharda School of Bio-Science & Technology, Sharda University, Greater Noida, India

A. K. Pandey (✉)
Department of Life Sciences, Sharda School of Bio-Science & Technology, Sharda University, Greater Noida, India
e-mail: Aashu.p20@gmail.com; Ashutosh.pandey1@sharda.ac.in



A. K. Pandey and E. Oksanen (eds.), *Tropospheric Ozone and Food Security*,
https://doi.org/10.1007/978-3-032-03608-7_6

apprehension. Ground level $O_3$ has been acknowledged as the primary air contaminant influencing of crop productivity losses in the majority of the universe, posing a serious risk to the world's ability to produce enough food to feed its expanding population. While $O_3$ source emissions have risen in Asia over the past few decades, $O_3$ quantities at the top have decreased in North America and Europe (Rai and Agarwal 2012). For much of eastern China, surface ozone pollution is a major air quality concern (Wang et al. 2020). Crop damage caused by ozone has been extensively documented in North America and Europe (Emberson et al. 2009; Rai and Agarwal 2012). The concentrations of $O_3$ at the ground level have considerably increased due to the rising emissions of nitrogen oxide and reactive hydrocarbons in urban areas. $O_3$ sources show that $O_3$ amounts are greater in rural than urban areas (Emberson et al. 2009; Rai and Agarwal 2012; Nguyen et al. 2022).

Ground level ozone ($O_3$) is a subordinate air pollutant, but contempt its damaging effects on health of human beings, different studies have confirmed that it reasons significant harm to plants and agricultural product (Amann et al. 2008; Fuhrer 2009). Many agricultural crops experience significant yield losses when exposed to higher ozone concentrations more than 40 ppb (part per billion) by volume (ppbv) (Fuhrer et al. 1997). Just exposure to $O_3$ damages more than 90% of the vegetation (Felzer et al. 2007). Elevated ozone concentrations have been shown in numerous biological and ecological studies to seriously impair vegetation and diminish crop productivity, which can have a negative financial impact and jeopardize food security (Li et al. 2022).

People have known for many years about the harmful consequences of hydrogen sulfide, a naturally occurring colorless, highly soluble, and flammable gas (Christou et al. 2013). Over the past ten years, a growing body of data has highlighted its biological roles in mammals. It is ranked third among gaseous transmitters, afterward carbon monoxide (CO) and nitric oxide (NO) (Jia et al. 2015).

The impact of significant geothermic development on productivity on agriculture may be significant due to the release of non-condensable stack airs, mainly carbon dioxide ($CO_2$) and hydrogen sulfide ($H_2S$). The most significant phytotoxic air pollutant, ozone ($O_3$), is found downwind of urban areas close to recognized geothermal zones. Prior research on the effects of $H_2S$ on plants has only examined visible signs of damage to a widespread range of yields, as well as the reduction of photosynthesis in spinach and lettuce. While levels between 0.03 and 0.10 ppm significantly stimulated growth in sugar beet, alfalfa, and lettuce, nonstop disinfection of multiple crop and different species of tree at 3.0 ppm $H_2S$ resulted in visible damage, growth reduction and demise of subtle species; no interaction with O a was investigated: As a result, a field fumigation study was conducted to examine the joint consequences of hydrogen sulfide and ozone on snap bean (*Phaseolus vulgaris* L.) yields (Coyne and Bingham 1978b).

## 6.2 Discovery of Ozone

In 1839, during an experiment involving the electrolysis of water, Christian Friedrich Schonbein made the initial discovery of Ozone. One of Europe's first measurements was carried out in Paris by Albert-Lévy. According to his demonstration, $O_3$ was 11 ± 2 ppbv between 1876 and 1910 (Tingey et al. 2004). The first ongoing measurements were conducted in 1956 at the German site of Arkona-Zingst. The results indicated that during the 1950s and 1960s, ozone was in the range of 15–20 ppbv (Lee et al. 2003). The first negative effects of $O_3$ were documented in Inland valleys surrounding Los Angeles in which frequently saw highest annual ozone levels of 300–400 ppbv during the 1970s. The number of days with ozone greater than 95 ppbv increased from 114 in 1963 to 163 days in 1978 as a result of air pollution (Gaudel et al. 2018). According to Zhang et al. (2015), there was an increase of 0.35 ppbv annually in Japan and South Korea between 2000 and 2014, and a rise of 2–4 ppbv per decade was advanced described in other places. Recent reports of inter annual and decadal variations have surfaced elsewhere. In the plains of North China, increases of 3–5 ppbv were observed between 2001 and 2006; these increases were primarily ascribed to variations in temperature and cloud cover, with afforestation and increased VOC production having a negligible impact of 1–2% (Zhang 2015). According to Li et al. (2014), there appears to be a larger concentration rise toward southern latitudes. Correspondingly, El Tololo Mountain in Chile has shown an increase of 0.66 ppbv every ten years.

## 6.3 Tropospheric Ozone Formation

About 90% of ozone layer is found in the stratosphere. Although ground level or surface $O_3$ accounts for about 10% of the entire $O_3$ column, its concentrations have been steadily rising over last century. Tropospheric $O_3$ production is the consequence of photochemical responses concerning carbon monoxide (CO), methane ($CH_4$), and other hydrocarbons in the existence of $NO_x$ ($NO + NO_2$). Ozone likewise can be destroyed through photochemical responses with NO, $HO_2$, or OH. 63% of $NO_x$ emissions come from burning fossil fuels; other small sources include lightning (10%), burning biomass (14%), soils (11%), and other sources (Wang et al. 2019).

In addition, burning fossil fuels, direct fuel evaporation, the use of solvents in chemical reactions, and natural vegetation can all produce hydrocarbons1. In out-of-town or rural areas with little $NO_x$ concentration, $O_3$ production is usually $NO_x$-limited; in urban areas with high $NO_x$ concentrations, it is VOC-limited. $O_3$ can also enter a region through local winds. Due to differing longitudinal disseminations of $NO_x$ and VOC production as well as NO destruction of $O_3$, the highest concentrations of $O_3$ are found windward of municipal centers rather than in urban areas. When the temperature is high and the sun is shining, O production occurs (Oltmans and Levy 1994).

1. **Formation of Ozone from $NO_x$**

$$NO + 1/2O_2 \rightarrow 2NO_2$$
$$\downarrow$$
$$NO_2 + h\nu \rightarrow NO + O$$
$$\downarrow$$
$$O + O_2 \rightarrow O_3$$

   $NO_2$ weakens when $O_3$ grasps its highest. $O_3$ concentration grasps during the late morning and early afternoon hours.

2. **Establishment of Ozone from Carbon Monoxide**

$$CO + 2O_2 + h\nu \rightarrow CO_2 + O_3$$

3. **Formation of Ozone from Methane**

$$CH_4 + 4O_2 + 2h\nu \rightarrow HCHO + H_2O + 2O_3$$
$$\downarrow$$
$$HCHO + h\nu \rightarrow H + HCO$$
$$\downarrow$$
$$HCO + h\nu \rightarrow H + CO$$
$$\downarrow$$
$$CO + 2O_2 + h\nu \rightarrow CO_2 + O_3$$

4. **Formation of Ozone from Methane Hydrocarbon**

$$RH + 4O_2 + 2h\nu \rightarrow RCHO + H_2O + 2O_3$$

### *6.3.1 Surface Ozone Levels in India*

India has been providing surface $O_3$ readings since 1989 from a number of locations, such as Delhi, Varanasi, Chandigarh, Ahmedabad, Agra, Bhubaneshwar, Berhampur, Cochin, Pune, and so on. Variations in surface $O_3$ amongst sites may be due to localized levels of human activity, different seasons for monitoring, or different methods. Bhatia et al. (2010) reported that $O_3$ concentration fluctuated annually between 24 and 63 ppbv in Varanasi and between 7 and 46 ppbv in Delhi. Oxygen fluctuates according to the season, reaching definite maxima in the summer and fall and minima in the winter and monsoon. At New Delhi, India, average $O_3$ concentrations varied from 10 to 22 ppb in the winter to 23– 44 ppb in the summer.

Delhi's surface $O_3$ values regularly surpassed the 80 parts per billion hourly averages required by the WHO for ambient air quality. There are very few reports on surface $O_3$ levels in rural and isolated areas, despite the high risk of $O_3$ damage to agricultural crops in these areas. Surface $O_3$ concentrations across India's regions often accumulated above the 40 ppb (AOT 40) threshold, often exceeding 3000 ppb, according to modeling analysis 45. It has been discovered that crop plants have an AOT40 as high as 5.3 ppm h; this value was measured over the course of three months during daylight hours (Fuhrer et al. 1997).

Research conducted by Emberson et al. (2009) found that mean 7 h $O_3$ concentrations (M 7) in many parts of South Asia can reach 50–90 ppb. Global photochemical models predict that $O_3$ concentrations in some parts of Asia will rise significantly more until 2030 under current legislative emission scenarios. According to Meehl et al. $O_3$ concentrations in Asia are expected to rise by 20–25% between 2015 and 2050 and by 40–60% by 2100. Despite ideal climate conditions for the formation of $O_3$, very little information from systematic monitoring of $O_3$ levels exists in India. Ozone concentrations across the whole Indian region ranged from 25 to 100 ppb, according to Mittal et al. (2007) study, which used the HANK model. The researchers make it abundantly evident that the concentration of $O_3$ increased steadily between 1992 and 2008, with higher peaks found in rural regions. $O_3$ concentrations ranged from 6 to 10.2 ppb at Varanasi urban sites in a field transect study conducted between 1989 and 1991 (Pandey and Agrawal 1992). At a municipal site in Delhi, hours of daylight $O_3$ concentrations (9 h) were reported to vary between 9.4 and 128.3 ppb during the same period. According to Singh et al. 1997 during the winter of 1993, Delhi's 10 h ground level mean $O_3$ concentrations ranged from 34 to 126 ppb. In Pune, India, from August 1991 to July 1992, the average daily $O_3$ concentration was 27 ppb, and the hourly concentration ranged from 2 to 69 ppb (Khemani et al. 1995). Lal et al. (2000) examined the outline of $O_3$ concentrations at a town site in Ahmadabad, India, from 1991 to 1995. They found that it was uncommon to find mean $O_3$ concentrations during the daytime that were higher than 80 ppb. According to Jain et al. (2005), at a town site in New Delhi, the monthly average $O_3$ concentrations varied from 50 to 82 ppb in the autumn (October–November) and from 62 to 95 ppb in the summer (April–June).

Amongst 2002 and 2006, the 7-hourly $O_3$ levels at Varanasi suburban sites ranged from 23.4 to 62.4 ppb. Seasonal fluctuations in ozone concentration were noted, through summer having the highest concentration, winter following, and the rainy season having the lowest. Mean $O_3$ concentrations at a 12 h period varied from 36.4 to 48 ppb at a residential site during the growth of wheat period (November–March) and from 23.4 to 44.4 ppb at a rustic site during the growth of rice period (July–October) (Rai et al. 2007). The mean $O_3$ concentrations measured in 2007–2008 and 2008–2009 at a rural Varanasi site by Sarkar and Agrawal (2010) were 45.3 and 47.3 ppb, correspondingly. $O_3$ concentrations ranged from 2.0 to 137 ppb in 2008–2009 and from 4.0 to 125 ppb in 2007–2008, respectively.

The highest concentrations of surface ozone over the contaminated Northern station Delhi typically occur in May, when levels surpass 60 parts per billion, and the lowest, observed in August, when levels fall below 20 parts per billion. Strong insolation, which facilitates the photochemical ozone formation in the presence of precursors, is the cause of the elevated levels of surface ozone detected in May. The variability of surface ozone over Varanasi, another polluted eastern station, is comparable to that of Delhi (apart from the winter minimum), peaking at 60 parts per billion in May and falling to a minimum of 20 parts per billion in August. At the Udaipur Western station, surface ozone levels range from 35 to 40 parts per billion to 15–20 parts per billion in April, with the lowest values recorded in September. Yadav et al. (2016) also reported similar maxima and minima in surface ozone over Udaipur.

The weekly, monthly, annually, yearly, periodic, and diurnal scales of the supreme daily 8-h average (MD8A) are defined from 2020 to 2022. As more than 60% of these cities have ozone levels above the World Health Organization (WHO) starting point (70 $\mu g\ m^{-3}$), elevated ozone levels remain alarm in the Northern, Central, Eastern, and Western sub-regions. In Northern, Western, and Central sub-regions, the average percentage of cities exceeding WHO grade (100 $\mu g\ m^{-3}$) is 15%. Between 2020 and 2021, there was a decline in the mean concentration of ozone, but in 2022, not so much. While the monthly variation in ozone levels in Southern India is relatively stable, it peaks in the Western, Northern, and Central sub-regions between April and May (Kumar et al. 2024).

In the Northern part of India, ozone production is restricted to $NO_x$ over Pantnagar, a semi-urban area only 225 km northeast of New Delhi, and VOC over the nation's capital, New Delhi (Chen et al. 2020).

## 6.4 Uptake of Ozone

This is caused by the complex process of plant uptake of $O_3$, which comes from micrometeorology and enters the plant canopy. $O_3$ is absorbed by plants from the unstable atmosphere overhead. After entering the covering, ozone can be engrossed by tissues, leaves, stems, and soil. Stomata principally absorb $O_3$ to the leaves. Reduced stomatal conductance is often observed following $O_3$ exposure. Various resistances at varying intensities affect the $O_3$ flux from the troposphere into leaves. Boundary layer and stomatal resistances are high due to high irradiance, high temperatures, and low wind speeds, which restricts $O_3$ uptake. Therefore, high $O_3$ levels are usually observed in the summer and spring (Zhao et al. 2024).

Therefore, the width of stomatal pores acts a major role in determining the uptake of $O_3$ (at a given atmospheric concentration), resulting in substantial uptake occurring under environmental conditions that favor stomatal opening and low uptake occurring under conditions that favor stomatal closure. The levels of ambient $O_3$ and stomatal conductance both affect how much $O_3$ is absorbed by leaves. The regulation of stomatal opening ensures that the plant gets enough carbon dioxide ($CO_2$) without experiencing a deficiency of water in its leaves (Ariura et al. 2023).

$O_3$ is phytotoxic because of its strong oxidant potential (redox latent = + 2.07 V) and ability to produce other lethal species like $OH^-$, $O^-$, and HO. The structure, composition, and purpose of the plasma membrane are attacked by these harmful substances, which allows $O_3$ to exert its main effects. Transportable across membranes, $H_2O_2$ can trigger the production of additional hazardous species and initiate a signaling chain involving messenger molecules. It has been suggested that $H_2O_2$, salicylic acid, Jasmonate, and calcium function as second messengers in signal pathways for plant defense reaction (Prophet et al. 2024).

## 6.5 Losses in Crop Yield Due to Ozone and Their Consequences for Global Agriculture

Increased $O_3$ concentrations during crop growth seasons result in significant reductions in global crop yields. Current and projected $O_3$ levels' impact on crop productivity in numerous portions of the biosphere have been estimated through a quantity of modeling researches. Studies that combine earth system models or atmospheric chemical transport models with data on world crop dispersal have found relationships statistically among crop yield and ozone, allowing for the determination of together current and potential agricultural yields (Van Dingenen et al. 2009; Avnery et al. 2011; Tang et al. 2013).

According to Burney and Ramanathan (2014) and McGrath et al. (2015), a number of studies have used statistical models to compute yield variations related to $O_3$ over the preceding 20–30 years by combining ancient state or county yield facts with data on air contamination and climate variability. From both approaches, it is evident that crop yields are being considerably reduced by the current levels of $O_3$. Global pollution from $O_3$ is estimated to lower harvests of 3–4% for rice, 2–4% for maize 4–15% for wheat, and 5–15% for soybeans (Van Dingenen et al. 2009; Avnery et al. 2011). A solitary median dose–response function is assigned to, respectively, crop species and then applied to all international locations.

Given that crop species differ greatly in their genotypic sensitivity to ozone, this is a known limitation. Furthermore, the $O_3$ relationship with dose in Asian germplasm is different from that of North American germplasm (Emberson et al. 2009; Osborne et al. 2016). Moreover, the $O_3$ response is influenced by variations in other environmental factors like temperature, drought, and nitrogen availability (Fuhrer and Booker 2003; Yendrek et al. 2013; McGrath et al. 2015).

Notwithstanding these doubts, modeling studies show unequivocally that the detrimental properties of $O_3$ on world agricultural production are expensive—at least as expensive as the predictable costs of the most current climate alteration to agriculture (Lobell and Field 2007).

Global estimates suggest that in 2000, crop yield losses resulting from elevated tropospheric $O_3$ accounted for approximately 40% of global economic losses, primarily in China and India. Short-lived pollutants such as black carbon also affect these areas, in addition to $O_3$ pollution. Because of climate change and brief air contaminants, wheat yields were predicted to reduce by 36% in 2010; air pollution was responsible for 90% of the yield loss (Burney and Ramanathan 2014).

Without stronger laws governing air contamination, it is estimated that by 2020, augmented ozone will supplementary lower wheat yields in China by 8.1–3.4% and in India by 5.4–7.7% (Tang et al. 2013). Further projections suggest that, by 2050, wheat production in South Asia may drop by 40% due to rising $O_3$ pollution (Tai et al. 2014). Based on Chuwah et al. (2015), this could result in an 8.9% increase in the amount of agricultural land required in Asia.

Crop yields can be significantly impacted by the combined effects between planting season temperature, water accessibility, and ozone, according to an analysis of past patterns in tropospheric and agricultural yields. According to McGrath et al. (2015), $O_3$ had no discernible impact on historically high US maize and soybean yields in unusually rainy years. On the other hand, background $O_3$ has been demonstrated to lower soybean and maize yields by 10–20% in dry years, contingent on the temperature of the growing season. The effects of $O_3$ may worsen as a result of warmer, drier growing seasons brought on by climate change.

Furthermore, research by McGrath et al. (2015) demonstrated that firmer regulations on air pollution can quickly offset the undesirable properties of $O_3$ on crop yields. Circumstantial $O_3$ has lessened its negative impact on US maize and soybean yields since 2000, when the Environmental Protection Agency in the United States imposed stronger regulations for $O_3$ pollution (Federal Register 1997). Increasing regulations on $O_3$ precursors in a quantitative and timely manner can significantly increase crop productivity; other countries should take note of this encouraging trend in the USA.

The impacts of climate change, both alone and in conjunction with ozone exposure, on crop yield were examined by Da et al. (2022). The yields of winter wheat, soybeans, corn, cotton, barley, spring wheat, sorghum, peanuts, rice, sorghum, and sunflower were all affected in this way in the US. All crops were found to be severely damaged by ozone exposure, with the exception of winter wheat and barley. Furthermore, for crops damaged by ozone, they found that for every 1 ppm h upsurge in AOT40, annual yields decrease by 0.3–0.2%. This translates into an annual loss of $6.03 billion (in US dollars) for these damaged crops, or 5.8% of the total revenue in 2018. They also found that, contrary to results from chamber experiments, wheat grown in winter is not susceptible to exposure to ozone. The researchers theorize that the discrepancy may be caused by the lower levels of ozone in main winter wheat-growing areas (like the Great Plains) as well as the chiller production seasons (early spring). They also corroborate experimental data that corn generally shows higher sensitivity than soybeans, which contradicts the findings of the chamber experiment (Da et al. 2022).

Estimates of crop loss based on surface ozone observations from 17 sites for rice and wheat by Shyam et al. (2017) indicate that crop loss occurs unquestionably in India. The range of rice loss is between 0.3 and 6.7%, whereas the range of wheat loss is higher, ranging from 4.2 to 15%. During the winter wheat cropping season, the average ozone levels are highest in the Northern and Western regions. Because of their higher yields, these two regions significantly contribute to India's overall wheat loss. In India, ozone levels are considerably lower during the summer monsoon and rice cropping seasons than they are during the winter. This guards against serious ozone-related damage to the rice crop throughout the kharif season. As was pointed out earlier, the kharif season accounts for 85–90% of India's annual harvest of rice.

These crop losses for India are compared with predictions from other sources. In 2015, Sinha et al. assessed the loss of rice and wheat in two Northern Indian states (Haryana and Punjab) to be between 19 and 21% and 11–41%, respectively, based on observational data from a single site. According to Debaje's (2014) estimation, 3–16% and 5–30% of India's wheat and rice, respectively, were lost. His predictions are based on information gathered from three sites in the Indian state of Maharashtra and six distinct places in various states and territories. In the Northern Indian region, predictions for wheat and rice currently range from 4 to 13 and 1–8%, correspondingly. Although the findings are based on a vast network of observation sites, they remain far less than the earlier estimations.

Yi et al. (2020) used a unique county-specific group to conduct a thorough investigation of the effect of ozone pollution on maize yield. The experimental approach included a spatial error model since the spatial connection. According to the researchers' analysis, there is a negative correlation between the amount of surface ozone and maize yield, which is dependable with preceding ecological and economic studies. Moreover, the outcome of the nonlinear simulation indicates that maize is most vulnerable to ozone during the second month of the growing period.

Earlier environmental research did not clearly show the effect of ozone on plant seedlings, but as the plants age, this effect will become more pronounced. Using the empirical data collected during this period, we demonstrate that between 2013 and 2015, the average yearly yield of maize was reduced by 4.234 million tons due to ozone pollution, or approximately 1.9% of the total domestic output. More than any other region, Shandong, being the principal producer of maize, lowers its annual loss of maize output by nearly one million tons (Yi et al. 2020).

The study conducted by Pellegrini et al. (2011) and Shakeri et al. (2016) found that the content of anthocyanins, phenolic acids, and tannins were all substantially greater in *Melissa officinalis*, a common herbal remedy used for treating anxiety, dementia, and diseases related to the central nervous system (CNS) with higher concentration of ozone.

Researchers from Brazil investigated how long-term exposure to ozone affected *Capsicum baccatum*. The pericarp of the ozone-exposed plants presented a 50% reduction in both capsaicin and dihydrocapsaicin compared to control plants, which showed no change in either compound. Additionally, according to Bortolin et al. (2016), the pericarp's total phenolic content and carotenoid increased by 17% and 52.8%, correspondingly.

There are several ways in which ozone can reduce crop yields. The initial instance is acute noticeable harm, which can cause a species whose market value is dependent on how they look, such as many horticultural crops, to clearly and immediately lose economic value. Several reports of ozone harm to crops have been made in Europe and North America; the most damage has been experienced by crops grown with irrigation in the Mediterranean province (Fumigalli et al. 2001).

In Europe and North America, crop ozone injury is well-documented, with watered crops in the Mediterranean region experiencing the worst effects (Fumigalli et al. 2001). For instance, in 1998, there was an ozone episode north of Athens that triggered *Lactuca sativa* and *Chicorium endivum* to have such significant necrosis and swelling that the crops could not be sold locally (Velissariou 1999), which had disastrous commercial effects on local manufacturers.

Numerous instances of ozone-induced apparent harm to crops outside of North America and Western Europe have been documented by Emberson et al. (2003) and (2001). When there isn't any visible harm, ozone can reduce the commercial yield of many different types of crops, primarily by speeding up the senescence of leaves and lowering photosynthetic rates. There is a wealth of evidence showing that ambient levels of ozone have a detrimental effect on crop yields in North America and Western Europe, but little is known about how ozone impacts crop yields in other regions. Wahid et al. (2005) carried out research in Pakistan which confirmed that the sifting of ambient air pollution at a Lahore suburb improved the yields of local wheat (*Triticum aestivum*) and rice (*Oryza sativa*) by about 40%.

This was at a location where sulfur dioxide ($SO_2$) concentrations were minimal but ozone and nitrogen dioxide ($NO_2$) levels were critical. Succeeding measured fumigation experiments with these cultivars revealed that no growth or yield was affected by $NO_2$ at levels found at the Lahore site, but ozone at these levels had very a major impact on native varieties of both rice and wheat (Maggs and Ashmore 1998).

Ozone can also significantly affect the productivity of *Phaseolus vulgaris* in the Valley of Mexico (de Bauer 2003), soybean (Glycine max) in Pakistan's Punjab (Wahid et al. 2001), radish and turnip yield at a suburban site in the Nile delta (Hassan et al. 1995), and *Lycopersicon esculentum* within and around New Delhi, according to studies conducted with ozone-protecting chemicals. Even though there isn't any hard evidence that ozone harms crop in mainland China, new research indicates that ozone concentrations in rural areas are sufficiently elevated to hypothetically affect the yields of other crops, including wheat grown in winter (Chameides et al. 1999; Zheng et al. 1998).

Black et al. (2000) described that ozone directly affects germination of pollen and tube growth, fertilization, and the abscission or abortion of flowers, pods, individual ovules, or seeds. Among these studies were those that Stewart et al. (1996), Bosac et al. (1994), and Schoene et al. (2004) published.

Total yield decreases (RYLs) for the eight-year average from 2010 to 2017 were as follows: 8.13–25.14% for wheat; 6.72–10.71% for double early rice; 7.11–8.53% for twice late rice; 6.79–10.22% for single rice; and 0.48–5.29% for northward maize, up to 94% for south maize, and up to 21.48% for soybeans, according to Avnery et al. (2011). The countrywide average ryls (ARYLs) for wheat were between 11 and 22%; for rice, they were between 8 and 9%; for maize, they were between 2 and 4%; for south maize, they were approximately 1%; and for soybeans, they were between 8.27 and 12.59%. The ARYLs consider the percentages of crop production loss relative to the total production that would be produced in a lack of ozone pollution.

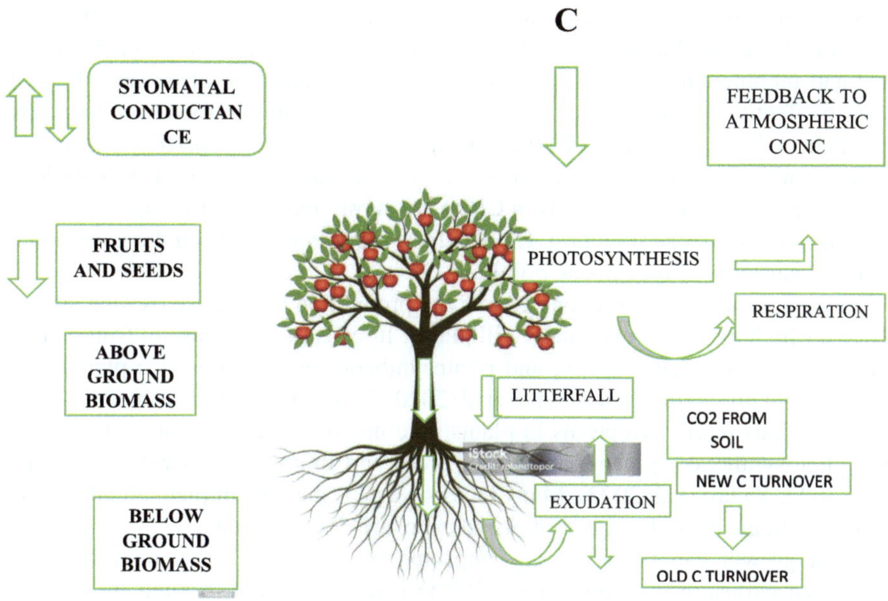


An illustration representing the effects of ozone on C-cycling in vegetation. *Source* Ciais et al. (2014)

Previous studies (Avnery et al. 2011; Lin et al. 2018; Feng et al. 2019; Zhao et al. 2021; Wang et al. 2022) produced comparable outcomes. Using yearly agricultural output data from the Statistical Yearbook of China, the researchers also discovered that, between 2010 and 2017, the country's total CPL for wheat fluctuated between 13.81 and 36.51 Mt. Single, early, double and late rice were included in the yearly CPLs for rice, which ranged from 16.89 2022Mt in 2010–20.03 Mt in 2014. The range of corn CPLs was 4.59–8.17 Mt, and the range of soybean CPLs was 1.09–1.84 Mt. In light of this, the economic impacts of surface ozone exposure varied widely: for wheat, they fluctuated from USD 3.86 billion to USD 14.29 billion; for double rice (early and late rice that led nearly similarly), they varied from USD 2.96 billion to USD 6.49 billion; for maize (with north maize accounting for over 90% of the total), they reached from USD 1.16 billion to USD 3.53 billion; and for soybeans, they ranged from USD 0.82 billion to USD 1.43 billion. Chinese crop productivity significantly decreased across 2010 and 2017 as a result of ozone depletion. The primary staple crops grown by the Indian population are rice and wheat, both of which are highly vulnerable to rising $O_3$ pollution. The results of the M7 and AOT40 indices indicate that the invention of agricultural crops in India may be highly vulnerable to rising $O_3$ concentrations brought on by an increase in anthropogenic pollution in both rural and urban areas (Avnery et al. 2011; Lin et al. 2018; Feng et al. 2019; Zhao et al. 2021; Wang et al. 2022).

The need for a more organized and balanced approach to controlling ozone precursors is underscored by the intersections between main crop growing regions, inhabited and manufacturing zones, and hotspots for concentration of ozone. This will help to effectively bound concentration of ozone and maintain food sanctuary. The researchers demonstrated that China's stringent air quality laws, particularly in rural areas, significantly decreased the losses in crop yield linked to high ground level ozone exposure. Still, crop yields in China are surprisingly heavily impacted by the current levels of ozone pollution. China's food security depends on the government working harder to reduce ozone pollution.

Being a strong oxidant, ozone causes the production of reactive oxygen species (ROS), which harm and eventually kill plants. It leads to a decrease in fruit quality, harvest, biomass accumulation, and photosynthetic rate. It stops rice plants from growing and producing yields (Shao et al. 2020; Seth et al. 2012). Elevated $O_3$ levels cause obvious injury symptoms in plants, like necrosis and chlorosis, which show that photosynthetic pigments are out of order and impede photosynthesis (Gupta et al. 2023; Yadav et al. 2022). Many factors, such as the type of plant and the ozone regime, influence the damage that ozone exposure does to plants.

Plant growth can be negatively impacted by oxidative stress resulting from prolonged exposure to ozone stress. Ozone can enter a leaf through its stomata, altering the chemical composition of its cells. The ozone-mediated oxidation process raises ROS activity. A plant's ability to withstand oxidative damage is more susceptible early in the vegetative development process than it is later. It has been observed that most plants under ozone stress exhibit reduced leaf area, biomass, and overall development. When exposed to high levels of ozone stress, some plants may show either positive or negative responses in their photopigment content (Han et al. 2020).

The projected amount of wheat loss owing to ozone in advanced nations, higher middle-income countries and regions, and lower middle-income countries and territories is 39.7%, 31.8%, and 26.8%, respectively (Lal et al. 2017).

Additionally, a rise in ozone concentration would modify the composition of chemical of the foliage, changing the way plants and insects connect and ultimately impacting the services provided by the ecosystem (Agathokleous 2020). When increased ozone pass in the leaf through stomata, it oxidizes the cellular membrane by changing mesophyll cells, which lowers carbon fixation and Rubisco activity. This is just one way that the elevated ozone may have destroyed photosynthesis. The level of chlorophyll is lowered as a consequence (Saxena and Kulshrestha 2016). Because photosynthesis is directly correlated with both net assimilation rate and chlorophyll content, declining photosynthetic rates have been linked to lower crop biomass as a result of elevated tropospheric ozone concentrations.

## 6.6 Hydrogen Sulfide

Compared to coal, natural gas emits approximately 50% less carbon dioxide into the atmosphere, making it a more environmentally friendly fuel. The trend is moving toward using gas extracted from unfavorable lands, such as shale gas basins, as a result of the growing demand for natural gas, which is depleting natural reservoirs (Goodwin et al. 2015; Brace and Engelberth 2020). The natural gas extracted from these reservoirs is referred to as sour gas because it is concentrated with $H_2S$ gas and extremely acidic. According to Goodwin et al. (2015), sour gas subsidized 21% of all-natural gas in 2004 and is expected to contribute 27% by 2030, raising serious health concerns.

$H_2S$ is a colorless gas that smells strongly of rotten eggs. It is produced within plant cells by the action of enzymes called desulfhydrase. It occurs naturally in the atmosphere. It was first identified as a phytotoxic material, but more recent studies have shown that it also functions as a signaling molecule within plant cells (Gupta et al. 2021). From its initial identification as a sewer gas in the 1700s, hydrogen sulfide ($H_2S$) has been historically painstaking a harmful gas (Da-silva 2023). Protein per sulfidation is how $H_2S$ works. It has a strong preference for heme-containing proteins like myoglobin, cytochrome oxidase, catalase, hemoglobin, and peroxidase. According to Liu et al. (2021) per sulfidation can alter the structure and function of proteins, which can cause problems with cellular metabolism or even cause a cell's death.

Yang et al. (2008) state that, after carbon monoxide and nitric oxide, hydrogen sulfide ($H_2S$) is one of the three utmost vital obviously occurring gaseous molecules recognized for its gesticulating roles in biology. $H_2S$ synthesis and release typically occur in plants at multiple stages of metabolic processes. According to Rennenberg et al. (1990), it usually starts in cut divisions, tissue cultures, and leaf discs before being released into the environment by plant green cells.

$H_2S$ can cause tissue necrosis, shedding, and growth restriction in leaves of plants. It can also prevent roots from taking up mineral elements like phosphate (Liu et al. 2021). According to a recent study by Zhang et al. (2017a, b), $H_2S$ can cause nitrosative and oxidative stress in *Arabidopsis* plants by generating reactive oxygen species, or ROS, and nitric oxide (NO), which may hinder taproot growth and control root system architecture.

Regarding the various ways that sulfur is supplied to the plants, it is important to comprehend the effects of hydrogen sulfide ($H_2S$), another gas that contains sulfur. High $H_2S$ dosages have been shown to lead to defoliation, leaf lesions, slowed expansion, and death of tissue in nearly plants (Montesinos-Pereira et al. 2016). On the other hand, Zhang et al. (2010) have documented that $H_2S$ is a vital chemical that plants produce and that controls how well they function. Furthermore, it has been shown that this signaling molecule boosts the antioxidant defenses of numerous plants against abiotic stressors.

Montesinos-Pereira et al. (2016) claim that applying $H_2S$ to some plants also decreased their quality and prospective as antioxidants.

Bronco cabbage (*Brassica oleracea*) was treated with higher concentrations of sodium hydrosulfide (an $H_2S$ donor) in order to monitor hemodynamic and antioxidative changes. Carotenoids, anthocyanins, flavonols, total phenolics, and sinigrin were found to be elevated at lower dose levels (Montesinos-Pereira et al. 2016). In *Nicotiana tabacum* plants grown at high temperatures, hydrogen sulfide has also been reported to mediate the biosynthesis of nicotine (Chen et al. 2016).

Many plants, such as cucumber, soybean, cantaloupe, cotton, pumpkin, squash, and squash, have been demonstrated to disperse hydrogen sulfide ($H_2S$) into the atmosphere around them when they expand normally. $H_2S$ production is known to improve sulfite and sulfate metabolic functions in addition to raising the enzymatic activity of vigorous cysteine (Rennenberg et al. 1983; Rennenberg 1984). Nonetheless, a number of studies have shown that $H_2S$ regulates plant physiological and biochemical features in ways that are incompatible. For example, it has been shown that $H_2S$ increases the detrimental impact on plant growth at higher doses and functions as a sulfur source at lower concentrations. (De Kok et al. 2002; Hancock and Whiteman 2016; Li et al. 2016). Most plants eventually lose their leaves when prolonged exposure to high $H_2S$ levels causes leaf damage, which can impede plant growth in general.

## 6.7 Effects of Hydrogen Sulfide on Crops

$H_2S$ has been demonstrated to positively impact plant growth at low concentrations. Mineral nutrition and exchange of gas parameters (photosynthesis, stomatal conductance, respiration, and relative water content) are components of physiological homeostasis. Similar to the growth and development of plants, small amounts of $H_2S$ are most effective for preserving physiological equilibrium (Gupta et al. 2021).

$H_2S$ was long considered to be a dangerous gas. Due to its harmful effects, $H_2S$ has definitely been demonstrated to be a phytotoxin when utilized at substantial amounts (e.g., in the mM range) (Koch et al. 1990; Koch and Erskine 2001). This is particularly valid in conjunction with other treatments such as salinity or high temperatures.

In addition to being extremely reactive, $H_2S$ is toxic to animals and fatal in extremely high levels. One effect, as demonstrated by Dorman et al. (2002), is the inhibition of cytochrome oxidase in the mitochondria, where concentrations above 30 ppm significantly reduced growth in specific tissues.

$H_2S$ was long considered to be a dangerous gas. Due to its harmful effects, $H_2S$ has definitely been demonstrated to be a phytotoxin when utilized at substantial amounts (e.g., in the mM range) (Koch et al. 1990; Koch and Erskine 2001). This is particularly valid in conjunction with other treatments such as salinity or high temperatures.

In addition to being extremely reactive, $H_2S$ is toxic to animals and fatal in extremely high levels. One effect, as demonstrated by Dorman et al. (2002), is the reserve of cytochrome oxidase in the mitochondria, where concentrations above 30 ppm significantly reduced growth in specific tissues.

But negative effects were also reported by other individuals. Thompson and Kats (1978) preserved a variety of plant species with persistent $H_2S$ fumigation. Medicago, grapes, pine, lettuce, sugar beets, pine, and fir all showed lesions of leaf, defoliation, and abridged plant growth when exposed to three thousand parts per billion (ppb) of $H_2S$. These results imply that a wide variety of plants are utilizing $H_2S$ as a phytotoxin. The evidence appeared to be mounting against $H_2S$ having a beneficial effect and having the ability to communicate with plants as a signaling molecule. It should be mentioned, though, that a lot of this kind of work has already been done.

If $H_2S$ were a signal to plants, then it ought to make sense to think that low concentrations would also be beneficial, rather than just a phytotoxin. As seen previously, plants are able to generate and release identifiable amounts of $H_2S$. This would suggest that, as has been projected for both RNS and ROS, $H_2S$ exists in both the environment and the cells themselves, allowing it to function as an intracellular and intercellular signal. $H_2S$ has benefits that cannot be disputed. Sugar beets, lettuce, and mesquite grew noticeably more when exposed to low fumigation levels, approximately 100 ppb, as reported by Thompson and Kats (1978). However, they asserted that the roots of sugar beets had a lower sugar content, which is bad for the crop. Interestingly—and perhaps a little off subject—they also noticed that fungal attack decreased following $H_2S$ treatment in their work with beets. This suggests that instead of stopping fungal growth, $H_2S$ may just have a lethal consequence. However, Bloem et al. (2004) showed that fungal infection, specifically with *Pyrenopeziza brassicae*, increased the amount of the $H_2S$-producing enzyme L-cysteine desulfhydrase, possibly indicating a higher potential for $H_2S$ release from the infected plant.

## 6.8 Sources of Hydrogen Sulfide for Plant Cells

For plant cells to be affected by $H_2S$, it must be present and in sufficient amounts. But that doesn't mean it does anything good for the plant just because it exists there. To put it plainly, a plant cell may wish to respond to two main sources of $H_2S$: internal and external sources. There are many different environmental factors, both natural and man-made, that can produce $H_2S$. Volcanic eruptions are one example of a natural source (Aiuppa et al. 2005). This suggests that plants living downstream of craters will be exposed to elevated levels of $H_2S$, to which they must either adapt or develop resistance.

$H_2S$ gas needs to be present in plant cells at a high concentration in order to be effective. The two types of sources of $H_2S$ emissions in the surroundings are natural and anthropogenic. The primary human activity responsible for $H_2S$ emissions is burning fuel. Agriculture, wastewater treatment, and the geothermal industries are examples of additional man-made sources. Near geothermal sites, there was 50 parts per billion of $H_2S$. $H_2S$ gas is also released by contemporary cars equipped with catalytic inverters (Kourtidis et al. 2008). Enough $H_2S$ gas is released during the drilling, purifying, coke oven, making of paper and treatment of waste processes. $H_2S$ is also produced during the mining procedure when xanthates break down in the presence of water (Bhomick and Rao 2014).

According to Aiuppa et al. (2005), volcanoes are thought to be a main natural source of $H_2S$ gas emissions. Plants raised in these types of environments would be resistant to $H_2S$. Additionally, groundwater, anoxic soils, marshlands, and sediments from the coast emit it. The atmospheric concentration of $H_2S$ is 5–30 $\mu g/m^3$ in areas close to the coast (Kourtidis et al. 2008).

### *6.8.1 Sulfur Dynamics and Synthesis of $H_2S$ in Soil*

In addition to volcanoes, anoxic marshland soils and coastal marine sediments are other organic sources of $H_2S$. Sulfur (S)-based compounds like sulfate ($SO_4^{2-}$) are produced by Archaea (e.g., *Archaeoglobus*) and bacteria (e.g., *Desulfovibrio, Desulfobacter*) that are found in damp soils and marshy areas. Sulfite ($SO_3^{2-}$) is the end product of this transformation. The genesis perspective view states that $H_2S$ primarily originates from four sources: inorganic volcanic sources, bacteria that reduce $SO_4^{2-}$, heat-induced breakdown of organic S substances, and thermochemical reactions.

The primary cause of $H_2S$ production in wet soils is $SO_4^{2-}$ reduction. The occurrence of anaerobic bacteria is the reason for the reduction that only happens in these soils. This is another important S cycling reaction that predominates in wet soils because plant remnants, like alfalfa (Medicago sativa), decompose quickly. The processes of assimilation ($SO_4^{2-}$ is decreased to the thiol) or destructive metabolism ($SO_4^{2-}$ decreases results in the production of $H_2S$) are involved in the $SO_4^{2-}$ reduction caused by bacteria. Since soil is the primary source of $H_2S$ production, information regarding the accessibility of S in the earth's crust—the primary component for $H_2S$ synthesis—must be reported. According to Ali Tabatabai (2005), the average S content of the earth's crust is between 0.06 and 0.10%, making it the thirteenth most prevalent element in nature.

Another common source of $H_2S$ is human activity. Waste treatment facilities (Zhang et al. 2008), various farming practices (Aneja et al. 2008), and the geothermal sector all release $H_2S$ into the atmosphere. The plant's aerial sections will be exposed to $H_2S$ due to the atmospheric release of $H_2S$ from these locations. Unexpectedly high concentrations of $H_2S$ may also be the cause of a similar situation in some urban settings. Catalytic converters, which have long been required for modern cars in many nations, have been proposed as one of the gas sources (Kourtidis et al. 2008). It seems that the accountable enzymes in plants are called desulfhydrases. L'eon et al. (2002) report that *Arabidopsis* contains a cysteine desulfhydrase located close to the plastids. Other groups have discovered a related enzyme that is present in mitochondria (Riemenschneider et al. 2005). It has been shown that the activities of these enzymes vary under specific conditions. For example, there has been evidence of increased activity following a pathogen challenge (Bloem et al. 2004).

It would be anticipated that naturally occurring $H_2S$ production could be quantified if $H_2S$ is indeed functioning as a signal in plants. Utilizing a sulfur-specific flame photometric detector, Wilson et al. (1978) demonstrated that plants such as cotton, cucumber, pumpkin squash, soybean, and squash could produce volatile sulfur compounds and $H_2S$. It was discovered that this activity was light-dependent. These emissions persisted for several hours when sulfate was given to the roots and the plants were lit. Furthermore, feeding sulfate to leaves through the petioles or mechanically damaging plant roots both markedly increased the rates of $H_2S$ emission. *Cucumis sativus* leaf discs treated with sulfate in a light source allowed Sekiya et al. (1982) to demonstrate that $H_2S$ could be released at 50–100 pmol $min^{-1}/cm^2$. Using cucumbers as their model species, they were able to demonstrate in this work that young leaves release a significant amount more $H_2S$ than mature leaves (Sekiya et al. 1982).

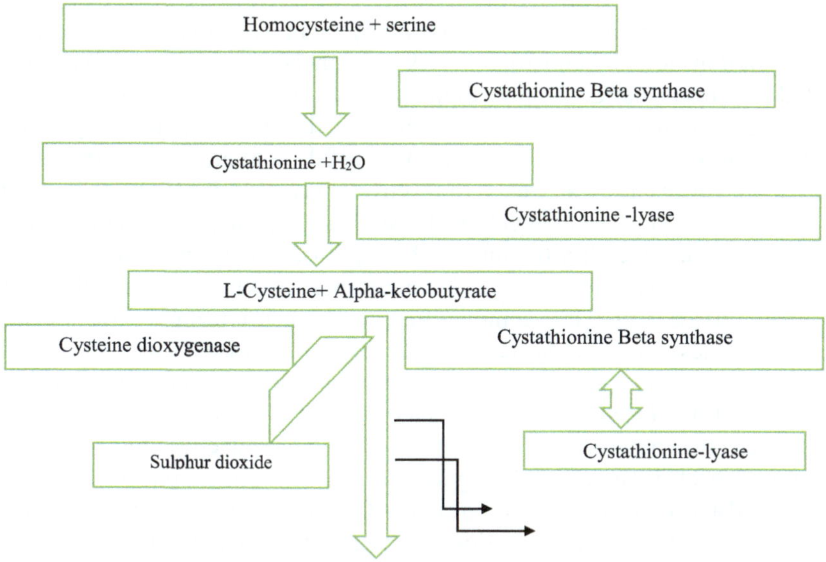


*Source* Arif et al. (2021)

## 6.8.2 *The Impact of Hydrogen Sulfur on Stressed Plants*

Regarding the various ways that sulfur is supplied to the plants, it is important to comprehend the effects of hydrogen sulfide ($H_2S$), another gas which includes sulfur. High $H_2S$ dosages have been shown to cause leaf lesions, defoliation, slowed growth, and death of tissue in some plants (Montesinos-Pereira et al. 2016). On the other hand, Zhang et al. (2010) have documented that $H_2S$ is a vital chemical that plants produce and that controls how well they function. Furthermore, it has been shown that this signaling molecule boosts the antioxidant response of numerous plants against environmental stressors. According to Montesinos-Pereira et al. (2016), the application of $H_2S$ also reduced the antioxidant likely and quality in certain plants. Increasing concentrations of sodium hydrosulfide (as an $H_2S$ donor) were practically applied to bronco cabbage (*Brassica oleracea*) in order to track physiological and antioxidative changes. At lesser behavior levels, there was an intensification in anthocyanin, carotenoids, flavanols, total phenolics, and sinigrin, according to Montesinos-Pereira et al. (2016). $H_2S$ has also been described to intermediate the biosynthesis of nicotine in nicotiana tabacum plants grown at high temperatures (Chen et al. 2016).

### *6.8.3 Combine Effects of Hydrogen Sulfide and Ozone*

Air pollutants, heavy metals, greenhouse gases, and trace gases have all seen rapid changes in their levels. Numerous of these chemical compounds connect to agricultural systems and affect crop yield in two ways: directly by changing the plant's capacity to withstand other stresses, both abiotic and biotic, or indirectly by changing the plant's growth and quality (Bender and Weigel 2011). Plants absorb ozone through their leaves, which causes an excess of reactive oxygen species (ROS) to build up in the mesophyll and protect their cell walls. ROS can impede photosynthetic electron transport and harm the ultrastructure of chloroplasts. Ozone has the ability to cause stomatal closure and modify stomatal conductance, which can impede the fixation of carbon dioxide ($CO_2$). It is common for ozone to cause leaf chlorosis. Under $O_3$ stress, all of these variables cause photosynthesis to decrease (Nowroz et al. 2024).

Plant physiological processes, such as water and nutrient uptake, respiration, and the translocation of assimilates and metabolites, are disrupted by prolonged exposure to high concentrations of $O_3$. Plant growth and reproductive efficiency are thus adversely impacted. Therefore, the biggest effects of $O_3$ stress on plants are a decrease in crop yield and a degradation in crop quality. Common markers of oxidative damage in plants under $O_3$ stress include elevated rates of ion loss, lipid peroxidation, and hydrogen peroxide buildup (Nowroz et al. 2024). Crop stress is caused by ozone, a transient climatic pollutant that, when exceeded, not only causes early senescence of the leaves, but also often inhibits photosynthesis, nutrient splitting, and the development of reproductive organs at concentrations of barely appropriate (40 ppb). Thiourea, a NO and $H_2S$ donor, enhances ozone recognition at the physiological and metabolic levels by raising glutathione reductase activity (Sahu 2022). Phaseolus lunatus benefited from NO exposure before ozone ($O_3$) stress, according to an investigation, by changing their antioxidant defense system and lowering the accumulation of volatile organic compounds in their bodies.

A significant gaseous signaling molecule, hydrogen sulfide ($H_2S$) is vital to numerous physiological and biological processes, such as fruit ripening, stomatal movement, photosynthesis, leaf senescence, seed germination, defense mechanisms against abiotic stresses, and stomatal movement (Shalaby et al. 2023). The atmosphere contains $H_2S$, which plants may absorb through their foliage. This has an adverse effect on the normal metabolism of sulfate in plants, which is absorbed by the roots of the plants (Ausma and De Kok 2019). Through stomata on the foliage, $H_2S$ is absorbed, serving as an important supply of sulfur (S) nutrients for plants.

Hydrogen sulfide ($H_2S$) at high concentrations is toxic to plants and stunts their growth. According to a study's findings, high $H_2S$ concentrations alter auxin transport, which modifies the root system architecture (RSA); the signaling system that underlie this process is still unknown (Zhang et al. 2017a, b).

The total amount of sulfur in petioles and leaf blades ranged from 0.05 to 0.25%, with averages of 0.10% in the $H_2S$ alone group, 0.15% in the control group, and 0.11°/5 in the $H_2S + O_3$ group. This observation included the variety of organic sulfur levels found in most plants. The expected and observed yield losses agreed with one another. Both theoretically and empirically, yield losses for $H_2S + O_3$ at 450 ppm h were always more than productivity losses for $H_2S$ alone (Coyne and Bingham 1978a, b). Field-grown bean plants were obviously affected by $H_2S$ with and without 03 at levels above 2.0 ppm when fumigated four hours a day. The degree of observable harm did not increase linearly with time. Injury grooves increased more during the early growth period (22–30 June) than during the later period (30 June–14 July). One part per million of $H_2S$ or $H_2S + O_3$ added after July 14 could harm more than 75% of the foliage, exceeding the 2.0 parts per million injury threshold (Coyne and Bingham 1978a, b).

For a long while, it was believed that $H_2S$ was a toxic gas. Undoubtedly, due to its harmful effects at high concentrations (such as in the milligram range), $H_2S$ has been classified as a phytotoxin. This is especially true when combined with other treatments like high salinity or temperature (Lisjak et al. 2010). Plant $H_2S$ emissions are highly significant ecologically because they have the potential to degrade toxic surface ozone. Over the past ten years, surface ozone concentrations have risen in rural areas.

## 6.9 Conclusion

In conclusion, the increasing levels of ground-level ozone present a significant menace to agricultural productivity and food security, particularly in regions like Asia where pollution levels are alarmingly high. The detrimental effects of ozone on key crops such as wheat, rice, maize, and soybeans have been well-documented, with studies indicating potential crop yield fatalities ranging from 4 to 15% depending on the crop and environmental conditions. The interaction between ozone and other abiotic issues, such as temperature, availability of water, further complicates the impact on crop yields, especially in the setting of climate change. Hydrogen sulfide ($H_2S$) emerges as a significant gaseous signaling molecule with profound implications for plant physiology and health. While it plays essential roles in various biological processes, including fruit ripening, stomatal movement, and defense mechanisms against abiotic stresses, elevated levels of $H_2S$ can lead to detrimental effects on plant growth and development. The dual nature of $H_2S$—acting as both a beneficial signaling compound and a potential stressor—highlights the complexity of plant responses to environmental challenges. Research indicates that while $H_2S$ can enhance antioxidant defenses and support metabolic functions, excessive exposure can result in tissue necrosis, impaired nutrient uptake, and overall growth inhibition. Understanding the balance between the beneficial and harmful effects of both ozone and $H_2S$ is crucial for developing strategies to optimize plant health and productivity, particularly in the context of increasing air pollution and changing environmental conditions. To mitigate these challenges, it is crucial for governments and policymakers to implement stricter air quality regulations and promote sustainable agricultural practices. Additionally, research into breeding ozone-tolerant crop varieties could provide a viable solution to enhance resilience against rising ozone levels. Ultimately, a coordinated approach that addresses both air pollution and agricultural productivity is essential to protection of food security and guarantee the agricultural sustainability system in the face of ongoing environmental changes.

Though there is limited research work on the combine effects of hydrogen sulfide and ozone, future studies should focus on the combine effects by which $H_2S$ and $O_3$ affect plant growth and productivity in agriculture to obtain more comprehensive and reliable data. Addressing these gaps will be crucial for advancing our understanding. By joining the positive aspects of $H_2S$ while modifying its negative impacts, we can improve crop resilience and contribute to sustainable agricultural practices in a rapidly changing world.

## References


Agathokleous E, Saitanis CJ, Feng Z, De Marco A, Araminiene V, Domingos M, Sicard P, Paoletti E (2020) Ozone biomonitoring: a versatile tool for science, education and regulation. Curr Opin Environ Sci Health 18:7–13

Aiuppa A, Inguaggiato S, McGonigle AJS, O'Dwyer M, Oppenheimer C, Padgett MJ, Rouwet D, Valenza M (2005) $H_2S$ fluxes from Mt. Etna, Stromboli, and Vulcano (Italy) and implications for the sulfur budget at volcanoes. Geochim Cosmochim Acta 69:1861–1871

Ali Tabatabai M (2005) Chemistry of sulfur in soils. Chem Process Soils 8:193–226

Amann M, Derwent D, Forsberg B, Hanninen O, Hurley F, Krzyzanowski M, de Leeuw F, Liu SJ, Mandin C, Schneider J, Schwarze P, Simpsonet D (2008) Health risks of ozone from longrange transboundary air pollution WHO: World Health Organization, Europe

Aneja VP, Blunden J, Roelle PA, Schlesinger WH, Knighton R, Niyogi D, Gilliam W, Jennings G, Duke CS (2008) Workshop on agricultural air quality: state of the science. Atmos Environ 42(14):3195–3208

Arif MS, Yasmeen T, Abbas Z, Ali S, Rizwan M, Aljarba NH, Alkahtani S, Abdel-Daim MM (2021) Role of exogenous and endogenous hydrogen sulfide ($H_2S$) on functional traits of plants under heavy metal stresses: a recent perspective. Front Plant Sci 11:545453. https://doi.org/10.3389/fpls.2020.545453

Ariura R, Matsumoto M, Li J, Fuse T, Aoki T, Zhang Y, Kinose Y, Yamaguchi M, Izuta T, Watanabe M (2023) Effects of elevated ozone and carbon dioxide on the dynamic photosynthesis of Fagus crenata seedlings under variable light conditions. Sci Total Environ 891:164398

Ashmore MR (2002) Effects of oxidants at the whole plant and community level. Air Pollut Plant Life 2

Ausma T, De Kok LJ (2019) Atmospheric H2S: impact on plant functioning. Front Plant Sci 10:743

Avnery S, Mauzerall DL, Liu J, Horowitz LW (2011) Global crop yield reductions due to surface ozone exposure: 2. Year 2030 potential crop production losses and economic damage under two scenarios of O3 pollution. Atmos Environ 45(13):2297–2309

Bender J, Weigel HJ (2011) Changes in atmospheric chemistry and crop health: a review. Agronomy Sust Dev 31:81–89. https://doi.org/10.1051/agro/2010013

Bhatia A, Singh S D, Banerjee B, Mina U (2010) Project completion of Impact of surface ozone on productivity of rice and maize, DST, New Delhi

Bhomick PC, Rao KS (2014) Sources and effects of hydrogen sulfide. J Appl Chem 3(3):914–918

Black VJ, Black CR, Roberts JA, Stewart CA (2000) Impact of ozone on the reproductive development of plants. New Phytol 147:421–447

Bloem E, Riemenschneider A, Volker J, Papenbrock J, Schmidt A, Salac I, Haneklaus S, Schnug E (2004) Sulphur supply and infection with Pyrenopeziza brassicae influence L-cysteine desulphydrase activity in Brassica napus L. J Exp Bot 55(406):2305–23012

Bortolin R, Caregnato F et al (2016) Chronic ozone exposure alters the secondary metabolite profile, antioxidant potential, anti-inflammatory property, and quality of red pepper fruit from *Capsicum baccatum*. Ecotoxicol Environ Saf 129:16–24

Bosac C, Roberts JA, Black VJ, Black CR (1994) Impact of $O_3$ and $SO_2$ on reproductive development in oilseed rape (*Brassica napus L.*) II. Reproductive site losses. New Phytol 126(1):71–79

Brace EC, Engelberth AS (2020) Assessing viability of soybean oils to remove hydrogen sulfide from natural gas. ACS Sustain Chem Eng 8(25):9377–9384

Burney J, Ramanathan V (2014) Recent climate and air pollution impacts on Indian agriculture. Proc Natl Acad Sci USA 111:16319–16324

Chameides WL, Xingsheng L, Xiaoyan T, Xiuji Z, Luo C, Kiang CS, St. John J, Saylor RD, Liu SC, Lam KS, Wang T (1999) Is ozone pollution affecting crop yields in China?. Geophys Res Lett 26(7):867–870

Chen S, Wang H, Lu K, Zeng L, Hu M, Zhang Y (2020) The trend of surface ozone in Beijing from 2013 to 2019: indications of the persisting strong atmospheric oxidation capacity. Atmos Environ 242:117801

Chen X, Chen Q, Zhang X, Li R, Jia Y, Ef AA, Jia A, Hu L, Hu X (2016) Hydrogen sulfide mediates nicotine biosynthesis in tobacco (*Nicotiana tabacum*) under high temperature conditions. Plant Physiol Biochem 104:174–179. https://doi.org/10.1016/j.plaphy.2016.02.033

Christou A, Manganaris GA, Papadopoulos I, Fotopoulos V (2013) Hydrogen sulfide induces systemic tolerance to salinity and non–ionic osmotic stress in strawberry plants through modification of reactive species biosynthesis and transcriptional regulation of multiple defence pathways. Journal of experimental botany 64(7): 1953–1966

Chuwah C, van Noije T, van Vuuren DP, Stehfest E, Hazeleger W (2015) Global impacts of surface ozone changes on crop yields and land use. Atmos Environ 106:11–23

Ciais P, Sabine C, Bala G, Bopp L, Brovkin V, House JI (2014) Carbon and other biogeochemical cycles. In: Climate change 2013: the physical science basis. Contribution of working group I to the fifth assessment report of the intergovernmental panel on climate change. Cambridge University Press, pp 465–570

Coyne PI, Bingham GE (1978a) Photosynthetic and stomatal light responses in snap beans exposed to hydrogen sulfide and ozone. University of California

Coyne PI, Bingham GE (1978b) Photosynthesis and stomatal light responses in snap beans exposed to hydrogen sulfide and ozone. J Air Pollut Control Assoc 28(11):1119–1123. https://doi.org/10.1080/00022470.1978.10470715

Da Y, Xu Y, McCarl B (2022) Effects of surface ozone and climate on historical (1980–2015) crop yields in the United States: implication for mid-21st century projection. Environ Resour Econ 81(2):355–378

da Silva LM, Mena IF, Montiel MA, Saez C, Motheo AJ, Rodrigo MA (2023) Electrochemical generation of ozone for application in environmental remediation. Results Eng 20:101436
De Bauer ML (2003) Air pollution impacts on vegetation in Mexico. In: Air pollution impacts on crops and forests: a global assessment, pp 263–286
De Kok LJ, Castro A, Durenkamp M, Stuiver CE, Westerman S, Yang L, Stulen I (2002) Sulphur in plant physiology. In: PROCEEDINGS-INTERNATIONAL FERTILISER SOCIETY IFS; 1999.
Debaje SB (2014) Estimated crop yield losses due to surface ozone exposure and economic damage in India. Environ Sci Pollut Res 21(12):7329–7338
Dorman DC, Moulin FJ, McManus BE, Mahle KC, James RA, Struve MF (2002) Cytochrome oxidase inhibition induced by acute hydrogen sulfide inhalation: correlation with tissue sulfide concentrations in the rat brain, liver, lung, and nasal epithelium. Toxicol Sci 65:18–25
Emberson LD, Ashmore MR, Cambridge H, Tuovinen J-P, Simpson D (2000) Modelling stomatal flux across Europe. Environ Pollut 109:403–413
Emberson LD, Ashmore MR, Simpson D, Tuovinen JP, Cambridge HM (2001) Modelling and mapping ozone deposition in Europe. Water Air Soil Pollut 130(1):577–582
Emberson L, Ashmore M, Murray F (2003) Air pollution impacts on crops and forests: a global assessment, pp xiii+–372
Emberson LD, Buker P, Ashmore MR, Mills G, Jackson LS, Agrawal M, Atikuzzaman MD, Cinderby S, Engardt M, Jamir C, Kobayshi K, Oanh OTR, Quadir QF, Wahid A (2009) A comparison of North-America and Asian exposure response data for ozone effects on crop yields. Atmos Environ 43:1945–1953
Federal Register (1997) Federal Register 62, pp 38856–38896, codified in 40 CFR part 50
Felzer BS, Cronin T, Reilly JM et al (2007) Impacts of ozone on trees and crops. CR Geosci 339:784–798
Feng Z, Kobayashi K, Li P, Xu Y, Tang H, Guo A, Paoletti E, Calatayud V (2019) Impacts of current ozone pollution on wheat yield in China as estimated with observed ozone, meteorology and day of flowering. Atmos Environ 217:116945
Fuhrer J (2009) Ozone risk for crops and pastures in present and future climates. Naturwissenschaften 96(2):173–194
Fuhrer J, Booker F (2003) Ecological issues related to ozone: agricultural issues. Environ Int 29(2–3):141–154
Fuhrer J, Skarby L, Ashmore MR (1997) Critical levels for ozone effects on vegetation in Europe. Environ Pollut 97(1–2):91–106
Fumigalli I, Gimeno B, Velissariou D, De Temmerman L, Mills G (2001) Evidence of ozone-induced adverse effects on crops in the Mediterranean region. Atmos Environ 35:2583–2587
Gaudel A, Cooper OR, Ancellet G, Barret B, Boynard A, Burrows JP, Clerbaux C, Coheur P-F, Cuesta J, Cuevas E (2018) Tropospheric ozone assessment report: present-day distribution and trends of tropospheric ozone relevant to climate and global atmospheric chemistry model evaluation. Elementa 6:39
Goodwin MJ, Musa OM, Steed JW (2015) Problems associated with sour gas in the oilfield industry and their solutions. Energy Fuel 29(8):4667–4682
Gupta A, Agrawal SB, Agrawal M (2023) Evaluation of toxicity of tropospheric ozone on tomato (*Solanum lycopersicum L.*) cultivars: ROS production, defense strategies and intraspecific sensitivity. J Plant Growth Regul 42(10):6444–6460
Gupta SK, Marwa N, Pandey AK, Zhang Y-B, Zhang J-L (2021) Chapter 17—Hydrogen sulfide homeostasis in plants: an overview. In: Hydrogen sulfide in plant biology. Academic Press, pp 341–363. https://doi.org/10.1016/B978-0-323-85862-5.00017-8
Han H, Liu J, Shu L, Wang T, Yuan H (2020) Local and synoptic meteorological influences on daily variability in summertime surface ozone in eastern China. Atmos Chem Phys 20(1):203–222
Hancock JT, Whiteman M (2016) Hydrogen sulfide signaling: interactions with nitric oxide and reactive oxygen species. Ann NY Acad Sci 1365(1):5–14
Hassan IA, Ashmore MR, Bell JNB (1995) Effect of ozone on radish and turnip under Egyptian field conditions. Environ Pollut 89:107–114

Hou X, Zhang Y, Lv X, Lee J (2023) The impact of meteorological conditions and emissions on tropospheric column ozone trends in recent years. Remote Sens 15(22):5293

Jain SL, Arya BC, Kumar A, Ghude SD, Kulkarni PS (2005) Observational study of surface ozone at New Delhi, India. Int J Remote Sensi 26(16):3515–3524

Jia H, Hu Y, Fan T, Li J (2015) Hydrogen sulfide modulates actin-dependent auxin transport via regulating ABPs results in changing of root development in Arabidopsis. Sci Rep 5:8251

Khemani LT, Momin GA, Rao PSP, Vijaykumar R, Safai PD (1995) Study of surface ozone behaviour at urban and forested sites in India. Atmos Environ 29:2021–2024

Koch MS, Erskine JM (2001) Sulfide as a phytotoxin to the tropical seagrass Thalassia testudinum: interactions with light, salinity and temperature. J Exp Mar Biol Ecol 266(1):81–95

Koch MS, Mendelssohn IA, McKee KL (1990) Mechanism for the hydrogen sulphide-induced growth limitation in wetland macrophytes. Limnol Oceanogr 35:399–408

Kourtidis K, Kelesis A, Petrakakis M (2008) Hydrogen sulfide ($H_2S$) in urban ambient air. Atmos Environ 42(32):7476–7482

Kumar C, Dogra A, Kumari N et al (2024) Multi-time-scale surface ozone exposure and associated premature mortalities over Indian cities in different climatological sub-regions. Air Qual Atmos Health. https://doi.org/10.1007/s11869-024-01547-w

Lal S, Venkataramani S, Naja M, Kuniyal JC, Mandal TK, Bhuyan PK, Kumari KM, Tripathi SN, Sarkar U, Das T, Swamy YV (2017) Loss of crop yields in India due to surface ozone: an estimation based on a network of observations. Environ Sci Pollut Res 24(26):20972–20981

Lal S, Naja M, Subbaraya BH (2000) Seasonal variations in surface ozone and its precursors over an urban site in India. Atmos Environ 34:2713–2724

L'eon S, Touraine B, Briat JF, Lobr'eaux S (2002) The AtNFS2 gene from Arabidopsis thaliana encodes a NifS-like plastidial cysteine desulphurase. Biochem J 366:557–564

Lee EH, Tingey DT, Hogsett WE, Laurence JA (2003) History of tropospheric ozone for the San Bernardino Mountains of Southern California, 1963–1999. Atmos Environ 37:2705–2717

Li D, Shindell D, Ding D, Lu X, Zhang L, Zhang Y (2022) Surface ozone impacts on major crop production in China from 2010 to 2017. Atmos Chem Phys 22:2625–2638. https://doi.org/10.5194/acp-22-2625-2022

Li JF, Lu K, Lv W, Li J, Zhong L, Ou Y, Chen D, Huang X, Zhang Y (2014) Fast increasing of surface ozone concentrations in Pearl River Delta characterized by a regional air quality monitoring network during 2006–2011. J Environ Sci 26:23–36

Li K, Jacob DJ, Shen L, Lu X, De Smedt I, Liao H (2020) Increases in surface ozone pollution in China from 2013 to 2019: anthropogenic and meteorological influences. Atmos Chem Phys 20:11423–11433. https://doi.org/10.5194/acp-20-11423-2020

Li ZG, Min X, Zhou ZH (2016) Hydrogen sulfide: a signal molecule in plant cross-adaptation. Front Plant Sci 7:1621

Lin Y, Jiang F, Zhao J, Zhu G, He X, Ma X, Li S, Sabel CE, Wang H (2018) Impacts of O3 on premature mortality and crop yield loss across China. Atmos Environ 194:41–47

Lisjak M, Srivastava N, Teklic T, Civale L, Lewandowski K, Wilson I, Wood ME, Whiteman M, Hancock JT (2010) A novel hydrogen sulfide donor causes stomatal opening and reduces nitric oxide accumulation. Plant Physiol Biochem 48(12):931–935

Liu H, Wang J, Liu J, Liu T, Xue S (2021) Hydrogen sulfide ($H_2S$) signaling in plant development and stress responses. aBIOTECH 2:32–63. https://doi.org/10.1007/s42994-021-00035-4

Liu Y, Wang T, Stavrakou T, Elguindi N, Doumbia T, Granier C, Bouarar I, Gaubert B, Brasseur GP (2021) Diverse response of surface ozone to COVID-19 lockdown in China. Sci Total Environ 789:147739

Lobell DB, Field CB (2007) Global scale climate-crop yield relationships and the impact of recent warming. Environ Res Lett 2:1

Lu X, Zhang L, Chen Y, Zhou M, Zheng B, Li K, Liu Y, Lin J, Fu TM, Zhang Q (2019) Exploring 2016–2017 surface ozone pollution over China: source contributions and meteorological influences. Atmos Chem Phys 19(12):8339–8361

Maggs R, Ashmore MR (1998) Growth and yield responses of Pakistan rice (*Oryza sativa L.*) cultivars to $O_3$ and $NO_2$. Environ Pollut 103(2–3):159–170

McGrath JM, Betzelberger AM, Wang S, Shook E, Zhu X-G, Long SP, Ainsworth EA (2015) An analysis of ozone damage to historical maize and soybean yields in the United States. Proc Natl Acad Sci USA 112:14390–14395

Mittal ML, Hess PG, Jain SL, Arya BC, Sharma C (2007) Surface ozone in the Indian region. Atmos Environ 41:6572–6584

Montesinos-Pereira D, Barrameda-Medina Y, Baenas N, Moreno DA, Sanchez-Rodriguez E, Blasco B, Ruiz JM (2016) Evaluation of hydrogen sulfide supply to biostimulate the nutritive and phytochemical quality and the antioxidant capacity of Cabbage (*Brassica oleracea L.* '*Bronco*'). J Appl Bot Food Qual 89

Nguyen DH, Lin C, Vu CT, Cheruiyot NK, Nguyen MK, Le TH, Lukkhasorn W, Vo TDH, Bui XT (2022) Tropospheric ozone and $NO_x$: a review of worldwide variation and meteorological influences. Environ Technol Innov 102809

Nowroz F, Hasanuzzaman M, Siddika A, Parvin K, Caparros PG, Nahar K, Prasad PVV (2024) Elevated tropospheric ozone and crop production: potential negative effects and plant defense mechanisms. Front Plant Sci 14:1244515. https://doi.org/10.3389/fpls.2023.1244515

Oltmans SJ, Levy H II (1994) Surface ozone measurements from a global network. Atmos Environ 28:9–24

Osborne SA, Mills G, Hayes F, Ainsworth EA, Büker P, Emberson L (2016) Has the sensitivity of soybean cultivars to ozone pollution increased with time? an analysis of published dose–response data. Glob Change Biol 22(9):3097–3111

Pallavi S, Anju S, Mritunjay T, Simran K (2019) Impact of tropospheric ozone on plant metabolism–a review. Pollut Res 38(1):175–180

Pandey J, Agrawal M (1992) Ozone: concentration variabilities in a seasonally dry tropical climate. Environ Int 18:515–520

Pellegrini E, Carucci G, Campanella A et al (2011) Ozone stress in *Melissa officinalis* plants assessed by photosynthetic function. Environ Exp Bot 73:94–101

Prophet AM, Polley K, Van Berkel GJ, Limmer DT, Wilson KR (2024) Iodide oxidation by ozone at the surface of aqueous microdroplets. Chem Sci 15(2):736–756

Rai R, Agrawal M, Agrawal SB (2007) Assessment of yield losses in tropical wheat using open top chambers. Atmos Environ 41:9543–9554

Rai R, Agrawal M (2012) Impact of tropospheric ozone on crop plants. Proceedings of the National Academy of Sciences, India Section B: Biological Sciences 82(2): 241–257.

Rennenberg H, Reski G, Polle A (1983) Hydrogen sulfide emission by cultured tobacco cells. Zeitschrift für Pflanzenphysiologie 111(3):189–202

Rennenberg H (1984) Fate of excess sulfur in higher plants. Annu Rev Plant Physiol (United States), 35

Rennenberg H, Huber B, Schroder P, Stahl K, Haunold W, Georgii HW, Slovik S, Pfanz H (1990) Emission of volatile sulfur compounds from spruce trees. Plant Physiol 92(3):560–564

Riemenschneider A, Nikiforova V, Hoefgen R, De Kok LJ, Papenbrock J (2005) Impact of elevated H2S on metabolite levels, activity of enzymes and expression of genes involved in cysteine metabolism. Plant Physiol Biochem 43(5):473–483

Sahu MP (2022) Thiourea can mitigate the adverse effect of ozone on crop productivity. Sustainable management of environmental contaminants: eco-friendly remediation approaches. Springer, Cham, pp 509–532

Sarkar A, Agrawal SB (2010) Elevated ozone and two modern wheat cultivars: an assessment of dose dependent sensitivity with respect to growth, reproductive and yield parameters. Environ Exp Bot 69(3):328–337

Saxena P, Kulshrestha U (2016) Biochemical effects of air pollutants on plants. In: Plant responses to air pollution (pp 59–70). Singapore: Springer Singapore

Schoene K, Franz JT, Masuch G (2004) The effect of ozone on pollen development in *Lolium perenne L.* Environ Pollut 131(3):347–354

Sekiya J, Schmidt A, Rennenberg H, Wilson LG, Filner P (1982) Hydrogen sulfide emission by cucumber leaves in response to sulfate in light and dark. Phytochemistry 21:2173–2178

Seth CS, Misra V, Chauhan LKS (2012) Accumulation, detoxification, and genotoxicity of heavy metals in Indian mustard (*Brassica juncea* L.). Int J Phytoremediation 14(1):1–13

Shakeri A, Sahebkar A, Javadi B (2016) Melissa officinalis L.—a review of its traditional uses, phytochemistry and pharmacology. J Ethnopharmacol

Shalaby OAES, Farag R, Ibrahim MFM (2023) Effect of hydrogen sulfide and hydrogen peroxide on growth, yield and nutrient content of broccoli plants grown under saline conditions. Scientia Horticulturae 316:112035. https://doi.org/10.1016/j.scienta.2023.11203

Shao Z, Zhang Y, Mu H, Wang Y, Wang Y, Yang L (2020) Ozone-induced reduction in rice yield is closely related to the response of spikelet density under ozone stress. Sci Total Environ 712:136560

Shyam L, Sethuraman V, Manish N, Jagdish CK, Tuhin KM, Pradip KB, Kandikonda MK, Sachchida N, Tripathi US, Trupti D, Yerramsetti VS, Kotalo RG, Harish G, Modathi K, Satheesh K (2017) Loss of crop yields in India due to surface ozone: an estimation based on a network of observations. Environ Sci Pollut Res 24:20972–20981. https://doi.org/10.1007/s11356-017-9729-3

Singh A, Sarin SM, Shanmugam P, Sharma N, Attri AK, Jain VK (1997) Ozone distribution in the urban environment of Delhi during winter months. Atmos Environ 31:3421–3427

Stewart CA, Black VJ, Black CR, Roberts JA (1996) Direct effects of ozone on the reproductive development of Brassica species. J Plant Physiol 148(1–2):172–178

Tai APK, Martin MV, Heald CL (2014) Threat to future global food security from climate change and ozone air pollution. Nat Clim Chang 4:817–821

Tang H, Takigawa M, Liu G, Zhu J, Kobayashi K (2013) A projection of ozone-induced wheat production loss in China and India for the years 2000 and 2020 with exposure-based and flux-based approaches. Glob Chang Biol 19:2739–2752

Thompson CR, Kats G (1978) Effects of continuous hydrogen sulfide fumigation on crop and forest plants. Environ Sci Technol 12(5):550–553

Tingey D, Hogsett W, Lee E, Laurence JA (2004) Stricter ozone ambient air quality standard has beneficial effect on ponderosa pine in California. Environ Manag 34:397–405

Van Dingenen R, Raes F, Krol MC, Emberson L, Cofala J (2009) The global impact of $O_3$ on agricultural crop yields under current and future air quality legislation. Atmos Environ 43:604–618

Velissariou D (1999) Toxic effects and losses of commercial value of lettuce and other vegetables due to photochemical air pollution in agricultural areas of Attica, Greece. In: Fuhrer J, Achermann B (eds) Critical levels for ozone—Level II. Swiss Agency for Environment, Forest and Landscape, Bern, Switzerland, pp 253–256

Wahid A, Milne E, Shamsi SR, Ashmore MR, Marshall FM (2001) Effects of oxidants on soybean growth and yield in the Pakistan Punjab. Environ Pollut 113(3):271–280

Wang P, Chen Y, Hu J, Zhang H, Ying Q (2019) Attribution of Tropospheric ozone to $NO_x$ and VOC emissions: considering ozone formation in the transition regime. Environ Sci Technol 53(3):1404–1412. https://doi.org/10.1021/acs.est.8b05981

Wang X, Jacob DJ, Fu X, Wang T, Breton ML, Hallquist M, Liu Z, McDuffie EE, Liao H (2020) Effects of anthropogenic chlorine on PM2. 5 and ozone air quality in China. Environ Sci Technol 54(16):9908–9916

Wang W, Parrish DD, Wang S, Bao F, Ni R, Li X, Yang S, Wang H, Cheng Y, Su H (2022) Long-term trend of ozone pollution in China during 2014–2020: distinct seasonal and spatial characteristics and ozone sensitivity. Atmos Chem Phys 22(13):8935–8949

Yadav M, Gupta P, Seth CS (2022) Foliar application of α-lipoic acid attenuates cadmium toxicity on photosynthetic pigments and nitrogen metabolism in *Solanum lycopersicum* L. Acta Physiol Plant 44(11):112

Yadav R, Sahu LK, Beig G, Jaaffrey SNA (2016) Role of long-range transport and local meteorology in seasonal variation of surface ozone and its precursors at an urban site in India. Atmos Res 176–177:96–107. https://doi.org/10.1016/J.ATMOSRES.2016.02.018

Yendrek CR, Leisner CP, Ainsworth EA (2013) Chronic ozone exacerbates the reduction in photosynthesis and acceleration of senescence caused by limited N availability in *Nicotiana sylvestris*. Glob Chang Biol 19:3155–3166

Yi F, Feng J, Wang Y, Jiang F (2020) Influence of surface ozone on crop yield of maize in China. J Integr Agric 19(2):578–589

Zhang H, Hu LY, Li P, Hu KD, Jiang CX, Luo JP (2010) Hydrogen sulfide alleviated chromium toxicity in wheat. Biol Plant 54(4):743–747

Zhang H, Jiao H, Jiang CX, Wang SH, Wei ZJ, Luo JP, Jones RL (2010) Hydrogen sulfide protects soybean seedlings against drought-induced oxidative stress. Acta Physiologiae Plant 32(5):849–857

Zhang L, De Schryver P, De Gusseme B, De Muynck W, Boon N, Verstraete W (2008) Chemical and biological technologies for hydrogen sulfide emission control in sewer systems: a review. Water Res 42:1–12

Zhang P, Li F, Wiegman CH, Zhang M, Hong Y, Gong J, Chang Y, Zhang J, Adcock I, Chung KF, Zhou X (2015) Inhibitory effect of hydrogen sulfide on ozone–induced airway inflammation, oxidative stress, and bronchial hyperresponsiveness. Am J Respir Cell Mol Biol 52(1):129–137

Zhang P, Luo Q, Wang R et al (2017a) Hydrogen sulfide toxicity inhibits primary root growth through the ROS-NO pathway. Sci Rep 7:868. https://doi.org/10.1038/s41598-017-01046-2

Zhang P, Luo Q, Wang R, Xu J (2017b) Hydrogen sulfide toxicity inhibits primary root growth through the ROS-NO pathway. Sci Rep 7:868. https://doi.org/10.1038/s41598-017-01046-2

Zhao F, Liu C, Cai Z, Liu X, Bak J, Kim J, Hu Q, Xia C, Zhang C, Sun Y, Wang W (2021) Ozone profile retrievals from TROPOMI: implication for the variation of tropospheric ozone during the outbreak of COVID-19 in China. Sci Total Environ 764:142886

Zhao H, Huang J, Zheng Y (2024) Observation of ozone deposition flux and its contribution to stomatal uptake over a winter wheat field in eastern China. Atmos Environ 326

Zheng Y, Stevenson KJ, Barrowcliffe R, Chen S, Wang H, Barnes JD (1998) Ozone levels in Chongqing: a potential threat to crop plants commonly grown in the region?. Environ Polluti 99(3):299–308

# Chapter 7
# Combined Effect of Drought Stress and High Tropospheric Ozone Conditions on Crop Yield

**Bhanu Pandey and Annesha Ghosh**

**Abstract** The combination of drought stress and high tropospheric ozone conditions poses a significant threat to crop yield and global food security. This chapter provides an in-depth analysis of the individual effects of drought stress and ozone on crop productivity, as well as the complex interactions and different responses (synergistic/additive/antagonistic) that occur when these stressors act together. The chapter discusses the physiological, biochemical, and morphological mechanisms underlying the combined effect, emphasizing the oxidative stress, water relations, and photosynthetic impairment induced by the interaction of drought stress and ozone. Furthermore, the chapter explores the implications for crop productivity and food security, highlighting the yield losses, vulnerability of certain regions, and the importance of adaptive strategies. The conclusion emphasizes the need for multidisciplinary approaches and the implementation of sustainable practices to mitigate the combined effect and ensure food security in a changing climate.

**Keywords** Abiotic stress · Agriculture · Antagonistic · Synergistic · Water deficit stress

## 7.1 Introduction

Agricultural production is vital for sustaining global food security, providing livelihoods, and supporting economic development. However, the ever-increasing challenges posed by climate change threaten the productivity and stability of crop systems worldwide. Two significant environmental stressors that have garnered increasing attention in recent years are drought stress and high tropospheric ozone conditions

B. Pandey
CSIR-Central Institute of Mining & Fuel Research, Dhanbad 826001, Jharkhand, India

A. Ghosh (✉)
Department of Life Science, School of Natural Sciences, Central University of Jharkhand, Cheri Manatu, Ranchi 835222, Jharkhand, India
e-mail: anne19901900@gmail.com; annesha.ghosh@cuj.ac.in

 
A. K. Pandey and E. Oksanen (eds.), *Tropospheric Ozone and Food Security*,
https://doi.org/10.1007/978-3-032-03608-7_7

(Otu-Larbi et al. 2020; Kask et al. 2021; Hoshika et al. 2022). These stressors, individually and in combination, can profoundly impact crop growth, development, and ultimately, yield (Tian et al. 2016; Grantz et al. 2019; Ghosh et al. 2020a; Feng et al. 2022; Prodhan et al. 2022; Broberg et al. 2023).

Drought stress, which occurs due to inadequate water supply, is a major abiotic factor that significantly impacts crop yields. As climate patterns shift, drought occurrences are likely to become more common and severe, creating a substantial risk to global food security (Singh et al. 2022).

Drought stress can interfere with numerous physiological and biochemical functions in plants, resulting in decreased photosynthesis rates, impaired nutrient uptake, hormonal imbalances, oxidative stress, and cellular damage. These physiological alterations ultimately manifest as reduced crop growth and yield losses (Zhang et al. 2019; Hussain et al. 2019; Zhao et al. 2020; Liang et al. 2020; Askarnejad et al. 2021; Zou et al. 2021; Kang et al. 2022; Wahab et al. 2022). Simultaneously, increased concentrations of tropospheric ozone—a harmful air pollutant generated through the reaction of sunlight with nitrogen oxides and volatile organic compounds—are emerging as a significant issue for agricultural systems (Ashmore 2005; Ainsworth et al. 2012; Ghosh et al. 2021b). Tropospheric ozone concentrations have risen due to human-driven activities like industrial emissions, agricultural practices, and fossil fuel combustion. Ozone enters plant tissues through stomatal pores, where it triggers a cascade of oxidative reactions that impede plant growth and development. The effects of ozone on crops include reduced photosynthetic capacity, accelerated leaf senescence, altered hormonal regulation, compromised nutrient uptake, and heightened vulnerability to pests and diseases to diseases and pests (Reich and Amundson 1985; Vandermeiren et al. 2005; Fiscus et al. 2005; Burkart et al. 2013; Yendrek et al. 2017; Romero et al. 2020; Hasan et al. 2022).

While drought stress and high tropospheric ozone conditions are distinct stressors, their combined effect on crops is a subject of growing interest and concern. A mounting collection of evidences suggest that the interaction between drought and ozone, can produce both positive and negative outcomes on plants. Research indicates that some crop species have developed mechanisms to tolerate these stresses better than others. Furthermore, when water is scarce, plants close their stomata, reducing the uptake of ozone and minimizing the damage caused by ozone (Herbinger et al. 2002; Khan and Soja 2003). Additionally, water scarcity has the potential to exacerbate the effects of ozone, impairing a plant's defenses and limiting its ability to fix carbon (Hayes et al. 2015; Cotrozzi et al. 2016; Ghosh et al. 2020a) and these effects may become worse due to the synergistic interplay between water shortage stress and ozone (Grulke et al. 2002).

While the effects of various stresses on plants are intricate, and the interplay of multiple factors occurring simultaneously requires more research. Similarly, the mechanisms underlying this combined effect are complex and multifaceted, involving intricate physiological, biochemistry, and genetic processes within plants. Understanding the combined effect of drought stress and high tropospheric ozone conditions on crop yield is of paramount importance for developing resilient agricultural systems that can cope with climate change. This knowledge can guide the establishment of strategies and technologies to alleviate the adverse effects of these stressors and improve crop yield under challenging conditions (Rivero et al. 2022). Additionally, unravelling the intricate interplay between drought stress and ozone can provide insights into the underlying molecular pathways involved in plant responses to multiple stresses, thus opening up avenues for targeted breeding and genetic engineering approaches to improve crops' resilience to stress (Iyer et al. 2013).

In this text, we seek to offer a thorough examination of how drought stress and ozone interact to affect crop yield. We will review and synthesize the current state of knowledge regarding the biochemical, physiological, and genetic responses of crops to individual and combined stressors. Furthermore, we will explore the underlying mechanisms and interactions that contribute to the observed synergistic effects. Finally, we will discuss possible approaches for alleviating the negative impacts of these stressors and improving crop resilience amidst climate change. Through this chapter, we hope to contribute to the broader understanding of the difficulties presented by drought stress and rising ozone conditions on agriculture. By shedding light on the combined effect of these stressors on crop yield, we seek to offer a foundation for upcoming research and stimulate discussions on the development of sustainable and adaptive agricultural practices.

## 7.2 Tropospheric Ozone—A Phytotoxic Agent

Tropospheric ozone, a secondary air pollutant created through intricate play of different photochemical reactions of precursor gases, is recognized as a significant phytotoxic agent with detrimental effects on plant growth and their advancement (Ghosh et al. 2020b; Gupta et al. 2023; Poornima et al. 2023). Rising ozone levels due to changing pattern of its precursors' emission have become a global concern, particularly in south east Asia, regions of intense industrial and agricultural activities (Li et al. 2023). The spring season in India has been associated with the greatest ozone concentrations, which can reach 100 ppb throughout the day as reported by Ainsworth (2017). According to estimates made by Fowler et al. (2008), South Asia's ambient ozone concentration may rise from its present level of 40–60 ppb by 2100. Tropospheric ozone ($O_3$) is extremely reactive composed of three oxygen atoms. It is predominantly found in the lower atmosphere, known as the troposphere, where it can persist for hours to weeks, posing a significant threat to plants. While stratospheric ozone is essential for protecting the Earth from damaging ultraviolet (UV) radiation, tropospheric ozone acts as a pollutant that harms both mankind and vegetation.

The formation, transport, and dispersion of atmospheric pollutants are all heavily influenced by meteorological factors (Garrido-perez et al. 2019). There are large geographical and temporal variations in ozone concentration due to changes in climatic circumstances such as solar radiation, temperature, relative humidity, wind speed and direction, and boundary layer height (Ghosh et al. 2018). Thus, tropospheric ozone concentrations vary spatially and temporally, with higher levels often observed in rural areas and regions downwind of industrial centers (Ghosh et al. 2018, 2021b). Seasonal variations of ozone are also prevalent with its level often tend to be higher during the warm season when solar radiation and the emission of ozone precursors are at their peak. Furthermore, the regional and local ozone concentration is determined by the across-the-border movement of these precursor gases. Although it has a very limited chemical lifetime on Earth's surface, once released into the troposphere it can remain for weeks or months (Dentener et al. 2010). Transport over such great distances is mostly controlled by the strength and direction of the wind (Zhang et al. 2015). Additionally, ozone concentrations can be elevated in rural agricultural areas due to the release of ozone precursors from agricultural practices, like the use of nitrogen-based fertilizers and crop residue burning (Chandra et al. 2002).

When tropospheric ozone encounters plant tissues, it can penetrate through the stomatal openings on leaves, leading to the formation of reactive oxygen species (ROS) and triggering a series of phytotoxic reactions (Ainsworth 2017). The phytotoxicity of ozone can be attributed to its high reactivity and ability to interact with cellular components, including proteinaceous molecules, lipids, and nucleic acids. One of the primary mechanisms through which ozone exerts phytotoxicity is oxidative stress. Ozone-induced ROS, such as superoxide radicals ($O_2$·-) and hydrogen peroxide ($H_2O_2$), disrupt the dynamic equilibrium between ROS production and detoxification mechanisms in plant cells. Excessive ROS accumulation can lead to oxidative harm to key cellular structures such as fatty acid components, proteinaceous compounds, and DNA structure, ultimately disrupting cellular functions and metabolic processes (Wedow et al. 2021).

Furthermore, ozone can disrupt photosynthesis, a crucial process for plant growth and yield. Ozone exposure hampers photosynthetic efficiency by impairing the functioning of photosystem II (PSII) and diminishing the function of crucial enzymes that facilitates carbon fixation. As a result, the net photosynthetic rate decreases, limiting the production of assimilates necessary for plant growth and advancement (Anderson 2003; Leisner and Ainsworth 2012). Ozone-induced phytotoxicity also affects stomatal conductance and water relations in plants. This closure also disrupts transpiration, causing alterations in water balance and potentially reducing and/or exacerbating the effects of drought stress (Herbinger et al. 2002; Khan and Soja 2003; Pellegrini et al. 2019). Furthermore, ozone can disrupt hormonal balance in plants, affecting various physiological processes. Ozone exposure alters the production and signaling of plant hormones, including abscisic acid (ABA), ethylene, and jasmonic acid, which play crucial roles in plant development, growth, and responses to stress (Hasan et al. 2021). These hormonal imbalances can influence plant growth, root development, and the activation of stress-responsive genes.

The impact of tropospheric ozone on plants is not uniform across different species, genotypes, and developmental stages (Morgan et al. 2003; Feng et al. 2008; Peng et al. 2020; Rathore and Chaudhary 2021). Some plant species and genotypes exhibit varying degrees of sensitivity or tolerance to ozone, with visible symptoms ranging from visible leaf injury and chlorosis to reduced growth and yield (Singh et al. 2018). Factors influencing the plant's sensitivity to ozone include the presence of natural defense mechanisms, antioxidant capacity, and the ability to repair ozone-induced damage. The harmful effects of tropospheric ozone in plants can be further exacerbated when they are simultaneously exposed to multiple stressors such as heat, drought, flooding, and radiation. In such cases, the negative impacts of ozone get intensified by compromising the plant's defense mechanisms, increasing its vulnerability to oxidative stress, and exacerbating ozone-induced damage. Thus, the combined effect of such stressors can lead to greater reductions in plant growth, yield, and overall crop productivity compared to the effects of either stressor alone (Bohler et al. 2015).

## 7.3 How Does Ozone Impacts Crop Yield?

Understanding the specific mechanisms through which tropospheric ozone impacts crop yield is essential for devising effective strategies to counteract its detrimental effects on agricultural productivity. In this here, we will explore the various pathways through which ozone influences crop yield, considering impacts on plant growth, both direct and indirect, their development, and reproductive processes.

### *7.3.1 Direct Effects on Plant Physiology*

The maximum ozone input into the apoplastic region was mostly regulated by stomatal pore opening and their dimension (Castagna and Ranieri 2009), while cuticular uptake of ozone has been observed to be negligible in several cases. Guard cell ozone stomatal flow is primarily driven by hormonal and chemical redox signaling (Kangasjärvi et al. 2005). However, it has been discovered that stomatal conductance decreases in ozone-fumigated plants as a protective strategy to reduce the amount of ozone entering the cells (Yadav et al. 2020; Hoshika et al. 2019; Kinose et al. 2020). Rapid diffusion of ozone into the sub-stomatal area initiates a cascade of events in the cell wall, plasma membrane, and apoplast that ultimately result in reactive oxygen species (ROS) being produced (Ainsworth 2017; Castagna and Raneiri 2009). Apoplastic ascorbate (ASA) removes reactive oxygen species; however, if ROS levels rise beyond the detoxifying threshold, an alarm signal is activated. Overproduction of ROS damages the permeability of photosynthetic organelle membranes, which ultimately impairs the operation of the photosynthetic machinery and stomata (Calatayud et al. 2011). In the symplastic area, ROS are propagated by plasma membrane-specific NADPH oxidase and peroxidase. The activation of plasma membrane-bound NADPH oxidase by a heterotrimeric G-protein results in a biphasic pattern of ROS burst. Ozone fumigations were found to cause the kinase protein OST1 to activate plasma membrane-bound NADPH oxidase, which in turn activated the ion channel in the guard cells that controls stomatal opening (Ainsworth 2017).

Ozone disrupts the photosynthetic machinery, particularly the functioning of photosystem II (PSII). According to a research by Thwe et al. (2014), severe ozone treatment of tomato reduced photosynthesis because it suppressed the highest carboxylation rate constrained by the rate of electron transport required for RuBP generation (Jmax), the use of triose phosphate, and the initial downward slope of the spectrum curve at low photosynthetic photon flux density. Due to heightened production of plant hormones such abscisic acid (ABA), ozone stimulates stomatal closure (McAdam et al. 2017) and induce declination of mesophyll conductance ($gmCO_2$) as reported in *Populus* (Xu et al. 2019), *Phaseolus vulgaris* (Flowers et al. 2007), and *Fagus sylvatica* (Hoshika et al. 2020). Although the underlying mechanisms of ozone's impacts on $gmCO_2$ have not yet been fully investigated in crops, numerous potential causes have been identified in trees (Nadal et al. 2021). In example, Hoshika et al. (2020) hypothesized that the mesophyll cells of beech leaves exposed to ozone underwent structural alterations concurrent with the decline in $gmCO_2$. The fact that ozone damages mesophyll cells or modifies the ultrastructure of mesophyll and other leaf morphological features, which may inhibit $CO_2$ diffusion inside leaves (Gao et al. 2017) supports the work of Hoshika et al. (2020). The activation of aquaporins is another process impacted by ozone, raising the resistance of mesophyll to $CO_2$ diffusion (Eichelmann et al. 2004).

The kinetics of chlorophyll fluorescence is a crucial non-destructive method for controlling a plant's photosynthetic efficiency (Kalaji et al. 2016). The decrease in $F_v/F_m$ under ozone may have been an indication of photoinhibition-induced damage to plants' photosystem II. It has been observed that the $F_v/F_m$ ratio decreases in many crop varieties, including wheat (Fatima et al. 2019; Ghosh et al. 2020c; Yadav et al. 2020), soybean (Singh et al. 2009), common bean (Kuang et al. 2013), rice (Sarkar and Agrawal 2010), and tomato (Degl'innocenti et al. 2007). Collectively, these consequences compromise plant growth, development, and reproduction, leading to lower crop output.

### 7.3.2 Indirect Effects on Plant-Environment Interactions

The indirect effects of ozone on plant-environment interactions can extend to the broader plant community. Ozone exposure can selectively impact certain plant species, favoring ozone-tolerant species over sensitive ones (Oksanen et al. 2009). This can lead to shifts in plant community composition and diversity, altering species interactions, competition for resources, and ecosystem functioning (Agathokleous et al. 2020). Changes in plant community dynamics can have cascading effects on other trophic levels, such as herbivores, pollinators, and predators (Couture et al. 2012; Cotrozzi et al. 2021) further influencing ecosystem structure and function. Ozone can indirectly influence plant–herbivore interactions through changes in plant chemistry (Khaling et al. 2015). Ozone exposure can alter the composition and concentration of secondary metabolites in plants, such as phenolics and terpenoids, which are often involved in plant defense against herbivores (Agathokleous et al.

2020). Changes in plant chemistry can affect herbivore feeding behavior, preference, and performance. For example, some studies have shown that ozone-exposed plants may become less palatable or more toxic to herbivores, leading to reduced herbivory (Khaling et al. 2020). Conversely, other studies have suggested that ozone-induced changes in plant chemistry may attract certain herbivores or alter their behavior, potentially disrupting natural plant–herbivore interactions.

Ozone exposure can also impact plant–microbe interactions indirectly. Plants rely on beneficial microbes, such as mycorrhizal fungi and rhizobacteria, for nutrient uptake, disease resistance, and overall plant health. Ozone can disrupt these interactions by altering the composition and activity of rhizospheric microbial communities (Andersen 2003; Chen et al. 2019; Zhang et al. 2019; Agathokleous et al. 2020). Changes in microbial diversity and abundance can affect nutrient cycling, soil structure, and plant–microbe signaling, ultimately influencing plant growth and performance. Additionally, ozone-induced changes in plant physiology and defense mechanisms can affect the recruitment and activity of beneficial microbes, potentially disrupting the balance of plant–microbe interactions.

### 7.3.3 *Crop-Specific Responses to Ozone*

It is worth noting that the sensitivity of crop species to ozone can be influenced by various factors such as ozone concentration, exposure duration, temperature, humidity, and interactions with other environmental stressors in addition to specific crop species, genotypes, and developmental stages. Different crops exhibit varying degrees of sensitivity or tolerance to ozone, influenced by genetic factors and natural defense mechanisms. Several studies have concentrated on examining how various crops, including *Oryza sativa* (Pandey et al. 2015; Ashrafuzzaman et al. 2017, 2020), wheat (Biswas et al. 2008; Saitanis et al. 2014; Singh et al. 2018), *Vigna radiata* (Chaudhary and Agrawal 2015), *Phaseolus vulgaris* (Flowers et al. 2007), *Glycine max* (Betzelberger et al. 2010) respond across different varieties to ozone stress. However, while taking into account the Indian situation, there is currently a lack of information on evaluating a large number of crops/or cultivars at once for their relative susceptibility to and resilience to ozone.

The variability in crop-specific responses to ozone underscores the importance of considering ozone tolerance and sensitivity in breeding programs and crop selection. Developing ozone-tolerant cultivars can mitigate the detrimental impacts of ozone on crop production and enhance the resilience of agricultural systems to this phytotoxic agent. Understanding the specific mechanisms through which ozone impacts crop yield is essential for implementing strategies to mitigate its negative effects and enhance agricultural sustainability.

## 7.4 Drought Stress—A Rising Abiotic Stress Around the Globe

Drought stress is a pervasive abiotic stress factor that poses significant challenges to global agricultural systems and food security. As climate change continues to unfold, alterations in rainfall distribution and the growing occurrence of severe weather incidents, and water scarcity have led to the rising prevalence and severity of drought stress worldwide (Pandey et al. 2019). Drought stress occurs when the availability of water falls below the critical threshold required by plants for normal growth, development, and physiological functions. It is characterized by a deficiency in soil moisture content, leading to limited water uptake by plant roots. Drought stress can manifest in different forms, including prolonged dry spells, inadequate rainfall, high temperatures, and excessive evapotranspiration rates (Wrey et al. 1994; Wilhite 2000). The severity and duration of drought stress can vary across different regions and cropping systems, determined by variables such as climate, soil composition, and agricultural practices. Both rainfed and irrigated agricultural systems are susceptible to drought stress, with rainfed systems being particularly vulnerable due to their reliance on natural precipitation.

### *7.4.1 Global Trends Contributing to Rising Drought Stress*

Several global trends contribute to the escalating prevalence and severity of drought stress in agricultural systems which are as follows.

#### 7.4.1.1 Climate Change

A major peril to human society and the Earth's ecosystems is climate change (IPCC 2007, 2013). The altering climate patterns associated with global warming have led to shifts in rainfall patterns, characterized by heightened variability and stronger rainfall events. This variability disrupts the regularity of water availability, causing drought periods to become more common and longer-lasting in many areas worldwide. The Intergovernmental Panel on Climate Change (IPCC) has concluded that increased temperatures and less precipitation are the primary causes of the more severe and longer droughts that have been recorded over larger areas over the twentieth century (IPCC 2007). Human activities have been exacerbating drought conditions since the turn of the twentieth century, as demonstrated by the research of Marvel et al. (2019).

#### 7.4.1.2 Water Scarcity

Growing water scarcity driven by population boom, urbanization, and competition for water resources exacerbates the impact of drought stress on agricultural systems. The term "water scarcity" is used to describe the widespread problem of insufficient freshwater resources around the globe (Alcamo et al. 2000; Vörösmarty et al. 2000). This is because human water over-consumption has a negative effect on the water supply and quality. The World Economic Forum ranks the water scarcity as the most pressing and impactful threat of our era in its Global Risks 2015 Report (World Economic Forum 2015). Water scarcity restricts irrigation capacity, reducing the ability to diminish the influence of drought stress on crop harvests.

#### 7.4.1.3 Land Degradation and Desertification

Climate change and drought exemplified how land degradation, such as soil erosion, salinization, and desertification, reduces water-holding capacity of soils and exacerbates the effects of drought stress, leading to a general decrease in vegetation and biomass (Webb et al. 2017; IPCC 2019). Degraded soils have reduced infiltration rates, increased runoff, and decreased water retention, further intensifying the vulnerability of crops to drought stress. Drought can significantly worsen the present scenario of desertification by putting more strain on the already depleted surface and groundwater resources.

#### 7.4.1.4 Changing Agricultural Practices

Shifts in agricultural practices, such as the expansion of monocultures and intensive irrigation, can increase the susceptibility of crops to drought stress. Monocultures, with their uniform genetic makeup, lack the genetic diversity necessary for adapting to changing environmental conditions. Intensive irrigation practices, if not managed efficiently, can deplete water resources and contribute to water scarcity during drought periods.

#### 7.4.1.5 Socioeconomic Factors

Socioeconomic factors, including poverty, limited access to technology, and inadequate infrastructure, can exacerbate the impact of drought stress on vulnerable communities. Lack of financial resources and adaptive capacity can hinder the implementation of drought mitigation strategies and reduce resilience to its effects.

### 7.4.2 *Impact of Drought Stress on Plants*

The effects of drought stress on growth and advancement of plants are substantial. The impact of drought stress on crop yield is multifaceted, affecting diverse biochemical and physiological activities within plants. Understanding the characteristics and impacts of drought stress is essential for creating effective approaches to address its negative effects on crop yield. The following are some key mechanisms through which drought stress influences crop performance:

#### 7.4.2.1 Stomatal Closure and Reduced Photosynthesis

To conserve water, plants react to drought stress by shutting their stomata, the minute openings on leaves that allow for gas exchange. Such phenomenon has been reported in several crops such as wheat (Saradadevi et al. 2017), maize (Benešová et al. 2012), rice (Hu et al. 2017; Song et al. 2022), soybean (Castro-Valdecantos et al. 2021), peanut (Jiang et al. 2021) which is observed to be a strategy to conserve water during periods of water deficit conditions. In addition to stomatal closure, stomatal size reduction was also seen during moderate drought circumstances (Farooq et al. 2012).

The negative effect on plants' capacity to assimilate carbon dioxide during drought is due not only to the stomatal limitation but also to other non-stomatal factors such as reduction in leaf area, premature leaf senescence, and inefficient performance of different photosynthetic machineries (Farooq et al. 2009). Inhibition of essential photosynthetic enzymes and ATP synthases (Zandalinas et al. 2017) and reduced internal availability of $CO_2$ are often cited as causes of this latter consequence. Shutting the stomata limits the influx of carbon dioxide ($CO_2$) necessary for photosynthesis, reducing the plant's ability to produce carbohydrates which eventually leads to lower crop production. Ghosh et al. (2020a) reported lowering of stomatal conductance contributed to bringing about lower levels of photosynthesis in wheat cultivar HD 2967 at 50% water shortage relative to well-irrigated plants. Such finding was corroborated with the research of Miyashita et al. (2005) who found that stomatal restriction decreased photosynthesis and transpiration rate in kidney beans (*Phaseolus vulgaris* L.).

The apparent quantum yield, the concentration of ribulose-1,5-bisphosphate (RuBP), and the activity of RuBP all play crucial roles in the assimilation of light energy via photosynthesis. The regeneration RuBP and synthesis of adenosine triphosphate (ATP) can be negatively affected by stomatal constraint during water scarce conditions, leading to light-induced inhibition and disturbance of standard photochemical processes (Pandey et al. 2018) of crop plants. Moreover, under water stress, the activity of synthetic enzymes is inhibited, which limits RuBP production, as described by Gunasekera and Berkowitz (1993) in tobacco. Ribulose-1,5-bisphosphate carboxylase/oxygenase (RuBisCo) is another chemical vital to plant photosynthesis during periods of water scarcity. RuBisCo enzyme activity was unaffected by or only mildly impacted by mild water constraint, but was reduced by severe drought (Galmés et al. 2013). Photochemical and metabolic processes, such as the rate of electron transport and photophosphorylation, can undergo noticeable shifts.

Chlorophyll is crucial for capturing, transmitting, and converting light energy during photosynthesis, which occurs in the chloroplast of green plant leaves. Continuous chlorophyll metabolism in plants is directly linked to photosynthesis and eventually to the plant yields. Under drought stress, chlorophyll content decreases and the ratio of chlorophyll a/b and carotenoid shifts, leading to modifications in photosynthetic performance. Drought may be directly responsible for the breakdown of chlorophyll, which may explain why there is less chlorophyll in the leaves. Chickpea's chlorophyll a, chlorophyll b, and total chlorophyll concentrations were observed to be drastically reduced by drought stress during vegetative growth and anthesis (Mafakheri et al. 2010). Different wheat cultivars responded differently to drought stress, as was observed by Zaefyzadeh et al. (2009) in a study of 13 native varieties of durum wheat from Iran and Azerbaijan. Under drought stress, the chlorophyll content of sensitive wheat cultivars dropped dramatically, but the chlorophyll levels in wheat varieties with resistance remained stable (Zaefyzadeh et al. 2009). Drought also makes it harder for plants to absorb nutrients, which can lead to nutritional shortage which can ultimately resulted to lower chlorophyll levels.

$C_4$ photosynthesis, $C_3$ photosynthesis, and the Crassulacean acid metabolism (CAM) routes are the three primary processes through which carbon dioxide is assimilated during photosynthesis in plants. The $C_4$ route outperformed the $C_3$ pathway under drought conditions. $C_4$ plants' leaves have a classic Kranz wreath form, and their water-use efficiency is much higher than that of plants that employ the $C_3$ route. When water is scarce, the plant's ability to ingest $CO_2$ and generate organic matter increases, helping it to better withstand drying out prematurely. At night, CAM plants allow $CO_2$ in through their stomata, convert it to malic acid via the enzyme phosphoenolpyruvate carboxylase (PEPC), and store it in vacuoles. Stomata are closed and carbon dioxide is released throughout the day due to MAL decarboxylation. The contradiction between transpiration through stomatal pores and $CO_2$ absorption during drought stress is resolved due to CAM's ability to fix $CO_2$ through stomatal opening at night (Yang et al. 2021).

#### 7.4.2.2 Reduced Water Uptake and Transpiration

Stomatal opening regulates $CO_2$ uptake for photosynthesis and water loss through transpiration (Shahinnia et al. 2016). Reduced leaf water potential, decreased photosynthetic rate, increased oxidative stress, and canopy temperature result from stomatal shutdown triggered by water scarcity (Ludlow and Muchow 1990; Yokota et al. 2002). Combating photosynthetic suppression under drought stress requires greater stomatal control over transpiration (Bota et al. 2004). Under drought stress, yield variation in many crops is mostly attributable to transpiration efficiency which forms a crucial component of water-use efficiency in plants. The effectiveness of transpiration is linked to a more extensive root system, therefore enhancing both characteristics is necessary. It has been observed that wheat varieties with both deep root biomass and transpiration efficiency characteristics will be better able to withstand drought stress (Pask and Reynolds 2013). Under drought stress, the transpiration rate fell by 53% in faba beans (Hao et al. 2013) and by 57% in soybeans (Mak et al. 2014). As xylem abscisic acid was measured to be 50 times higher during drought stress (Liu et al. 2005), with a considerable rise as the severity of stress increased (Liu et al. 2004), this reduction was caused by the drop of stomatal conductance, governed by root-based abscisic acid.

ERECTA, found in Arabidopsis, is a potential leucine-rich repeat receptor-like kinase (LRR-RLK) that controls stomatal count, epidermal cell growth and organization, mesophyll cell division, and cell-to-cell interaction, making it a prototypical example of a gene responsible for controlling transpiration efficiency (Masle et al. 2005). Reduced stomatal density and increased stomata size in Arabidopsis mutants improved drought tolerance by decreasing transpiration and increasing biomass build-up under drought stress (Pillitteri and Torii 2012). GTL2-LIKE1 (GTL1), in contrast to the ERECTA gene in Arabidopsis, represses the STOMATAL DENSITY AND DISTRIBUTION 1 (SDD1) gene to regulate stomata count, transpirational rate, and water use efficiency (Yoo et al. 2010).

#### 7.4.2.3 Oxidative Stress and Induction of Antioxidative Defense System

Drought stress triggers the accumulation of reactive oxygen species (ROS) within plant cells. These ROS, including superoxide radicals ($O_2 \cdot^-$) and hydrogen peroxide ($H_2O_2$), induce oxidative damage in cellular structures, leading to impaired cellular metabolism, protein degradation, and DNA damage. This oxidative stress contributes to reduced crop yield by disrupting cellular processes critical for growth and advancement. Under conditions of water deprivation stress, ROS act as signaling chemicals at low concentrations. The intrinsic properties of biomolecules including proteins, fatty, and nucleic acids can be altered by oxidative stress induced by ROS, resulting to cell death when the ROS level surpasses the protective mechanism (Kurutas 2016). The cellular defense against ROS is controlled by both enzymatic and non-enzymatic components, and keeping a substantial amount of these radical scavenging compounds has been shown to be an adaptive response under drought stress (Sahitya et al. 2018). Antioxidants can be either enzymatic (such as superoxide dismutase (SOD), catalase (CAT), ascorbate peroxidase (APX), dehydroascorbate reductase (DHAR), monodehydroascorbate reductase (MDHAR), glutathione reductase (GR), peroxidase (POD), glutathione peroxidase (GPX)) or non-enzymatic (such as ascorbate, glutathione, tocopherols, phenolics, carotenoids, and ascorbic acid) (Ghosh et al. 2020a). Superoxide dismutase's primary role is to detoxify $O_2 \cdot^-$ by converting it to $H_2O_2$. In plants, SOD is the first line of defense in the elimination mechanism for ROS. In plants, removing $H_2O_2$ is primarily the job of CAT and POD. In addition, the $H_2O_2$ scavenging enzymes GR, APX, DHAR, and MDHAR are of critical importance. When combined, they serve as a second line of defense in plants' ROS removal mechanisms. Scavenging ROS is a three-pronged defense, with GPX being a key player in the oxidative processing of lipids and alkyl peroxides. As seen in in different crop plants, antioxidant concentrations may be elevated higher in the recovery period versus the stress phase when experiencing oxidative stress (Patel et al. 2011; Osman 2015; Guler and Pehliban 2016).

Many different metabolic processes in plants result in the production of ROS. In green plants, chloroplasts are the primary sites of ROS production (Dietz 2016). Plants' ability to efficiently absorb light energy declines in water-deficient conditions. When plants are unable to fix carbon dioxide, their NADP+ levels drop, and the rate at which they transfer electrons from NADP+ to $O_2$ rises relatively. Oxygen is being utilized as electron acceptor and eventually give rise to $O_2^-$. In turn, $O_2^-$ can set off a chain reaction in plants that results in a high concentration of ROS (Pospíšil 2012). In plants, ROS participates in both defense against external stress and regulation of advancement and growth, as well as the harmful effects of plant damage against any abiotic stress. In plants, ROS play a crucial role in the oxidative burst, a defense mechanism against biotic and abiotic stress factors (Yang et al. 2021).

#### 7.4.2.4 Secondary Metabolite Production and Hormonal Regulation

Drought stress often induces the synthesis of secondary metabolites, including phenolics, flavonoids, and terpenoids. These compounds serve as signaling molecules and participate in plant defense mechanisms against various stresses. They contribute to antioxidant activity, regulate gene expression, and enhance stress tolerance, ultimately influencing crop yield. Compatible solutes, such as proline, glycine betaine, and sugars, aquaporins, dehydrins accumulate in plant cells under drought stress. These solutes contribute to osmotic adjustment, maintaining cellular turgor and protecting cellular structures from dehydration and damage. The proline content of rice was found to be considerably higher when the concentration of polyethylene glycol (PEG) was 30% (Patmi et al. 2020). Furthermore, to better withstand drought, the Lanzhou lily alters its osmotic regulation and secondary metabolite content, decreasing polysaccharide, soluble sugar, and fructose while boosting glucose, proline, and trehalose (Li et al. 2020). Khoyerdi et al. (2016) also found that numerous pistachio genotypes showed significant increases in proline, glycine betaine, total soluble carbohydrate, and sucrose content in response to drought stress. Additionally, compatible solutes can act as antioxidants, scavenging reactive oxygen species (ROS) and mitigating oxidative stress.

Drought stress has been reported to alter the balance of plant hormones, particularly abscisic acid (ABA) and cytokinins, which play essential roles in regulating plant growth, development, and stress responses. Plants produce the hormone ABA in response to drought, which stimulates stomatal closure to lessen transpiration and thus water loss. $Ca^{2+}$ acts as a second messenger in osmotic stress response (Case et al. 2007), and its content in the cytoplasm rises during stomatal closure. Furthermore, the imbalance in hormone levels influences various physiological processes, including seed germination, root growth, flowering time, and fruit development, ultimately impacting crop yield.

#### 7.4.2.5 Changes in Carbon and Nitrogen Metabolism

Drought stress alters carbon and nitrogen metabolism in plants, impacting the availability and allocation of these essential nutrients. Reduced carbon availability impairs energy production and affects biosynthetic processes, while altered nitrogen metabolism affects protein synthesis and enzymatic activity. These changes have cascading effects on plant growth, development, and ultimately crop yield. According to research by Ghosh et al. (2020a), water deficit stress condition has significantly affect the elemental stoichiometry (C:N:P:K) in soil, leading to an imbalanced stoichiometry inside the wheat cultivar HD 2967 which in turn, causing yield loss.

#### 7.4.2.6 Root Morphology and Depth

Drought stress often induces changes in root morphology, including increased root branching, elongation, and proliferation of root hairs (Xiao et al. 2020). These adaptations enhance the exploration of soil volume, enabling plants to access water from deeper soil layers. Additionally, changes in root architecture can increase the root-to-shoot ratio, influencing resource allocation and overall plant growth.

#### 7.4.2.7 Flowering Time and Reproductive Structures

Drought stress can influence flowering time, causing delayed flowering or even complete floral abortion (Mwanamwenge et al. 1999). These alterations in reproductive structures impact crop yield by reducing the number of flowers, impairing pollination, and affecting seed set and fruit development. The timing of drought stress during critical reproductive stages can have significant consequences for crop productivity.

## 7.5 Combined Effect of Ozone and Drought Stress on Food Security

Co-occurrence of elevated ozone concentrations and water deficit stress is frequently observed during summer heat (Wilkinson and Davies 2010). Extensive investigations have been conducted to elucidate the impacts of ozone and drought on crop plants, particularly in tropical and sub-tropical regions witnessing an escalating trend in ozone levels and exacerbating drought conditions (Wittig et al. 2009). Notably, under water deficit conditions, plants exhibit enhanced tolerance to ozone stress by regulating stomatal closure, resulting in reduced stomatal ozone uptake (Herbinger et al. 2002; Harmens et al. 2019) (Fig. 7.1). Harmen et al. (2019) proposed that deliberate water restriction could serve as a practical management approach for mitigating high ozone concentrations. However, these intricate interactions display considerable variability across studies, as evidenced by divergent findings (Ghosh et al. 2018; Pellegrini et al. 2019).

Studies have indicated the occurrence of additive and synergistic responses, in addition to the concept of drought-induced protection against ozone uptake, in the context of ozone absorption and drought (Alonso et al. 2014; Cotrozzi et al. 2016; Ghosh et al. 2020a). These responses involve a reduction in carbon fixation caused by low photosynthetic rates and limited availability to the overall defense mechanisms of plants. Nutrient absorption and translocation to different plant parts are hindered, resulting in high yield loss due to decreased ROS detoxification capacity. These combined factors lead to an abnormal shift in the plant's normal physiological functions.

The regulation of stomatal conductance in response to ozone stress and drought involves the interplay of two signaling molecules, ethylene and abscisic acid (ABA). Previous research by Wilkinson and Davies (2010) has demonstrated that ABA

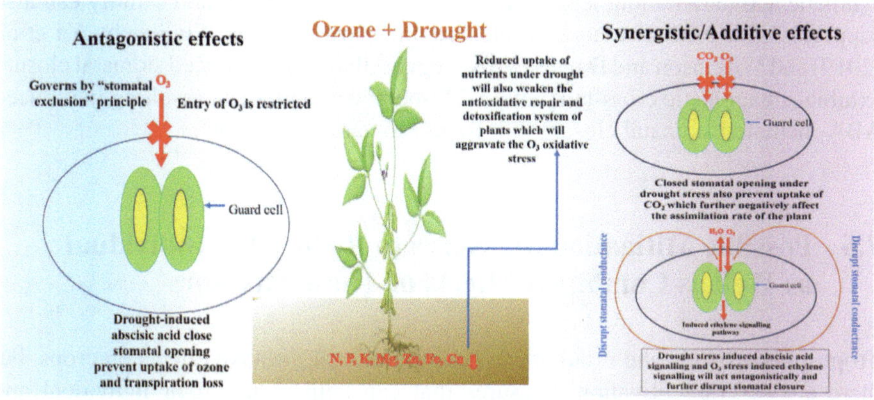


**Fig. 7.1** Pictorial diagram showing the possible interactive effect of tropospheric ozone and drought stress in plants. Under conditions of water scarcity, plants display increased tolerance to ozone stress by controlling stomatal closure, which results in decreased stomatal ozone uptake, exhibiting an antagonistic interaction. On the other hand, two signaling molecules, ethylene and abscisic acid, interact to control stomatal conductance in response to ozone stress and drought. Stomatal closure is primarily triggered by ABA when there is a water shortage. Additionally, in response to both ozone exposure and water deficit stress, ethylene synthesis rises. However, when it comes to modifying stomatal sensitivity in response to these stressors, ethylene and abscisic acid signaling are hostile to one another. Even in the presence of abscisic acid, which is produced when under the stress of a water deficit, ozone-induced ethylene stimulation impairs stomatal closure. As a result, by preventing stomatal closure, ethylene reduces the sensitivity of stomata to abscisic acid. Stomatal closure is hampered under such condition, which increases stomatal conductance and prolongs water loss through transpiration. Hence, in such cases plant may show synergistic or additive effect

primarily triggers stomatal closure during water deficit conditions. Furthermore, ethylene production increases in response to both ozone stress and water deficit stress, as shown in studies by Rao et al. (2002), Voisin et al. (2006), and Sobeih et al. (2004). However, an antagonistic relationship exists between ethylene and ABA signaling when it comes to modulating stomatal sensitivity under these stressors. Ozone-induced ethylene stimulation hampers stomatal closure, even in the presence of ABA, which is induced during water deficit stress. Consequently, the presence of ethylene diminishes the sensitivity of stomata to ABA by inhibiting stomatal closure.

Under conditions of water deficit and elevated ozone levels, stomatal closure is hindered, leading to increased stomatal conductance and continued water loss through transpiration. Wilkinson and Davies (2010) proposed a hypothetical mechanism suggesting that in the absence of either hormone, stomatal closure is induced through $H_2O_2$ signaling. However, when both hormones are present, the authors predict a complete failure of stomatal closure due to excessive $H_2O_2$ accumulation and nitric oxide (NO). Ethylene may interfere with ABA-induced $H_2O_2$ signaling by acting on the ETR1 gene, but the role of NO in these signaling pathways remains uncertain. These findings challenge the concept of the stomatal ozone exclusion principle, which suggests partial protection against ozone uptake under water deficit stress (Fig. 7.1). Proper water management in agricultural systems becomes crucial under changing climatic conditions. Other climate change factors such as high photosynthetic active radiation, temperature, vapor pressure deficit, and salinity can also suppress ABA-mediated stomatal closure under ozone stress. Studies by Yu et al. (2019) and Wilkinson and Davies (2009) suggest that ozone-induced stomatal closure exhibits antagonistic cross-talk with ABA-mediated regulation, resulting in reduced ABA-mediated stomatal closure in response to various stressors.

## 7.6 Possible Mitigation Measures to Reduce the Individual as Well as Combined Effects of These Stressors

Tropospheric ozone and drought can have significant negative effects on crops, but there are several mitigation measures that can help reduce their individual and combined impacts. However, it is important to note that these measures should be implemented in combination, considering the specific local conditions and crop requirements. Integrated approaches that address both ozone and drought stress will likely yield the best results in safeguarding crop productivity and resilience. Here are some possible strategies which has been discussed below and has been presented in Fig. 7.2.

**Fig. 7.2** Diagram showing the possible mitigation approaches to reduce the individual as well as combined effects of tropospheric ozone and drought-induced phytotoxicity in crop plants

### *7.6.1 Stress-Resistant Crop Varieties*

Breeders of plants can create and advocate for the usage of more ozone/drought-resistant crop varieties. These cultivars may feature genes that protect them from ozone-related toxicity. It has been proposed that selecting crops resistant to ozone/drought could help mitigate yield losses caused by these stresses (Blum and Didyk 2007). It has also been suggested (Sarkar and Agrawal 2010; Chaudhary and Agrawal 2015) that farmers could consider the present local ozone concentrations when deciding which crop varieties to use. Therefore, it is essential to tailor the screening and propagation of a crop to a given region. Numerous studies have looked into the sensitivity and tolerance of various cultivars and varieties to ozone in various plants, including *Triticum aestivum* (Saitanis et al. 2014; Singh et al. 2018; Begum et al. 2020), *Oryza sativa* (Frei et al. 2008; Sarkar and Agrawal 2010; Ashrafuzzaman et al. 2017), *Phaseolus vulgaris* (Guidi et al. 2010; Borowiak 2013; Leung et al. 2020), *Zea rays* (Singh et al. 2014; Yendrek et al. 2017; Gupta et al. 2020); *Amaranthus hypochondriacus* (Yadav et al. 2020), *Glycine max* (Jiang et al. 2018), *Arachis hypogaea* (Chaudhary and Rathore 2021). Additionally, biotechnological methods have been utilized to enhance crop resistance to a variety of abiotic challenges, including drought, salt, and high temperatures (Hirel et al. 2007). The long-term adaptation strategy of breeding or constructing novel plant types impervious to ozone can reduce ozone damage (Fuhrer 2009). However, research into ozone-tolerant genotypes has received less focus as compared to the development of drought-tolerant crop varieties. Therefore, we urgently require the development of ozone-resistant cultivars, which can be accomplished through either traditional breeding practices or the application of cutting-edge biotechnological technologies.

### *7.6.2 Irrigation Management*

Implementing efficient irrigation practices can help mitigate the effects of drought on crops. Techniques such as drip irrigation, precision irrigation, and soil moisture sensors can optimize water use and minimize water stress on plants. More widespread consensus on key ideas and performance indicators will be helpful in developing effective irrigation management strategies to address water scarcity. Supply management and demand management are two possible strategies for effective farm management (Pereira et al. 2002). Improved irrigation conveyance and distribution systems increase delivery flexibility and decrease system water wastages as part of supply management strategies. Farmers' participation and the training of irrigation agents and farmers are also important aspects of supply management. Finally, the development of new sources of water supplies is another component of supply management (Pereria et al. 2002). Better farm irrigation systems and deficit irrigation can help reduce demand for water. Increased irrigation uniformity is intimately tied to the development of better irrigation systems. This necessitates more thorough field appraisal, attentive upkeep, and improved design for irrigation systems.

### *7.6.3 Crop Rotation and Diversification*

Rotating crops and diversifying the types of crops grown in a particular area can help break the cycle of disease and pests associated with ozone and drought. Different crops have varying sensitivities to these stressors, so diversification can reduce overall vulnerability. Crop rotation (the pattern of crops planted through time) is one example of temporal crop diversity that has been employed by farmers for centuries to boost yields by restoring soil health and disrupting the life cycles of pests, weeds, and diseases (Tiemann et al. 2015). In the United States, a rotation of soybean and maize results in average yield increases of 5–10% compared to a monoculture system, even though the monoculture system typically requires more inputs (Gentry et al. 2013). Improved soil qualities, such as greater soil water uptake and storage and the number of beneficial soil bacteria, may contribute to improved production resilience to drought and other forms of harsh growth conditions in diverse rotations. One earlier research by Gaudin et al. (2015) found that a five-crop rotation, rather than a two-crop rotation consisting solely of maize and soybeans, increased maize yields by 7 percentage points during hot and dry years. It is unclear, however, whether this is a site-specific consequence due to unique environmental variables or crop-rotation composition, or a more general consequence of rotation diversification.

### 7.6.4 Mulching and Conservation Tillage

Mulching involves covering the soil around plants with organic materials like straw or wood chips. This practice helps retain soil moisture, suppress weeds, and moderate soil temperature, thus reducing drought stress. Conservation tillage techniques, such as no-till or reduced-till, can also help conserve soil moisture and enhance soil health. Yadav et al. (2018) found that using no-till techniques and applying mulch improved water storage of the soil profile, decreased evaporation, and boosted water productivity and net returns while decreasing weed density and biomass. Rice yield characteristics were improved by no-till and mulches due to increased soil moisture storage in the root zone via profile recharge. Another study by Yang et al. (2020) suggests that using no-tillage in conjunction with a maize straw mulch can increase grain yield, water use efficiency, and nitrogen use efficiency by retaining soil moisture for wheat tillering on dry days.

Many different types of tillage have been used to examine the emissions of oxides of nitrogen, methane, and carbon dioxide from farmland (Menéndez et al. 2008; Plaza-Bonilla et al. 2014), some of which are regarded as precursors of ozone. The positive effect of N fertilization on carbon sequestration may be nullified by improper tillage practices, which might increase $N_2O$ emissions (Snyder et al. 2009). Soil carbon retention and green-house gases emission reduction are both enhanced by the adoption of no-till farming techniques (Cole et al. 1997). Therefore, considering the development of ozone at ground level, it is suggested that various tillage strategies be investigated for their potential to mitigate the negative impacts of ozone on crop yields.

### 7.6.5 Improved Nutrient Management Through Soil Amendment Practices

Proper nutrient management, including balanced fertilization and soil testing, can optimize crop health and resilience to ozone and drought. Adequate nutrient supply can help plants withstand stress and recover more effectively. It is feasible to mitigate ozone's deleterious effects on crops by adjusting the use of nitrogen, phosphorus, and potassium (NPK) fertilizers, as well as manure and bio fertilizers (Singh et al.

2005, 2011, 2015; Calvo et al. 2009; Pandey et al. 2018). Maintaining photosynthetic efficiency through increased leaf area is a direct result of nitrogen's effect on dry matter synthesis (Dordas and Sioulas 2008). Schulte auf'mErley et al. (2007) found that increased N availability to plants could postpone leaf senescence in response to ozone stress. To some extent, yield losses can be avoided in ozone-stressed wheat cultivars like Sonalika and HD 2967, as demonstrated by research from Pandey et al. (2018). Total biomass in wheat cultivar HD 2967 increased after receiving 1.5 times the recommended N dose. *Beta vulgaris* L. was grown in an experiment by Singh et al. (2005) and found to benefit from an increase in NPK dose over the recommended level (1.5 times) when exposed to air pollutants such as $SO_2$ (2.5–42.5 ppb), $NO_2$ (10.6–65 ppb), and ozone (3.5–30.8 ppb). Manure application, in addition to inorganic fertilizers, is an effective farming method. For example, Calvo et al. (2009) found that applying organic manure at rates of 3 and 6 kg $m^2$ to the soil mitigated the potato crop's losses caused by a mean ozone exposure of 57 ppb for 8 h. Biochar fertilizer also showed promise in protecting wheat plants from the effects of ozone stress. Biochar addition to the soil at a rate of 2.5% was also found to be beneficial in mitigating the negative effects that ozone had on the growth characteristics of the test plants (Ghosh et al. 2021a). Numerous studies have also documented the positive impacts of biochar application under drought conditions (e.g., Artiola et al. 2012; Paneque et al. 2016). The photosynthesis of drought-stressed plants has been shown to be enhanced by biochar application (Akhtar et al. 2014; Paneque et al. 2016; Xiao et al. 2016). Biochar boosted the relative water content of drought-stressed tomato leaves and improved their chlorophyll content, stomatal conductance, assimilation rate, water usage efficiency, and stomatal density (Akhtar et al. 2014).

### *7.6.6 Education and Outreach*

Promoting awareness and providing education to farmers about the risks associated with tropospheric ozone and drought are crucial. Sharing information on best management practices, technological advancements, and available resources can empower farmers to make informed decisions and adopt appropriate mitigation measures.

### *7.6.7 Policy Support*

Governments can play a role by implementing policies that encourage sustainable farming practices and support farmers in adopting mitigation measures. This can include financial incentives, research funding, and regulations aimed at reducing ozone precursor emissions and promoting water conservation. Implementing advanced monitoring systems for ozone levels and drought conditions can provide early warnings and allow farmers to take timely action. This can include using weather stations, satellite data, and predictive models to track and predict ozone concentrations and drought severity.

## 7.7 Conclusion

The combined effect of drought stress and high tropospheric ozone conditions on crop yield presents a complex and multifaceted challenge to global food security. This chapter has shed light on the individual impacts of drought stress and ozone on crop productivity, as well as the interactive responses that occur when these stressors act together which can be either synergistic/additive/antagonistic. The physiological, biochemical, and morphological mechanisms underlying the combined effect have been explored, emphasizing the role of oxidative stress, water relations, and photosynthetic impairment.

Addressing the combined effect of drought stress and ozone requires a comprehensive and multidisciplinary approach that encompasses plant breeding, agronomic practices, and policy interventions. Developing stress-tolerant crop varieties, improving irrigation efficiency, and implementing ozone pollution control measures are crucial steps toward mitigating the combined impact. Moreover, promoting sustainable agricultural practices and adopting integrated approaches that consider both stressors simultaneously are essential for enhancing crop resilience and ensuring food security in a changing climate.

By understanding the complex interactions between drought stress and high tropospheric ozone levels, researchers, policymakers, and farmers can work together to develop strategies that minimize the negative impact on crop yield and global food security. The combined effect of these stressors cannot be overlooked, and concerted efforts are needed to adapt and mitigate their impact to safeguard the availability and affordability of food resources for future generations. Through collaboration and innovation, we can pave the way for a more sustainable and resilient agricultural system that can withstand the challenges posed by drought stress and high tropospheric ozone conditions.

**Acknowledgements** AG is grateful to the Head of Department of Life Science, Central University of Jharkhand, Ranchi, India. DST is being acknowledged for providing the financial assistance to AG as DST-INSPIRE Faculty Fellowship (IFA22-LSPA 154). BP is thankful to the Director, Central Institute of Mining and Fuel Research, Dhanbad, India. BP also acknowledges the Science and Engineering Research Board, Department of Science & Technology (SERB DST; File No.: CRG/2023/003085).

## References


Agathokleous E, Feng Z, Oksanen E, Sicard P, Wang Q, Saitanis C J, Paoletti E et al (2020) Ozone affects plant, insect, and soil microbial communities: A threat to terrestrial ecosystems and biodiversity. Sci Adv 6(33), eabc1176

Ainsworth EA (2017) Understanding and improving global crop response to ozone pollution. Plant J 90(5):886–897

Ainsworth EA, Yendrek CR, Sitch S, Collins WJ, Emberson LD (2012) The effects of tropospheric ozone on net primary productivity and implications for climate change. Annu Rev Plant Biol 63:637–661

Akhtar SS, Li G, Andersen MN, Liu F (2014) Biochar enhances yield and quality of tomato under reduced irrigation. Agric Water Manag 138:37–44

Alcamo J, Henrichs T, Rosch T (2000) World water in 2025. World water series report, 2

Alonso R, Elvira S, González-Fernández I, Calvete H, García-Gómez H, Bermejo V (2014) Drought stress does not protect Quercus ilex L. from ozone effects: Results from a comparative study of two subspecies differing in ozone sensitivity. Plant Biol 16(2):375–384

Andersen CP (2003) Source–sink balance and carbon allocation below ground in plants exposed to ozone. New Phytol 157(2):213–228

Artiola JF, Rasmussen C, Freitas R (2012) Effects of a biochar-amended alkaline soil on the growth of romaine lettuce and bermudagrass. Soil Sci 177(9):561–570

Ashmore MR (2005) Assessing the future global impacts of ozone on vegetation. Plant, Cell Environ 28(8):949–964

Ashrafuzzaman M, Lubna FA, Holtkamp F, Manning WJ, Kraska T, Frei M (2017) Diagnosing ozone stress and differential tolerance in rice (Oryza sativa L.) with ethylenediurea (EDU). Environ Pollut 230:339–350

Ashrafuzzaman M, Ueda Y, Frei M (2020) Natural sequence variation at the OsORAP1 locus is a marker for ozone tolerance in Asian rice. Environ Exp Bot 178:104153

Askarnejad MR, Soleymani A, Javanmard HR (2021) Barley (Hordeum vulgare L.) physiology including nutrient uptake affected by plant growth regulators under field drought conditions. J Plant Nutr 44(15):2201–2217

Begum H, Alam MS, Feng Y, Koua P, Ashrafuzzaman MD, Shrestha A, Frei M et al (2020) Genetic dissection of bread wheat diversity and identification of adaptive loci in response to elevated tropospheric ozone. Plant, Cell Environ 43(11):2650–2665

Benešová M, Hola D, Fischer L, Jedelský PL, Hnilička F, Wilhelmová N, Hniličková H et al (2012) The physiology and proteomics of drought tolerance in maize: early stomatal closure as a cause of lower tolerance to short-term dehydration?. PLoS ONE 7(6):e38017

Betzelberger AM, Gillespie KM, Mcgrath JM, Koester RP, Nelson RL, Ainsworth EA (2010) Effects of chronic elevated ozone concentration on antioxidant capacity, photosynthesis and seed yield of 10 soybean cultivars. Plant, Cell Environ 33(9):1569–1581

Biswas DK, Xu H, Li YG, Liu MZ, Chen YH, Sun JZ, Jiang GM (2008) Assessing the genetic relatedness of higher ozone sensitivity of modern wheat to its wild and cultivated progenitors/relatives. J Exp Bot 59(4):951–963

Blum O, Didyk N (2007) Study of ambient ozone phytotoxicity in Ukraine and ozone protective effect of some antioxidants. J Hazard Mater 149(3):598–602

Bohler S, Cuypers A, Vangronsveld J (2015) Interactive effects between ozone and drought: sorrow or joy? In: Combined stresses in plants: physiological, molecular, and biochemical aspects, pp 147–157

Borowiak K (2013) Effect of tropospheric ozone on photosynthetic activity of ozone-resistant and ozone-sensitive test plants (Nicotiana tabacum L., Phaseolus vulgaris L.). Polish J Ecol 61(4):693–703

Bota J, Medrano H, Flexas J (2004) Is photosynthesis limited by decreased Rubisco activity and RuBP content under progressive water stress? New Phytol 162(3):671–681

Broberg MC, Hayes F, Harmens H, Uddling J, Mills G, Pleijel H (2023) Effects of ozone, drought and heat stress on wheat yield and grain quality. Agr Ecosyst Environ 352:108505

Burkart S, Bender J, Tarkotta B, Faust S, Castagna A, Ranieri A, Weigel HJ (2013) Effects of ozone on leaf senescence, photochemical efficiency and grain yield in two winter wheat cultivars. J Agron Crop Sci 199(4):275–285

Calatayud V, Cerveró J, Calvo E, García-Breijo FJ, Reig-Armiñana J, Sanz MJ (2011) Responses of evergreen and deciduous Quercus species to enhanced ozone levels. Environ Pollut 159(1):55–63

Calvo E, Calvo I, Jimenez A, Porcuna JL, Sanz MJ (2009) Using manure to compensate ozone-induced yield loss in potato plants cultivated in the east of Spain. Agr Ecosyst Environ 131(3–4):185–192

Case RM, Eisner D, Gurney A, Jones O, Muallem S, Verkhratsky A (2007) Evolution of calcium homeostasis: from birth of the first cell to an omnipresent signalling system. Cell Calcium 42(4–5):345–350

Castagna A, Ranieri A (2009) Detoxification and repair process of ozone injury: from $O_3$ uptake to gene expression adjustment. Environ Pollut 157(5):1461–1469

Castro-Valdecantos P, Puértolas J, Albacete A, Dodd IC (2021) Girdling changes root and shoot hormonal balance but does not alter drought-induced stomatal closure in soybean. Environ Exp Bot 192:104657

Chandra S, Ziemke JR, Bhartia PK, Martin RV (2002) Tropical tropospheric ozone: implications for dynamics and biomass burning. J Geophys Res Atmosp 107(D14):ACH-3

Chaudhary IJ, Rathore D (2021) Assessment of dose–response relationship between ozone dose and groundnut (Arachis hypogaea L) cultivars using Open Top Chamber (OTC) and Ethylenediurea (EDU). Environ Technol Innov 22:101494

Chaudhary N, Agrawal SB (2015) The role of elevated ozone on growth, yield and seed quality amongst six cultivars of mung bean. Ecotoxicol Environ Saf 111:286–294

Chen Z, Zhuang Y, Xie X, Chen D, Cheng N, Yang L, Li R (2019) Understanding long-term variations of meteorological influences on ground ozone concentrations in Beijing During 2006–2016. Environ Pollut 245:29–37

Cole CV, Duxbury J, Freney J, Heinemeyer O, Minami K, Mosier A, Zhao Q et al (1997) Global estimates of potential mitigation of greenhouse gas emissions by agriculture. Nutr Cycl Agroecosyst 49:221–228

Cotrozzi L, Conti B, Lorenzini G, Pellegrini E, Nali C (2021) In the tripartite combination ozone-poplar-Chrysomela populi, the pollutant alters the plant-insect interaction via primary metabolites of foliage. Environ Res 201:111581

Cotrozzi L, Remorini D, Pellegrini E, Landi M, Massai R, Nali C, Lorenzini G et al (2016) Variations in physiological and biochemical traits of oak seedlings grown under drought and ozone stress. Physiol Plant 157(1):69–84

Couture JJ, Meehan TD, Lindroth RL (2012) Atmospheric change alters foliar quality of host trees and performance of two outbreak insect species. Oecologia 168:863–876

Degl'Innocenti E, Guidi L, Soldatini G F. (2007) Effects of elevated ozone on chlorophyll a fluorescence in symptomatic and asymptomatic leaves of two tomato genotypes. Biologia Plantarum 51:313–321

Dentener F, Keating T, Akimoto H, ECE U (2010) Hemispheric transport of air pollution 2010. Part A, Ozone and particulate matter/edited by Frank Dentener, Terry Keating and Hajime Akimoto; prepared by the task force on hemispheric transport of air pollution acting within the framework of the convention on long-range transboundary air pollution

Dietz KJ (2016) Thiol-based peroxidases and ascorbate peroxidases: why plants rely on multiple peroxidase systems in the photosynthesizing chloroplast? Mol Cells 39(1):20

Dordas CA, Sioulas C (2008) Safflower yield, chlorophyll content, photosynthesis, and water use efficiency response to nitrogen fertilization under rainfed conditions. Ind Crops Prod 27(1):75–85

Eichelmann H, Oja V, Rasulov B, Padu E, Bichele I, Pettai H, Laisk A et al (2004) Photosynthetic parameters of birch (Betula pendula Roth) leaves growing in normal and in $CO_2$-and $O_3$-enriched atmospheres. Plant, Cell Environ 27(4):479–495

Farooq M, Hussain M, Wahid A, Siddique KHM (2012) Drought stress in plants: an overview. In: Plant responses to drought stress: from morphological to molecular features, pp 1–33

Farooq M, Wahid A, Kobayashi NSMA, Fujita DBSMA, Basra SMA (2009) Plant drought stress: effects, mechanisms and management. Sustain Agric 153–188

Fatima A, Singh AA, Mukherjee A, Dolker T, Agrawal M, Agrawal SB (2019) Assessment of ozone sensitivity in three wheat cultivars using ethylenediurea. Plants 8(4):80

Feng Z, Kobayashi K, Ainsworth EA (2008) Impact of elevated ozone concentration on growth, physiology, and yield of wheat (Triticum aestivum L.): a meta-analysis. Global Change Biol 14(11):2696–2708

Feng Z, Xu Y, Kobayashi K, Dai L, Zhang T, Agathokleous E, Yue X et al (2022) Ozone pollution threatens the production of major staple crops in East Asia. Nat Food 3(1):47–56

Fiscus EL, Booker FL, Burkey KO (2005) Crop responses to ozone: uptake, modes of action, carbon assimilation and partitioning. Plant, Cell Environ 28(8):997–1011

Flowers MD, Fiscus EL, Burkey KO, Booker FL, Dubois JJB (2007) Photosynthesis, chlorophyll fluorescence, and yield of snap bean (Phaseolus vulgaris L.) genotypes differing in sensitivity to ozone. Environ Exp Bot 61(2):190–198

Fowler D, Amann M, Anderson R, Ashmore M, Cox P, Depledge M,... & Stevenson D (2008) Ground-level ozone in the 21st century: future trends, impacts and policy implications

Frei M, Tanaka JP, Wissuwa M (2008) Genotypic variation in tolerance to elevated ozone in rice: dissection of distinct genetic factors linked to tolerance mechanisms. J Exp Bot 59(13):3741–3752

Fuhrer J (2009) Ozone risk for crops and pastures in present and future climates. Naturwissenschaften 96(2):173–194

Galmés J, Aranjuelo I, Medrano H, Flexas J (2013) Variation in Rubisco content and activity under variable climatic factors. Photosynth Res 117:73–90

Gao F, Li P, Feng ZZ (2017) Interactive effects of ozone and drought stress on plants: a review. Chin J Plant Ecol 41:252–268

Garrido-Perez JM, Ordóñez C, García-Herrera R, Schnell JL (2019) The differing impact of air stagnation on summer ozone across Europe. Atmos Environ 219:117062

Gaudin AC, Tolhurst TN, Ker AP, Janovicek K, Tortora C, Martin RC, Deen W (2015) Increasing crop diversity mitigates weather variations and improves yield stability. PLoS ONE 10(2):e0113261

Gentry LF, Ruffo ML, Below FE (2013) Identifying factors controlling the continuous corn yield penalty. Agron J 105(2):295–303

Ghosh A, Agrawal M, Agrawal SB (2020a) Effect of water deficit stress on an Indian wheat cultivar (Triticum aestivum L. HD 2967) under ambient and elevated level of ozone. Sci Total Environ 714:136837

Ghosh A, Agrawal M, Agrawal SB (2021a) Examining the effectiveness of biomass-derived biochar for the amelioration of tropospheric ozone-induced phytotoxicity in the Indian wheat cultivar HD 2967. J Hazard Mater 408:124968

Ghosh A, Pandey AK, Agrawal M, Agrawal SB (2020b) Assessment of growth, physiological, and yield attributes of wheat cultivar HD 2967 under elevated ozone exposure adopting timely and delayed sowing conditions. Environ Sci Pollut Res 27:17205–17220

Ghosh A, Pandey B, Agrawal M, Agrawal SB (2020c) Interactive effects and competitive shift between Triticum aestivum L.(wheat) and Chenopodium album L.(fat-hen) under ambient and elevated ozone. Environ Pollut 265:114764

Ghosh A, Singh AA, Agrawal M, Agrawal SB (2018) Ozone toxicity and remediation in crop plants. Sustain Agric Rev 27:129–169

Ghosh A, Singh P, Agrawal M, Agrawal S (2021b) Chapter eleven response of crop plants to tropospheric ozone under different agronomic practices: An insight into the mechanisms. In: Tropospheric ozone: a hazard for vegetation and human health, p 380

Grantz DA, Paudel R, Shrestha A (2019) Tolerance of ozone and drought in common waterhemp (Amaranthus tuberculatus). J Crop Improv 33(2):236–253

Grulke NE, Preisler HK, Rose C, Kirsch J, Balduman L (2002) $O_3$ uptake and drought stress effects on carbon acquisition of ponderosa pine in natural stands. New Phytol 154(3):621–631

Guidi L, Degl'Innocenti E, Giordano C, Biricolti S, Tattini M (2010) Ozone tolerance in Phaseolus vulgaris depends on more than one mechanism. Environ Pollut 158(10):3164–3171

Guler NS, Pehlivan N (2016) Exogenous low-dose hydrogen peroxide enhances drought tolerance of soybean (Glycine max L.) through inducing antioxidant system. Acta Biol Hung 67:169–183

Gunasekera D, Berkowitz GA (1993) Use of transgenic plants with ribulose-1, 5-bisphosphate carboxylase/oxygenase antisense DNA to evaluate the rate limitation of photosynthesis under water stress. Plant Physiol 103(2):629–635

Gupta A, Yadav DS, Agrawal SB, Agrawal M (2023) Individual effects of high temperature and tropospheric ozone on tomato: a review. J Plant Growth Regul 42(3):1421–1443

Gupta SK, Sharma M, Majumder B, Maurya VK, Deeba F, Zhang JL, Pandey V (2020) Effects of ethylenediurea (EDU) on regulatory proteins in two maize (Zea mays L.) varieties under high tropospheric ozone phytotoxicity. Plant Physiol Biochem 154:675–688

Hao X, Li P, Feng Y, Han X, Gao J, Lin E, Han Y (2013) Effects of fully open-air [$CO_2$] elevation on leaf photosynthesis and ultrastructure of Isatis indigotica Fort. PLoS ONE 8(9):e74600

Harmens H, Hayes F, Sharps K, Radbourne A, Mills G (2019) Can reduced irrigation mitigate ozone impacts on an ozone-sensitive African wheat variety? Plants 8(7): 220

Hasan MM, Rahman MA, Skalicky M, Alabdallah NM, Waseem M, Jahan MS, Fang XW et al (2021) Ozone induced stomatal regulations, MAPK and phytohormone signaling in plants. Int J Mol Sci 22(12):6304

Hasan MM, Uddin MN, Ara-Sharmeen IF, Alharby H, Alzahrani Y, Hakeem KR, Zhang L (2019) Assisting phytoremediation of heavy metals using chemical amendments. Plants 8(9):295

Hasan MN, Hossain MA, Burritt DJ, Ashrafuzzaman M (2022) Cereals and phytohormones under ozone stress. In: Sustainable remedies for abiotic stress in cereals. Springer Nature Singapore, Singapore, pp 443–467

Hayes F, Williamson J, Mills G (2015) Species-specific responses to ozone and drought in six deciduous trees. Water Air Soil Pollut 226(5):156

Herbinger K, Tausz M, Wonisch A, Soja G, Sorger A, Grill D (2002) Complex interactive effects of drought and ozone stress on the antioxidant defence systems of two wheat cultivars. Plant Physiol Biochem 40(6–8):691–696

Hirel B, Le Gouis J, Ney B, Gallais A (2007) The challenge of improving nitrogen use efficiency in crop plants: towards a more central role for genetic variability and quantitative genetics within integrated approaches. J Exp Bot 58(9):2369–2387

Hoshika Y, De Carlo A, Baraldi R, Neri L, Carrari E, Agathokleous E, Paoletti E et al (2019) Ozone-induced impairment of night-time stomatal closure in $O_3$-sensitive poplar clone is affected by nitrogen but not by phosphorus enrichment. Sci Total Environ 692:713–722

Hoshika Y, Haworth M, Watanabe M, Koike T (2020) Interactive effect of leaf age and ozone on mesophyll conductance in Siebold's beech. Physiol Plant 170(2):172–186

Hoshika Y, Paoletti E, Centritto M, Gomes MTG, Puértolas J, Haworth M (2022) Species-specific variation of photosynthesis and mesophyll conductance to ozone and drought in three Mediterranean oaks. Physiol Plant 174(1):e13639

Hu Y, Wu Q, Peng Z, Sprague SA, Wang W, Park J, Park S et al (2017) Silencing of OsGRXS17 in rice improves drought stress tolerance by modulating ROS accumulation and stomatal closure. Sci Rep 7(1):15950

Hussain S, Rao MJ, Anjum MA, Ejaz S, Zakir I, Ali MA, Ahmad S et al (2019) Oxidative stress and antioxidant defense in plants under drought conditions. In: Plant abiotic stress tolerance: agronomic, molecular and biotechnological approaches, pp 207–219

IPCC (2007) Climate change 2007: the physical science basis; Contribution of working group I to the fourth assessment report of the intergovernmental panel on climate change. Accessed: 22 June 2023 from http://www.ipcc.ch/report/ar5/wg1/

IPCC (2013) Climate change 2013: the physical science basis IPCC working group I contribution to AR5. Accessed: 18 June 2023, from http://www.ipcc.ch/report/ar5/wg1/

IPCC (2019) Climate change and land: an IPCC special report on climate change, desertification, land degradation, sustainable land management, food security, and greenhouse gas fluxes in terrestrial ecosystems. In: International consensus regarding land system impacts of climate change as of class

Iyer NJ, Tang Y, Mahalingam R (2013) Physiological, biochemical and molecular responses to a combination of drought and ozone in Medicago truncatula. Plant, Cell Environ 36(3):706–720

Jiang C, Li X, Zou J, Ren J, Jin C, Zhang H, Jin H et al (2021) Comparative transcriptome analysis of genes involved in the drought stress response of two peanut (Arachis hypogaea L.) varieties. BMC Plant Biol 21(1):1–14

Jiang L, Feng Z, Dai L, Shang B, Paoletti E (2018) Large variability in ambient ozone sensitivity across 19 ethylenediurea-treated Chinese cultivars of soybean is driven by total ascorbate. J Environ Sci 64:10–22

Kalaji HM, Jajoo A, Oukarroum A, Brestic M, Zivcak M, Samborska IA, Ladle RJ et al (2016) Chlorophyll a fluorescence as a tool to monitor physiological status of plants under abiotic stress conditions. Acta Physiol Plant 38:1–11

Kang J, Peng Y, Xu W (2022) Crop root responses to drought stress: molecular mechanisms, nutrient regulations, and interactions with microorganisms in the rhizosphere. Int J Mol Sci 23(16):9310

Kangasjärvi J, Jaspers P, Kollist H (2005) Signalling and cell death in ozone-exposed plants. Plant, Cell Environ 28(8):1021–1036

Kask K, Kaurilind E, Talts E, Kännaste A, Niinemets Ü (2021) Combined acute ozone and water stress alters the quantitative relationships between $O_3$ uptake, photosynthetic characteristics and volatile emissions in Brassica nigra. Molecules 26(11):3114

Khaling E, Agyei T, Jokinen S, Holopainen JK, Blande JD (2020) The phytotoxic air-pollutant $O_3$ enhances the emission of herbivore-induced volatile organic compounds (VOCs) and affects the susceptibility of black mustard plants to pest attack. Environ Pollut 265:115030

Khaling E, Papazian S, Poelman EH, Holopainen JK, Albrectsen BR, Blande JD (2015) Ozone affects growth and development of Pieris brassicae on the wild host plant Brassica nigra. Environ Pollut 199:119–129

Khan S, Soja G (2003) Yield responses of wheat to ozone exposure as modified by drought-induced differences in ozone uptake. Water Air Soil Pollut 147:299–315

Khoyerdi FF, Shamshiri MH, Estaji A (2016) Changes in some physiological and osmotic parameters of several pistachio genotypes under drought stress. Sci Hortic 198:44–51

Kinose Y, Fukamachi Y, Watanabe M, Izuta T (2020) Ozone-induced change in the relationship between stomatal conductance and net photosynthetic rate is a factor determining cumulative stomatal ozone uptake by Fagus crenata seedlings. Trees 34:445–454

Kuang T, Lu C, Zhang L, Scheepers CC, Strasser RJ, Krüger GH (2013) Effect of ozone on photosynthesis and seed yield of sensitive (S156) and resistant (R123) Phaseolus vulgaris L. genotypes in open-top chambers. In: Photosynthesis research for food, fuel and the future: 15th international conference on photosynthesis. Springer, Berlin, pp 608–611

Kurutas EB (2015) The importance of antioxidants which play the role in cellular response against oxidative/nitrosative stress: current state. Nutr J 15(1):1–22

Leisner CP, Ainsworth EA (2012) Quantifying the effects of ozone on plant reproductive growth and development. Glob Change Biol 18(2):606–616

Leung F, Pang JY, Tai AP, Lam T, Tao DK, Sharps K (2020) Evidence of ozone-induced visible foliar injury in Hong Kong using Phaseolus vulgaris as a bioindicator. Atmosphere 11(3):266

Li S, Yang Y, Wang H, Li P, Li K, Ren L, Liao H et al (2023) Rapid increase in tropospheric ozone over Southeast Asia attributed to changes in precursor emission source regions and sectors. Atmos Environ 304:119776

Li W, Wang Y, Zhang Y, Wang R, Guo Z, Xie Z (2020) Impacts of drought stress on the morphology, physiology, and sugar content of Lanzhou lily (Lilium davidii var. unicolor). Acta Physiol Plant 42:1–11

Liang G, Liu J, Zhang J, Guo J (2020) Effects of drought stress on photosynthetic and physiological parameters of tomato. J Am Soc Hortic Sci 145(1):12–17

Liu F, Andersen MN, Jacobsen SE, Jensen CR (2005) Stomatal control and water use efficiency of soybean (Glycine max L. Merr.) during progressive soil drying. Environ Exp Bot 54(1):33–40

Liu F, Jensen CR, Andersen MN (2004) Pod set related to photosynthetic rate and endogenous ABA in soybeans subjected to different water regimes and exogenous ABA and BA at early reproductive stages. Ann Bot 94(3):405–411

Ludlow MM, Muchow RC (1990) A critical evaluation of traits for improving crop yields in water-limited environments. Adv Agron 43:107–153

Mafakheri A, Siosemardeh AF, Bahramnejad B, Struik PC, Sohrabi Y (2010) Effect of drought stress on yield, proline and chlorophyll contents in three chickpea cultivars. Aust J Crop Sci 4(8):580–585

Mak M, Babla M, Xu SC, O'Carrigan A, Liu XH, Gong YM, Chen ZH et al (2014) Leaf mesophyll $K^+$, $H^+$ and $Ca^{2+}$ fluxes are involved in drought-induced decrease in photosynthesis and stomatal closure in soybean. Environ Exp Bot 98:1–12

Marvel K, Cook BI, Bonfils CJ, Durack PJ, Smerdon JE, Williams AP (2019) Twentieth-century hydroclimate changes consistent with human influence. Nature 569(7754):59–65

Masle J, Gilmore SR, Farquhar GD (2005) The ERECTA gene regulates plant transpiration efficiency in Arabidopsis. Nature 436(7052):866–870

McAdam EL, Brodribb TJ, McAdam SA (2017) Does ozone increase ABA levels by non-enzymatic synthesis causing stomata to close? Plant, Cell Environ 40(5):741–747

Menéndez S, López-Bellido RJ, Benítez-Vega J, González-Murua C, López-Bellido L, Estavillo JM (2008) Long-term effect of tillage, crop rotation and N fertilization to wheat on gaseous emissions under rainfed Mediterranean conditions. Eur J Agron 28(4):559–569

Miyashita K, Tanakamaru S, Maitani T, Kimura K (2005) Recovery responses of photosynthesis, transpiration, and stomatal conductance in kidney bean following drought stress. Environ Exp Bot 53(2):205–214

Morgan PB, Ainsworth EA, Long SP (2003) How does elevated ozone impact soybean? A meta-analysis of photosynthesis, growth and yield. Plant, Cell Environ 26(8):1317–1328

Mwanamwenge JWAU, Loss SP, Siddique KHM, Cocks PS (1999) Effect of water stress during floral initiation, flowering and podding on the growth and yield of faba bean (Vicia faba L.). Eur J Agron 11(1):1–11

Nadal M, Carriquí M, Flexas J (2021) Chapter 3 Mesophyll conductance to $CO_2$ diffusion in a climate change scenario: effects of elevated $CO_2$, temperature and water stress. In: Photosynthesis, respiration, and climate change. Springer International Publishing, Cham, pp 49–78

Oksanen E, Manninen S, Vapaavuori E, Holopainen T (2009) Near-ambient ozone concentrations reduce the vigor of Betula and Populus species in Finland. AMBIO: J Hum Environ 38(8):413–417

Osman HS (2015) Enhancing antioxidant–yield relationship of pea plant under drought at different growth stages by exogenously applied glycine betaine and proline. Annal Agric Sci 60(2):389–402

Otu-Larbi F, Conte A, Fares S, Wild O, Ashworth K (2020) Current and future impacts of drought and ozone stress on Northern Hemisphere forests. Glob Change Biol 26(11):6218–6234

Pandey AK, Ghosh A, Agrawal M, Agrawal SB (2018) Effect of elevated ozone and varying levels of soil nitrogen in two wheat (Triticum aestivum L.) cultivars: Growth, gas-exchange, antioxidant status, grain yield and quality. Ecotoxicol Environ Saf 158:59–68

Pandey AK, Ghosh A, Rai K, Fatima A, Agrawal M, Agrawal SB (2019) Abiotic stress in plants: a general outline. In: Approaches for enhancing abiotic stress tolerance in plants. CRC Press, pp 1–46

Pandey AK, Majumder B, Keski-Saari S, Kontunen-Soppela S, Mishra A, Sahu N, Oksanen E et al (2015) Searching for common responsive parameters for ozone tolerance in 18 rice cultivars in India: results from ethylenediurea studies. Sci Total Environ 532:230–238

Paneque M, José M, Franco-Navarro JD, Colmenero-Flores JM, Knicker H (2016) Effect of biochar amendment on morphology, productivity and water relations of sunflower plants under non-irrigation conditions. CATENA 147:280–287

Pask AJD, Reynolds MP (2013) Breeding for yield potential has increased deep soil water extraction capacity in irrigated wheat. Crop Sci 53(5):2090–2104

Patel PK, Hemantaranjan A, Sarma BK, Radha S (2011) Growth and antioxidant system under drought stress in chickpea (Cicer arietinum L.) as sustained by salicylic acid. J Stress Physiol Biochem 7(4):130–144

Patmi YS, Pitoyo A (2020) Effect of drought stress on morphological, anatomical, and physiological characteristics of Cempo Ireng cultivar mutant rice (Oryza sativa L.) strain 51 irradiated by gamma-ray. J Phys Conf Ser 1436(1):012015

Pellegrini E, Hoshika Y, Dusart N, Cotrozzi L, Gérard J, Nali C, Paoletti E et al (2019) Antioxidative responses of three oak species under ozone and water stress conditions. Sci Total Environ 647:390–399

Peng, J., Shang, B., Xu, Y., Feng, Z., & Calatayud, V. (2020). Effects of ozone on maize (Zea mays L.) photosynthetic physiology, biomass and yield components based on exposure-and flux-response relationships. Environ Pollut 256:113466

Pereira LS, Oweis T, Zairi A (2002) Irrigation management under water scarcity. Agric Water Manag 57(3):175–206

Pillitteri LJ, Torii KU (2012) Mechanisms of stomatal development. Annu Rev Plant Biol 63:591–614

Plaza-Bonilla D, Álvaro-Fuentes J, Arrúe JL, Cantero-Martínez C (2014) Tillage and nitrogen fertilization effects on nitrous oxide yield-scaled emissions in a rainfed Mediterranean area. Agr Ecosyst Environ 189:43–52

Poornima R, Dhevagi P, Ramya A, Agathokleous E, Sahasa RGK, Ramakrishnan S (2023). Protectants to ameliorate ozone-induced damage in crops—a possible solution for sustainable agriculture. Crop Protection 106267

Pospíšil P (2012) Molecular mechanisms of production and scavenging of reactive oxygen species by photosystem II. Biochimica et Biophysica Acta (BBA)-Bioenergetics 1817(1):218–231

Prodhan FA, Zhang J, Sharma TPP, Nanzad L, Zhang D, Seka AM, Mohana HP et al (2022) Projection of future drought and its impact on simulated crop yield over South Asia using ensemble machine learning approach. Sci Total Environ 807:151029

Rao MV, Lee HI, Davis KR (2002) Ozone-induced ethylene production is dependent on salicylic acid, and both salicylic acid and ethylene act in concert to regulate ozone-induced cell death. Plant J 32(4):447–456

Rathore D, Chaudhary IJ (2021) Effects of tropospheric ozone on groundnut (Arachis hypogea L.) cultivars: Role of plant age and antioxidative potential. Atmosp Pollut Res 12(3):381–395

Reich PB, Amundson RG (1985) Ambient levels of ozone reduce net photosynthesis in tree and crop species. Science 230(4725):566–570

Rivero RM, Mittler R, Blumwald E, Zandalinas SI (2022) Developing climate-resilient crops: improving plant tolerance to stress combination. Plant J 109(2):373–389

Romero AM, Menéndez AI, Folcia AM, Martínez-Ghersa MA (2020) Tolerance to ozone might impose restrictions to plant disease management in tomato. Plant Biol 22(1):47–54

Sahitya UL, Krishna MSR, Deepthi R, Prasad GS, Kasim D (2018) Seed antioxidants interplay with drought stress tolerance indices in chilli (Capsicum annuum L) seedlings. BioMed Res Int

Saitanis CJ, Bari SM, Burkey KO, Stamatelopoulos D, Agathokleous E (2014) Screening of Bangladeshi winter wheat (Triticum aestivum L.) cultivars for sensitivity to ozone. Environ Sci Pollut Res 21:13560–13571

Saradadevi R, Palta JA, Siddique KH (2017) ABA-mediated stomatal response in regulating water use during the development of terminal drought in wheat. Front Plant Sci 8:1251

Sarkar A, Agrawal SB (2012) Evaluating the response of two high yielding Indian rice cultivars against ambient and elevated levels of ozone by using open top chambers. J Environ Manage 95:S19–S24

Sarkar A, Agrawal SB (2010) Elevated ozone and two modern wheat cultivars: an assessment of dose dependent sensitivity with respect to growth, reproductive and yield parameters. Environ Exp Bot 69(3):328–337

Schulte auf'm Erley G, Wijaya KA, Ulas A, Becker H, Wiesler F, Horst WJ (2007) Leaf senescence and N uptake parameters as selection traits for nitrogen efficiency of oilseed rape cultivars. Physiologia Plantarum 130(4):519–531

Shahinnia F, Le Roy J, Laborde B, Sznajder B, Kalambettu P, Mahjourimajd S, Fleury D et al (2016) Genetic association of stomatal traits and yield in wheat grown in low rainfall environments. BMC Plant Biol 16(1):1–14

Singh AA, Agrawal SB, Shahi JP, Agrawal M (2014) Assessment of growth and yield losses in two Zea mays L. cultivars (quality protein maize and nonquality protein maize) under projected levels of ozone. Environ Sci Pollut Res 21:2628–2641

Singh AA, Fatima A, Mishra AK, Chaudhary N, Mukherjee A, Agrawal M, Agrawal SB (2018) Assessment of ozone toxicity among 14 Indian wheat cultivars under field conditions: growth and productivity. Environ Monit Assess 190:1–14

Singh A, Agrawal SB, Rathore D (2005) Amelioration of Indian urban air pollution phytotoxicity in Beta vulgaris L. by modifying NPK nutrients. Environ Pollut 134(3):385–395

Singh J, Ashfaq M, Skinner CB, Anderson WB, Mishra V, Singh D (2022) Enhanced risk of concurrent regional droughts with increased ENSO variability and warming. Nat Clim Chang 12(2):163–170

Singh P, Agrawal M, Agrawal SB (2011) Differences in ozone sensitivity at different NPK levels of three tropical varieties of mustard (Brassica campestris L.): photosynthetic pigments, metabolites, and antioxidants. Water Air Soil Pollut 214:435–450

Singh P, Agrawal M, Agrawal SB, Singh S, Singh A (2015) Genotypic differences in utilization of nutrients in wheat under ambient ozone concentrations: growth, biomass and yield. Agr Ecosyst Environ 199:26–33

Snyder CS, Bruulsema TW, Jensen TL, Fixen PE (2009) Review of greenhouse gas emissions from crop production systems and fertilizer management effects. Agr Ecosyst Environ 133(3–4):247–266

Sobeih WY, Dodd IC, Bacon MA, Grierson D, Davies WJ (2004) Long-distance signals regulating stomatal conductance and leaf growth in tomato (Lycopersicon esculentum) plants subjected to partial root-zone drying. J Exp Bot 55(407):2353–2363

Song G, Son S, Lee KS, Park YJ, Suh EJ, Lee SI, Park SR (2022) OsWRKY114 negatively regulates drought tolerance by restricting stomatal closure in rice. Plants 11(15):1938

Thwe AA, Vercambre G, Gautier H, Gay F, Phattaralerphong J, Kasemsap P (2014) Response of photosynthesis and chlorophyll fluorescence to acute ozone stress in tomato (Solanum lycopersicum Mill.). Photosynthetica 52:105–116

Tian H, Ren W, Tao B, Sun G, Chappelka A, Wang X, Reilly J et al (2016) Climate extremes and ozone pollution: a growing threat to China's food security. Ecosyst Health Sustain 2(1):e01203

Tiemann LK, Grandy AS, Atkinson EE, Marin-Spiotta E, McDaniel MD (2015) Crop rotational diversity enhances belowground communities and functions in an agroecosystem. Ecol Lett 18(8):761–771

Vandermeiren K, Black C, Pleijel H, De Temmerman L (2005) Impact of rising tropospheric ozone on potato: effects on photosynthesis, growth, productivity and yield quality. Plant, Cell Environ 28(8):982–996

Voisin AS, Reidy B, Parent B, Rolland G, Redondo E, Gerentes D, Muller B et al (2006) Are ABA, ethylene or their interaction involved in the response of leaf growth to soil water deficit? An analysis using naturally occurring variation or genetic transformation of ABA production in maize. Plant, Cell Environ 29(9):1829–1840

Vorosmarty CJ, Green P, Salisbury J, Lammers RB (2000) Global water resources: vulnerability from climate change and population growth. Science 289(5477):284–288

Wahab A, Abdi G, Saleem MH, Ali B, Ullah S, Shah W, Marc RA et al (2022) Plants' physio-biochemical and phyto-hormonal responses to alleviate the adverse effects of drought stress: a comprehensive review. Plants 11(13):1620

Webb NP, Marshall NA, Stringer LC, Reed MS, Chappell A, Herrick JE (2017) Land degradation and climate change: building climate resilience in agriculture. Front Ecol Environ 15(8):450–459

Wedow JM, Ainsworth EA, Li S (2021) Plant biochemistry influences tropospheric ozone formation, destruction, deposition, and response. Trends Biochem Sci 46(12):992–1002

Wery J, Silim SN, Knights EJ, Malhotra RS, Cousin R (1994) Screening techniques and sources of tolerance to extremes of moisture and air temperature in cool season food legumes. In: Expanding the production and use of cool season food legumes: a global perspective of peristent constraints and of opportunities and strategies for further increasing the productivity and use of pea, lentil, faba bean, chickpea and grasspea in different farming systems, pp 439–456

Wilhite DA (2000) Drought as a natural hazard: concepts and definitions

Wilkinson S, Davies WJ (2009) Ozone suppresses soil drying-and abscisic acid (ABA)-induced stomatal closure via an ethylene-dependent mechanism. Plant, Cell Environ 32(8):949–959

Wilkinson S, Davies WJ (2010) Drought, ozone, ABA and ethylene: new insights from cell to plant to community. Plant, Cell Environ 33(4):510–525

Wittig VE, Ainsworth EA, Naidu SL, Karnosky DF, Long SP (2009) Quantifying the impact of current and future tropospheric ozone on tree biomass, growth, physiology and biochemistry: a quantitative meta-analysis. Glob Change Biol 15(2):396–424

World Economic Forum (2015) Global risks 2015, 10th edn. World Economic Forum, Geneva

Xiao Q, Zhu LX, Shen YF, Li SQ (2016) Sensitivity of soil water retention and availability to biochar addition in rainfed semi-arid farmland during a three-year field experiment. Field Crop Res 196:284–293

Xiao S, Liu L, Zhang Y, Sun H, Zhang K, Bai Z, Li C et al (2020) Fine root and root hair morphology of cotton under drought stress revealed with RhizoPot. J Agron Crop Sci 206(6):679–693

Xu Y, Feng Z, Shang B, Dai L, Uddling J, Tarvainen L (2019) Mesophyll conductance limitation of photosynthesis in poplar under elevated ozone. Sci Total Environ 657:136–145

Yadav GS, Das A, Lal R, Babu S, Meena RS, Patil SB, Datta M et al (2018) Conservation tillage and mulching effects on the adaptive capacity of direct-seeded upland rice (Oryza sativa L.) to alleviate weed and moisture stresses in the North Eastern Himalayan Region of India. Arch Agron Soil Sci 64(9):1254–1267

Yadav P, Mina U, Bhatia A (2020) Screening of forty Indian Amaranthus hypochondriacus cultivars for tolerance and susceptibility to tropospheric ozone stress. The Nucleus 63:281–291

Yang H, Wu G, Mo P, Chen S, Wang S, Xiao Y, Fan G et al (2020) The combined effects of maize straw mulch and no-tillage on grain yield and water and nitrogen use efficiency of dry-land winter wheat (Triticum aestivum L.). Soil Tillage Res 197:104485

Yang X, Lu M, Wang Y, Wang Y, Liu Z, Chen S (2021) Response mechanism of plants to drought stress. Horticulturae 7(3):50

Yendrek CR, Erice G, Montes CM, Tomaz T, Sorgini CA, Brown PJ, Ainsworth EA et al (2017) Elevated ozone reduces photosynthetic carbon gain by accelerating leaf senescence of inbred and hybrid maize in a genotype-specific manner. Plant, Cell Environ 40(12):3088–3100

Yokota A, Kawasaki S, Iwano M, Nakamura C, Miyake C, Akashi K (2002) Citrulline and DRIP-1 protein (ArgE homologue) in drought tolerance of wild watermelon. Ann Bot 89(7):825–832

Yoo CY, Pence HE, Jin JB, Miura K, Gosney MJ, Hasegawa PM, Mickelbart MV (2010) The Arabidopsis GTL1 transcription factor regulates water use efficiency and drought tolerance by modulating stomatal density via transrepression of SDD1. Plant Cell 22(12):4128–4141

Yu Y, Wang J, Li S, Kakan X, Zhou Y, Miao Y, Huang R et al (2019) Ascorbic acid integrates the antagonistic modulation of ethylene and abscisic acid in the accumulation of reactive oxygen species. Plant Physiol 179(4):1861–1875

Zaefyzadeh M, Quliyev RA, Babayeva S, Abbasov MA (2009) The effect of the interaction between genotypes and drought stress on the superoxide dismutase and chlorophyll content in durum wheat landraces. Turkish J Biol 33(1):1–7

Zandalinas SI, Sales C, Beltrán J, Gómez-Cadenas A, Arbona V (2017) Activation of secondary metabolism in citrus plants is associated to sensitivity to combined drought and high temperatures. Front Plant Sci 7:1954

Zhang F, Wang P, Zou YN, Wu QS, Kuča K (2019) Effects of mycorrhizal fungi on root-hair growth and hormone levels of taproot and lateral roots in trifoliate orange under drought stress. Arch Agron Soil Sci 65(9):1316–1330

Zhang H, Xu T, Zong Y, Tang H, Liu X, Wang Y (2015) Influence of meteorological conditions on pollutant dispersion in street canyon. Procedia Eng 121:899–905

Zhao W, Liu L, Shen Q, Yang J, Han X, Tian F, Wu J (2020) Effects of water stress on photosynthesis, yield, and water use efficiency in winter wheat. Water 12(8):2127

Zou YN, Wu QS, Kuča K (2021) Unravelling the role of arbuscular mycorrhizal fungi in mitigating the oxidative burst of plants under drought stress. Plant Biol 23:50–57

# Chapter 8
# Supplemental Nitrogen-Amendment Might Mitigate Higher Ozone Stress on Okra Plants (*Abelmoschus Esculentus* (L.) Moench) at Juvenile Stage

**Mamun Mandal, Shreya Roy, Anamika Roy, and Abhijit Sarkar**

**Abstract** Rising Ozone ($O_3$) concentrations in agricultural areas have been identified as a significant threat to Asian crop production, including India. This study investigated whether recommended NPK (control) and 1.5 times N fertilizer can potentially maintain Okra (*Abelmoschus esculentus* (L.) Moench) growth and development under $O_3$ stress under *in-vitro* conditions. We analyzed trends of chlorophyll (chl-A and chl-B), protein, amino acid content, enzymatic (such as ascorbic acid and proline) and non-enzymatic antioxidants (such as superoxide dismutase, catalase, peroxidase, and malondialdehyde) growth-related parameters, and total biomass production of the Okra plant. $O_3$ stress reduced plant growth-related parameters, biomass production, and lateral root and leaf numbers. The $O_3$-stressed plants showed decline in the contents of chl-A (0.1 μg/gm) and chl-B (0.35 μg/gm), protein (0.88 μg/gm), amino acid (16.42 μg/gm), and pH (6.1). On the contrary, the enzymatic and non-enzymatic antioxidants enzymatic activity increased. The application of recommended NPK plus 1.5 times N (1.5 N + $O_3$) reduced the impact of $O_3$ and maintained plants' higher growth and physiochemical traits. The 1.5 N-treated plants under $O_3$ stress conditions maintained higher chlorophyll, proteins, and enzymatic and non-enzymatic activities over the control treatment. These results suggest that 1.5 times N with the recommended NPK can sustain growth and development of Okra under $O_3$ stress by maintaining physiochemical activities.

**Keywords** Ozone stress · Okra plants · Juvenile stage · Nitrogen amendments

M. Mandal · S. Roy · A. Roy · A. Sarkar (✉)
Laboratory of Applied Stress Biology, Department of Botany, University of Gour Banga, Malda 732 103, West Bengal, India
e-mail: abbhijitbot@ugb.ac.in; abhijitbhu@gmail.com



A. K. Pandey and E. Oksanen (eds.), *Tropospheric Ozone and Food Security*,
https://doi.org/10.1007/978-3-032-03608-7_8

## 8.1 Introduction

A surge in industrialization, urbanization, and anthropogenic activities has resulted in a marked decline in ambient air quality with a range of secondary air pollutants impacting our environment (Sarkar and Agrawal 2010a, b). Tropospheric Ozone ($O_3$), a secondary air pollutant with phytotoxic properties, poses a significant threat to plant growth and human health (Tarasick et al. 2019). $O_3$ is produced through photochemical reactions involving nitrogen oxides ($NO_X$), carbon monoxide (CO), and volatile organic compounds (VOCs) emissions from vehicular sources and industrial activities (Shi et al. 2009). The photochemical $O_3$ formation requires sunlight, and its rate is notably higher during summer as opposed to winter. However, the photochemical lifetime of $O_3$ depends on the intensity of solar light, so it lasts longer in the winter than in the summer. Consequently, $O_3$ is transported over significantly greater distances in winter than in summer (Staehelin 2003).

Tropospheric $O_3$, a significant constituent of photochemical smog, poses a threat to crop plants and food security (Mukherjee et al. 2021; Puckette et al. 2007). Due to the negligible entry of $O_3$ through the leaf cuticle, stomata are a crucial factor in determining the $O_3$ flux into the apoplastic region (Sarkar et al. 2017). Upon penetration, $O_3$ reacts with cell walls and membranes, causing membrane disruption, and rapidly breaks into various reactive oxygen species (ROS) at the cell wall interface, inducing oxidative stress. Consequently, it leads to cell death, characterized by leaf bronzing, necrotic spots, and chlorotic flecking (Thwe et al. 2015; Puckette et al. 2007). The $O_3$-induced ROS can cause damage to plants through two mechanisms: the oxidation of DNA, proteins, and lipids or activation of the programmed cell death pathway. The second mechanism involves necrotic cell death that can trigger the programmed cell death pathway in surrounding tissues (Puckette et al. 2007). Furthermore, the physiological processes of plants can be affected directly by leaf injury, impairing photosynthesis, and indirectly by a change in the carbohydrate allocation pattern. This change results in a reduction in non-photosynthetic organ growth (Thwe et al. 2015). Exposure to higher $O_3$ concentration leads to visible injury in plants, which can be attributed to the collapse of mesophyll cells (Puckette et al. 2007). In addition, exposure to $O_3$ has a visible impact on the reproductive properties of plants. $O_3$ exposures alter flowering patterns and the phenology of flowers, leading to accelerated or delayed flowering and a reduction in flower numbers (Duque et al. 2021). Moreover, ambient $O_3$ reduces seed number, fruit number, and fruit weight, with some distinct reproductive growth stages being impacted by $O_3$ (Fuhrer et al. 2016). The relationship between plants and $O_3$ is known to be intricate, and the impact of $O_3$ on plant growth and metabolisms depends on the level and duration of $O_3$ exposure and the type of plant (Han et al. 2023).

*Abelmoschus esculentus* L. Moench, commonly known as Okra, is a highly nutritious vegetable crop cultivated mainly in tropical and subtropical regions. During the harvest season of 2018–2019, India produced almost 6 million tons of Okra from 0.5 million hectares (Singh et al. 2022). Fresh Okra pods can be consumed as a vegetable in salads, soups, and stews as a thickener (Gemede et al. 2015). Furthermore, Okra's leaves, buds, and flowers are edible (Adekiya et al. 2019). Additionally, the mucilage found in Okra has been used as a plasma transplant or blood volume expander due to its ability to bind to cholesterol and bile acids that carry toxins from the liver (Gemede et al. 2015). However, Okra is susceptible to $O_3$, as highlighted by Bulbovas et al. (2008), which can cause physiological disturbances and decrease the levels of antioxidants. Consequently, developing strategies to mitigate the unconditional consequences of $O_3$ on worldwide crop production deserves more attention to ensure food security.

Fertilizers, growth regulators, and antioxidants can alleviate the detrimental impact of $O_3$. Nutrients like phosphorus (P), potassium (K), and nitrogen (N) are fundamental in enhancing crop productivity and yield (Singh et al. 2015). Managing nutrient availability is crucial for plants to counteract the oxidative stress of $O_3$ exposure. Nutrients facilitate assimilation, influence carbohydrate allocation, and activate antioxidant enzymes in $O_3$-exposed plants, stimulating plant defense mechanisms (Podda et al. 2019). Effective nutrient management can significantly reduce ROS concentrations and activate antioxidant enzymes in $O_3$-exposed plants, boosting plant defense mechanisms under stress (Poornima et al. 2023). Carbon allocation can aid $O_3$-stressed plants to sustain yield upon nutrient amendments (Harmens et al. 2017). NPK is essential for maintaining protein levels, delaying leaf senescence, and alleviating the adverse effects of $O_3$ on growth, yield, and physiology (Poornima et al. 2023). NPK applications improve the plant's photosynthetic potential, which can be commonly associated with stomatal conductance (Sahoo et al. 2023). Apart from ensuring an adequate supply of nitrogen, maintaining optimal levels of K is equally essential for promoting cell membrane integrity and reducing ROS generation (Poornima et al. 2023).

Keeping this in view, the present study investigates all possible effects of $O_3$ on Okra, an important commercial vegetable crop in pot trials, with a particular focus on evaluating the potential benefits of recommended NPK and 1.5 times N fertilizer in promoting plant growth and assessing the associated physiological and biochemical responses of plants under conditions of $O_3$-induced stress. The study provides valuable insights into the optimal strategies for mitigating the detrimental impact of $O_3$ on crop yield and quality, which is critical for ensuring global food security and sustainability in the face of ongoing environmental challenges.

## 8.2 Materials and Methods

### *8.2.1 Experimental Design*

The experiment was conducted at the University of Gour Banga in Malda, West Bengal, situated in the northern region of the Indian subcontinent. The study area is located in the geographical coordinates of 24°58′59.60″ N, 88°08′17.98″ E, at an altitude of 20 m above sea level. The experiments were performed on a commercial vegetable crop, *Abelmoschus esculentus* L. Moench, commonly known as Okra, which belongs to the family Malvaceae. Prior to the experiment, the Okra seeds (Chiranjeevi F1) were sterilized with 80% alcohol, distilled water, and sodium hypochlorite (0.05%). Subsequently, an equal number of sterilized seeds were sowed in glass pots containing sterilized soil, and the most robust plantlet was selected after germination. Following that, the plantlet was transplanted into plastic pots containing sterilized soil, maintaining a consistent distance between each plant to prevent overcrowding. Pots were watered regularly throughout the growing period to ensure adequate hydration.

The experiment consisted of four treatments, each with three replicates: (i) Recommended dose was 80 kg $ha^{-1}$ N, 40 kg $ha^{-1}$ P, and 40 kg $ha^{-1}$ K added as fertilizer in the soil that was designated as the "*control*" treatment, (ii) recommended NPK and 1.5 times N was added in this set of treatments where the NPK was 120, 40, 40 kg $ha^{-1}$ that indicated as "*1.5 N*," (iii) "*control* + $O_3$," and (iv) "*1.5 N* + $O_3$." Furthermore, NPK fertilizers and N were added to the soil of pots at the experiment's beginning.

### 8.2.2 *Ozone Exposure*

A plastic closed-top chamber (1 m × 1 m × 1 m) was employed for $O_3$ exposure to plants. After 15 days of germination, plants are subjected to $O_3$ exposure for 14 days in an environmentally controlled and completely sealed $O_3$ chamber. $O_3$ was generated through an $O_3$ generator (Model: KENT 11,023 Car Ozone Air Disinfectant, India). At the time of the experiment, the ambient average $O_3$ concentration was 17.40 μg/$m^3$. The "control + $O_3$" and "1.5 N + $O_3$" plant sets were exposed to an average ambient air $O_3$ concentration plus 25% more $O_3$ for 1 h per day. The "control + $O_3$" and "1.5 N + $O_3$" sets were kept in chambers from 11:00 am to 12:00 pm throughout the experiment. The $O_3$ levels within the chambers were periodically monitored using the "gaseous pollutants sampler" (Model: APM 433, Envirotech, India).

### 8.2.3 *Plant Sampling and Analysis*

Three plants from each replicate plot were selected randomly for further investigation after 14 days of $O_3$ exposure.

#### 8.2.3.1 Morphological Characteristics

Random samples were taken for the determination of growth parameters. The study encompassed various growth aspects, including root length, shoot length, leaf number, fresh weight, dry weight, lateral root number, and leaf injury index. For dry weight, three samples were dug out carefully, thoroughly washed, and subsequently oven-dried at 80 °C till a constant weight was achieved.

#### 8.2.3.2 Biochemical Assays

Leaf extract pH was determined following Prasad and Rao (1982) with minor modifications. pH was measured with a digital pH meter (Model: pH tutor, EuTech) by homogenizing a 0.5 g sample of fresh leaves with 50 mL distilled water (Li et al. 2009).

Total chlorophyll content was determined using the modified Arnon method (1949). Chlorophyll was extracted from fresh leaf tissue weighing 500 mg using 2 mL of 80% acetone. The extracted solution was then subjected to spectrophotometric analysis, with absorbance values being measured at wavelengths of 645 and 663 nm.

Ascorbic acid was estimated using Keller and Schweiger (1977) DCPIP dye reduction method. The plant leaf material of about 0.05 g was macerated with 5 ml of trichloroacetic acid (6% w/v) and centrifuged for 10 min at 10,000 rpm. The optical density of the sample before and after bleaching was measured at 520 nm.

The protein content was determined following the method of Lowry et al. (1951). The supernatant of the centrifuged plant sample was measured against the Bovine Serum Albumin (BSA) standard curve. Proline content was estimated using the method described by Bates et al. (1973). The phenol content was measured using the Bray and Thorpe (1954) method. Amino acids were estimated using the Moore and Stein (1948) method.

#### 8.2.3.3 Antioxidant Analysis

Superoxide dismutase (SOD) activity was measured following the method described by Fridovich (1974) using the nitro blue tetrazolium (NBT), in which a 50% reduction of NBT was taken as the measure of SOD activity. Following the Aebi method, catalase (CAT) activity was estimated by adding $H_2O_2$ to the supernatant of the plant sample (Aebi, 1984). CAT activity was determined by measuring $O_2$ release from enzymatic dissociation of $H_2O_2$. Malondialdehyde (MDA) content, a lipid peroxidation product, was assessed using thiobarbituric acid (TBA), as described by Heath and Packer (1968). Peroxidase (POD) contents were measured following Chance and Mehley (1955).

### 8.2.4 Statistical Analysis

Data of physiological and biochemical parameters were analyzed through one-way analysis of variance (ANOVA) tests to examine the significance of changes in the treatments. Duncan's multiple range ($p \leq 0.05$) tests were performed post hoc on parameters subjected to various ANOVA tests. All the statistical tests were performed using SPSS software. The variability of the mean is indicated by the standard error.

## 8.3 Results

### 8.3.1 Effect on Morphological Parameters

Upon $O_3$ exposure, Okra undergoes significant morphological changes. Shoot and root length *were adversely affected by* $O_3$ exposure; however, recommended NPK amendments with 1.5 times N (1.5 N) had been found to promote shoot and root length compared to control and $O_3$ (control + $O_3$), suggesting that 1.5 times N amendment can ameliorate adverse impacts of $O_3$ (Fig. 8.1a and b).

$O_3$-exposed plants showed a significant reduction in fresh weight compared to the control. However, 1.5 N-treated plants increased fresh weight even after $O_3$ exposure, highlighting that nutrients are critical in mitigating $O_3$ stress (Fig. 8.1c). In addition, dry weight was also reduced significantly in the $O_3$-treated plants (control + $O_3$) compared to the control. In this case, dry weight decreased by 54 and 6.5% in control + O3 and 1.5 N + O3 treatments, respectively, compared to control (Fig. 8.1d).

Leaves injury, in the form of interveinal yellowing and chlorotic stippling, was first observed in control + $O_3$ plants after the 7 days of $O_3$ treatments and after the 14 days of $O_3$ treatments in all the replicates (Fig. 8.1e), initially showing minor injury, but later, it became prominent and appeared in 1.5 N + $O_3$ successively. $O_3$-treated plants exhibited a decrease in leaf numbers (Fig. 8.1f). The highest reduction was observed in control + $O_3$ plants but was not significantly different from the other treatments. The $O_3$ exposure significantly decreased lateral root numbers. The highest number of lateral roots (Fig. 8.1g) was noticed in 1.5 N (15), followed by 1.5 N + $O_3$ (11), control (10), and control + $O_3$ (8).

### 8.3.2 Effect on Biochemical Parameters

The $O_3$ stress decreased chl-a and chl-b concentrations in the Okra plant leaves (Fig. 8.2a and b). The plants with 1.5 N applications showed less decrease in chl-a content due to $O_3$ stress than those with control + $O_3$. Chl-b content decreased by 19% in plants 1.5 N + $O_3$ than the 1.5 N treatments, whereas the plants of 1.5 N +

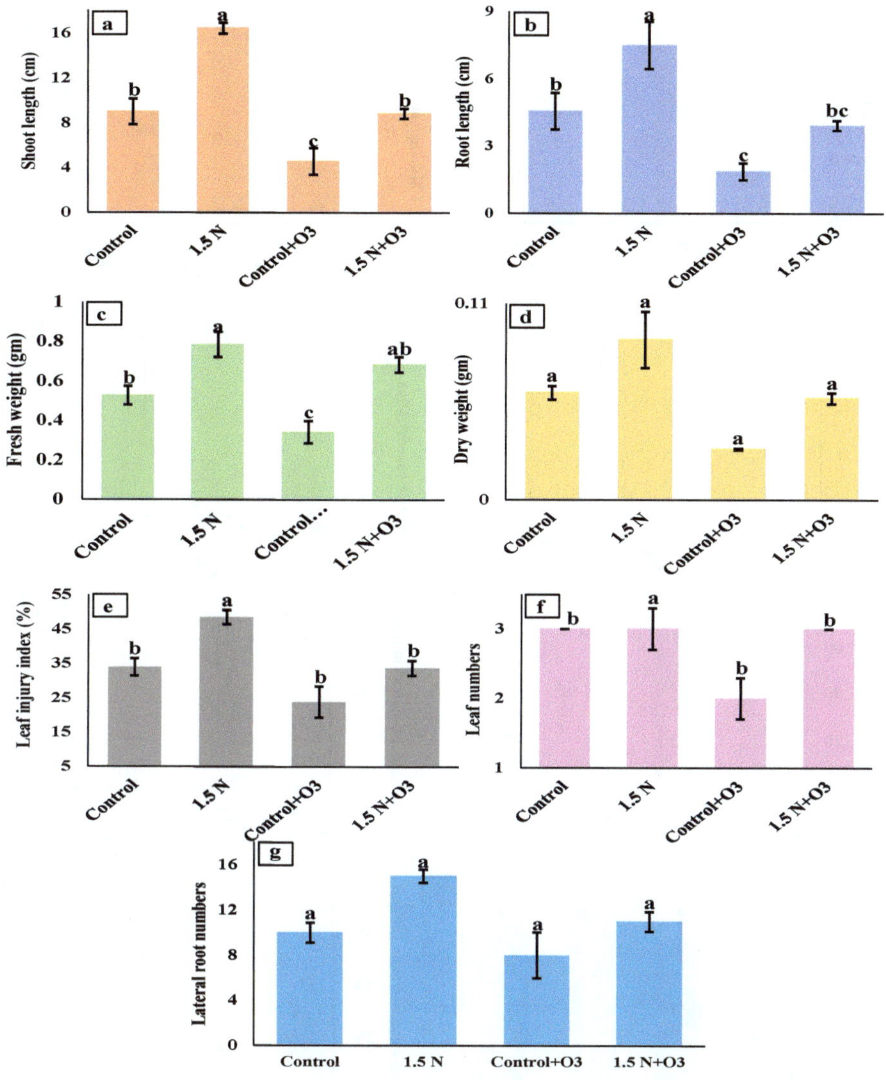


**Fig. 8.1** Shoot length (**a**), root length (**b**), fresh weight (**c**), dry weight (**d**), leaf injury index (**e**), leaf numbers (**f**), and lateral root numbers (G) of Okra plants after the 14 days of $O_3$ treatments. Values are mean ± SE (standard error). According to Duncan's test at $p < 0.05$, bars showing different letters indicate significant differences

$O_3$ exhibited a rise in chl-b content by 46% compared to the control + $O_3$ plant's treatment.

The amount of amino acid decreased in the presence of $O_3$ stresses in Okra plants (Fig. 8.2c). Under $O_3$ stress, the plants showed 71.81% lower in the contents of amino acids as compared to the control plants but were significantly not different. Due to

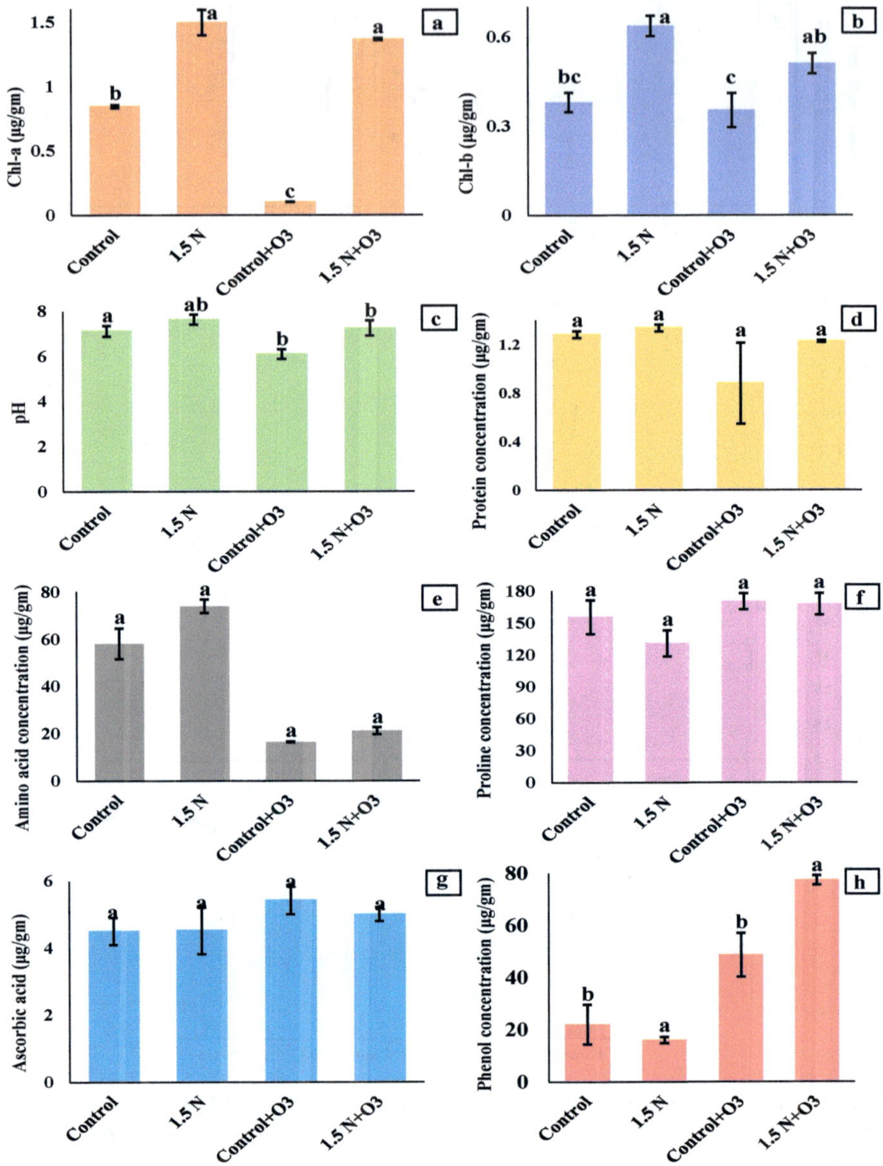


**Fig. 8.2** Chl-a (**a**), Chl-b (**b**), pH (**c**), protein (**d**), amino acid (**e**), proline (**f**), ascorbic acid (**g**), and phenol (**h**) concentration of Okra plants after the 14 days of $O_3$ treatments. Values are mean ± SE. According to Duncan's test at $p < 0.05$, bars showing different letters indicate significant differences

$O_3$ stresses, the protein content decreased in 31.25% of Okra plants (Fig. 8.2d). Moreover, more protein production was observed in plants treated with 1.5 N than in control. The protein contents improved up to 40% in 1.5 N + $O_3$ treatments compared to control + $O_3$ plants. $O_3$ stress conditions also decreased the pH content in plants.

Additionally, $O_3$ stresses increased production of proline, ascorbic acid, and phenol concentration (Fig. 8.2). The changes in ROS, including SOD, POD, CAT, and MDA, were determined due to $O_3$ in the plants of Okra, as shown in Fig. 8.3.

The $O_3$ stresses increased the production of SOD (Fig. 8.3a). Under $O_3$ stress, the plants with 1.5 N displayed less increase in SOD than control + $O_3$ treatments. During $O_3$ stress, SOD increased by 134 and 48.9% in control + $O_3$ and 1.5 N + $O_3$ treatments, respectively, as compared to control conditions. CAT activity increased due to $O_3$ stresses in Okra plants (Fig. 8.3b). CAT activity increased more than 100% in all treatments compared to the control conditions, but these are significantly different from each other. The $O_3$ stress resulted in an increase in lipid peroxidation

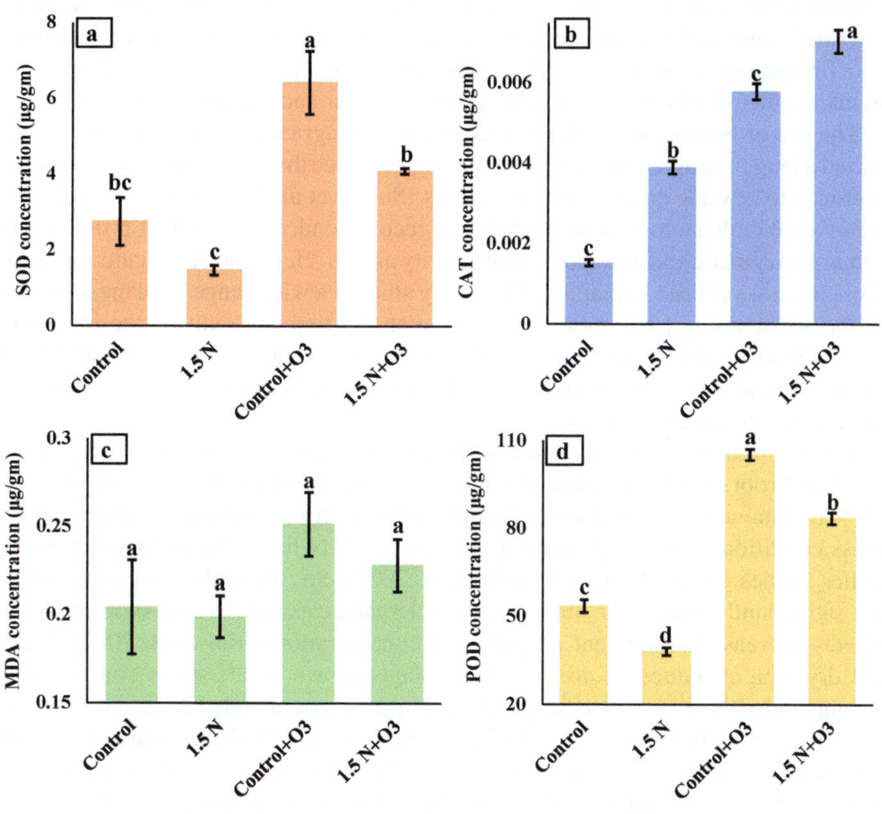


**Fig. 8.3** Effect of $O_3$ on SOD (**a**), CAT (**b**), MDA (**c**), and POD (**d**) concentration of Okra plants after the 14 days of $O_3$ treatments. Values are mean ± SE. According to Duncan's test at $p < 0.05$, bars showing different letters indicate significant differences

measured by the production of MDA (Fig. 8.3c). However, plants with 1.5 N treatments showed lower MDA production than control plants. The plants of 1.5 N + $O_3$ showed a lower amount of MDA than control + $O_3$. Similarly, the activity POD increased (Fig. 8.3d) under the $O_3$ stress condition. The plants with 1.5 N showed lower POD activity than the control + $O_3$ plants.

## 8.4 Discussion

$O_3$, the most significant abiotic stress in agriculture, is especially relevant to Okra (*Abelmoschus esculentus* (L.) Moench) due to its susceptibility to $O_3$ stress conditions, and which impact the crop's physicochemical characteristics significantly diminishing overall production. The $O_3$ existence during the juvenile stage causes a considerable decrease in plant growth and production associated with plant physiological and biochemical activities due to disturbance in functionality. Investigation on $O_3$ tolerance in Okra has not yet been reported; to the best of our knowledge, we have first reported about it. But in the last few years, many studies have been done on many other plants, such as wheat and rice (Sarkar and Agrawal 2010a, b).

The use of beneficial fertilizers such as NPK in agricultural production systems started a long time ago, and there is increasing evidence that NPK can enhance plants' tolerance to adverse environmental stresses (Sarkar et al. 2010; Singh et al. 2009). However, the physiological significance of recommended NPK with 1.5 times N in terms of enzymatic and non-enzymatic activity and its effects on Okra plants under $O_3$ stress has not yet been explained. This study studied a wide range of plant strategies by the application of recommended NPK with 1.5 times N reduced the impact of $O_3$ on growth and physiochemical traits of the plants and improved the ability of Okra plants to maintain advantageous physiological, biochemical, and agronomic characteristics under $O_3$ stress.

The $O_3$ treatments applied during the juvenile stage significantly reduced the shoot and root length and plant dry matter production (Fig. 8.1). The plants with 1.5 N maintained greater shoot and root length and dry matter production under $O_3$ stress conditions. Growth-promoting effects of 1.5 N have also been described in earlier studies under $O_3$ stress (Singh et al. 2009). So, the higher nutrient content can significantly induce the abiotic stress tolerance capacity. Wheat shoot and root lengths decreased at ambient air pollutant concentrations (Rai et al. 2007). Fresh and dry weight reduced considerably in plants grown in $O_3$ stress conditions at recommended NPK, whereas at 1.5 times N plus recommended NPK can increase the biomass weight (such as fresh and dry weight) where the dry weight did not vary significantly, but the fresh weight varied significantly. This suggests that the more biomass in the plant under $O_3$ stress may be maintained at a higher N level.

On the other hand, no significant changes were found in lateral root numbers. However, in the case of leaf numbers, significant differences were found between treatments. NPK levels suggest that plants can keep the photosynthate for its repair and maintain leaf formation and expansion. Although, the production of more leaves numbers with smaller areas is reported in a polluted environment (Singh et al. 2009). One of the main signs of $O_3$-induced stress on plants is thought to be visible damage in the form of necrotic patches on leaves (Sarkar and Agrawal 2010a). This particular leaf damage index has been used to measure and evaluate the effect of $O_3$ on plants as a fundamental part for the past few years. On the other hand, the type of plant used determines how severe the $O_3$-induced leaf damage would be, and various agricultural plants have developed distinct defense mechanisms against $O_3$ exposure (Sarkar et al. 2010). In the present experiment, initiation of leaf injury was observed after 7 days of $O_3$ exposure, where the control + $O_3$ treated plants were more affected than the other treatments.

$O_3$ may bring on early plant metabolic changes, albeit the exact type of plant will determine how much of a shift occurs. A few publications suggest these metabolic alterations may occur before phenotypic manifestations such as damage, apparent lesions, or early senescence (Sarkar and Agrawal 2010a). First and foremost, $O_3$ causes plants to lose all of their chlorophyll by either producing SOD or inhibiting the production of this pigment (Han et al. 2023). After exposure to $O_3$ concentrations of around 50 ppb for one month, wheat leaves showed a notable decrease in total chlorophyll, according to Pleijel et al. (2006). Our current results have also demonstrated a similar pattern. Under $O_3$ stress, Okra showed a considerable decrease in chl-a and chl-b. A one-way ANOVA revealed that the primary factor responsible for the substantial change ($p < 0.05$) in both parameters for chl-a and chl-b is the $O_3$ treatment (Fig. 8.2). Previous investigation has indicated that while fewer light-harvesting antennae complexes shield PSII from photoinhibition, a decrease in plant pigment content may act as an adaptation against $O_3$ stress (Han et al. 2023). Moreover, 1.5 N + $O_3$ treatments improve the chl-a and chl-b contents compared to the control + $O_3$. Similarly, a previous study reported that double the recommended NPK showed better results than recommended NPK due to nitrogen, which helps photosynthesis by increasing RuBP carboxylase activity. Usually, stomata allow $O_3$ to enter the intercellular spaces. Once inside, it interacts with the cell wall and/or cell membrane elements to break down and move from the stomata to the mesophyll regions. $O_3$ generated superoxide, hydroxyl radical, peroxyl radical, and $H_2O_2$, exposing the biological system to increased oxidative stress in the intercellular gaps in an aqueous environment (Sarkar et al. 2010). One of ROS's primary functions is to structurally change cellular proteins by changing their secondary and tertiary structures, which increases the proteins' vulnerability to proteolysis and lowers the quantity of total protein present. Our data also demonstrated a decrease in the amount of protein under $O_3$ stress. A notable reduction in the overall protein concentration under ambient $O_3$ pollution has also been documented (Han et al. 2023). However, 1.5 times N enhances the protein concentration in the Okra plants in the present study.

Since $O_3$ typically causes oxidative stress, we mainly aimed to evaluate the Okra plants' antioxidative defense mechanism. The most prevalent antioxidant in plants, ascorbic acid is a vital defense against ROS produced by $O_3$ (Sarkar et al. 2017). Rai et al. (2007) reported an 11.2% rise in ascorbic acid and a 46.2% rise in proline content in wheat in $O_3$ stress conditions. We also show an increment in both cases of ascorbic acid and proline under elevated $O_3$ concentration (Fig. 8.2). A similar trend was found for proline. Proline acts as an osmolyte, and under stress conditions, higher accumulation of proline protects cells from desiccation by maintaining high intracellular osmotic concentration. However, the 1.5 times N can reduce these concentrations under $O_3$ stress conditions. Total phenol also showed an increasing trend, not only in $O_3$ exposure but also in all treatments (Fig. 8.2). As phenols efficiently scavenge the ROS within the cell, their increment under higher $O_3$ exposure indicates the plant's adaptation against higher oxidative stress. Rai et al. (2007) also observed a 38.5% increment in total phenol under ambient $O_3$ pollution at 60 days after germination. One-way ANOVA showed that proline and phenol content significantly varied due to $O_3$ treatment, but the ascorbic acid was not varied.

It is widely known that a plant's capacity to detoxify $O_3$ and $O_3$-generated oxidative radicals is the factor that primarily determines its tolerance to $O_3$. Plants have a variety of enzymes at their disposal to handle this kind of stress. The degree of $O_3$ stress, as well as the suggested NPK and 1.5 times N, determined how much ROS (such as SOD, CAT, POD, and MDA) and MDA accumulated in the leaves. The level of oxidative stress is indicated by the MDA concentration between cells. Damage from MDA may set off a series of developmental signals that include early senescence, increased permeability of cell membranes, loss of chlorophyll, macromolecule disintegration, and massive remobilization of nutrients, all of which shorten the period needed for crop growth. The increased levels of MDA and ROS in control plants under $O_3$ stress suggest that the subcellular antioxidant system is less effective, and that ROS accumulates more readily in these plants. It is believed that suppressing the photosynthesis process also causes the buildup of ROS in the cell organelles. Tight control of ROS synthesis and accumulation in plant cells is necessary because excessive ROS production may damage cells by slowing down plant development and ultimately resulting in a loss of economic output. Compared to control + $O_3$ plants, 1.5 N + $O_3$ plants had reduced MDA and ROS levels, suggesting they were better at scavenging ROS under $O_3$ conditions.

Plants with 1.5 N exhibited enhanced redox defense status against ROS damage, as evidenced by the decreased enzymatic antioxidant activities of SOD, POD, and MDA levels. This was achieved by downregulating the peroxidation of cell membrane lipids and structural and functional proteins under $O_3$ stress (Fig. 8.3). According to our findings, the $O_3$-stressed plants at 1.5 N may have protected the photosynthetic process by developing a beneficial ROS-detoxifying antioxidant system. Similar results were reported by Singh et al. (2009), who proposed that increased N provided sustenance for plants' $O_3$ tolerance by shielding cell membranes (Han et al. 2023).

## 8.5 Conclusions

Plant growth and physiochemical functions of Okra plants were highly influenced by Ozone stress. However, the 1.5 times higher N treatment (1.5 N + $O_3$) benefited the plants by allowing them to maintain their development process and functionality efficiently, ultimately promoting growth under $O_3$ stress conditions. The plants with 1.5 times higher N treatment sustained higher dry matter production, fresh weight, and leaf and root number increase compared to control treatments. Plants from the 1.5 N + $O_3$ treatment showed a lower pH, chlorophyll, amino acid, and protein content decrease than the control treatment. However, this set also showed less MDA production than the $O_3$-treated plants. They maintained higher enzymatic activity of SOD and POD except CAT, with increased production of total proteins, phenol, amino acid, and proline except proline compared to the $O_3$-treated plants. The present study gives new insights into various effects of recommended NPK and 1.5 times N on morphological and physiochemical traits in response to $O_3$ stress in Okra. Such information can further provide practical guidelines for research efforts to improve $O_3$ tolerance and yield sustainability in Okra crops.

**Acknowledgements** M.M. acknowledges the University Grants Commission (UGC, New Delhi, Govt. of India) as the author receives fellowship in the form of Senior Research Fellowship (CSIR-UGC SRF; Award No. 16-6(DEC. 2018)/2019(NET/CSIR); UGC-Ref. No.: 900/(CSIR-UGC NET DEC. 2018); dated: 02.04.2019). A.R. acknowledges the Council of Scientific Research (CSIR), New Delhi, for providing fellowship in the form of Junior Research Fellowship from GoI. All authors also acknowledge the infrastructural support in the form of the DBT-BOOST program, Department of Science and Technology and Biotechnology, GoWB, GoI (vide Ref. No. 1089/BT(Estt)/1P-07/2018; dated: 24.01.2019) to the Department of Botany, University of Gour Banga, Malda, West Bengal, India.

## References

Adekiya AO, Agbede TM, Aboyeji CM, Dunsin O, Ugbe JO (2019) Green manures and NPK fertilizer effects on soil properties, growth, yield, mineral and vitamin C composition of Okra (*Abelmoschus esculentus* (L.) Moench). J Saudi Soc Agric Sci 18(2):218–223. https://doi.org/10.1016/j.jssas.2017.05.005

Aebi H (1984) Catalase in vitro. Methods Enzymol 105:121–126. https://doi.org/10.1016/s0076-6879(84)05016-3

Arnon DI (1949) Copper enzymes in isolated chloroplasts: Polyphenol oxidise in *Beta vulgaris.* Plant Physiol 2:1–15. https://doi.org/10.1104/pp.24.1.1

Bates LS, Waldren RP, Teare ID (1973) Rapid determination of free proline for water-stress studies. Plant Soil 39:205–207. https://doi.org/10.1007/BF00018060

Bray HG, Thorpe WY (1954) Analysis of phenolic compounds of interest in metabolism. Methods Biochem Anal 1:27–52. https://doi.org/10.1002/9780470110171.ch2

Bulbovas P, Sant'Anna SMR, Moraes RMD, Lima EDS, Pina JM, Esposito MP, Ferreira M, Spielmann AA, Sollito CM, Rinnert CH, Peralta DF (2008) Assessment of the sensitivity of Okra (*Abelmoschus esculentus* (L.) Moench.-Malvaceae) to ozone. Hoehnea 35:359–366

Chance B, Maehly AC (1955) Assay of catalase and peroxidase. Methods Enzymol 2:764–775. https://doi.org/10.1016/S0076-6879(55)02300-8

Duque L, Poelman EH, Steffan-Dewenter I (2021) Effects of ozone stress on flowering phenology, plant-pollinator interactions and plant reproductive success. Environ Pollut 272:115953. https://doi.org/10.1016/j.envpol.2020.11595

Fridovich I (1974) Superoxide dismutase. Adv Enzymol Relat Areas Mol Biol 41:35–97. https://doi.org/10.1002/9780470122860.ch2

Fuhrer J, Val Martin M, Mills G, Heald CL, Harmens H, Hayes F, Sharps K, Bender J, Ashmore MR (2016) Current and future Ozone risks to global terrestrial biodiversity and ecosystem processes. Ecol Evol 6(24):8785–8799. https://doi.org/10.1002/ece3.2568

Gemede HF, Ratta N, Haki GD, Woldegiorgis AZ, Beyene F (2015) Nutritional quality and health benefits of Okra (*Abelmoschus esculentus*): a review. J Food Process Technol 6(458):2

Han YJ, Beck W, Mewis I, Förster N, Ulrichs C (2023) Effect of ozone stresses on growth and secondary plant metabolism of *Brassica campestris* L. ssp. *chinensis*. Horticulturae 9:966. https://doi.org/10.3390/horticulturae9090966

Harmens H, Hayes F, Sharps K, Mills G, Calatayud V (2017) Leaf traits and photosynthetic responses of Betula pendula saplings to a range of ground-level ozone concentrations at a range of nitrogen loads. J Plant Physiol 211:42–52. https://doi.org/10.1016/j.jplph.2017.01.002

Heath RL, Packer L (1968) Photoperoxidation in isolated chloroplasts. I. Kinetics and stoichiometry of fatty acid peroxidation. Arch Bioch Biophys 125:189–198. https://doi.org/10.1016/0003-9861(68)90654-1

Keller T, Schwager H (1977) Air pollution and ascorbic acid. Eur J Forest Pathol 7(6):338–350. https://doi.org/10.1111/j.1439-0329.1977.tb00603.x

Li FL, Bao WK, Wu N (2009) Effects of water stress on growth, dry matter allocation and water-use efficiency of a leguminous species *Sophora davidii*. Agrofor Syst 77:193–201. https://doi.org/10.1007/s10457-008-9199-1

Lowry OH, Rosenbourgh NJ, Farr AL, Randall RJ (1951) Protein measurement with the Folin phenol reagent. J Biol Chem 193(1):265–275

Moore S, Stein WH (1948) Photometric nin-hydrin method for use in the chromatography of amino acids. J Biol Chem 176:367–388. https://doi.org/10.1016/S0021-9258(18)51034-6

Mukherjee A, Yadav DS, Agrawal SB, Agrawal M (2021) Ozone a persistent challenge to food security in India: current status and policy implications. Curr Opin Environ Sci Health 19:100220. https://doi.org/10.1016/j.coesh.2020.10.008

Pleijel H, Eriksen AB, Danielsson H, Bondesson N, Selldén G (2006) Differential ozone sensitivity in an old and a modern Swedish wheat cultivar—grain yield and quality, leaf chlorophyll and stomatal conductance. Environ Exp Botany 56(1):63–71. https://doi.org/10.1016/j.envexpbot.2005.01.004

Podda A, Pisuttu C, Hoshika Y, Pellegrini E, Carrari E, Lorenzini G, Nali C, Cotrozzi L, Zhang L, Baraldi R, Neri L (2019) Can nutrient fertilization mitigate the effects of ozone exposure on an ozone-sensitive poplar clone? Sci Total Environ 657:340–350. https://doi.org/10.1016/j.scitotenv.2018.11.459

Poornima R, Dhevagi P, Ramya A, Agathokleous E, Sahasa RGK, Ramakrishnan S (2023) Protectants to ameliorate Ozone-induced damage in crops—a possible solution for sustainable agriculture. Crop Prot 106267. https://doi.org/10.1016/j.cropro.2023.106267

Prasad BJ, Rao DN (1982) Relative sensitivity of a leguminous and a cereal crop to sulphur dioxide pollution. Environ Pollut Series A Ecol Bio 29(1):57–70. https://doi.org/10.1016/0143-1471(82)90054-X

Puckette MC, Weng H, Mahalingam R (2007) Physiological and biochemical responses to acute ozone-induced oxidative stress in *Medicago truncatula*. Plant Physiol Biochem 45(1):70–79. https://doi.org/10.1016/j.plaphy.2006.12.004

Rai R, Agrawal M, Agrawal SB (2007) Assessment of yield losses in tropical wheat using open top chambers. Atmos Environ 41(40):9543–9554. https://doi.org/10.1016/j.atmosenv.2007.08.038

Sahoo A, Madheshiya P, Mishra AK, Tiwari S (2023) Combating ozone stress through N fertilization: a case study of Indian bean (*Dolichos lablab* L.). Front Plant Sci 14:1125529. https://doi.org/10.3389/fpls.2023.1125529

Sarkar A, Agrawal SB (2010a) Identification of ozone stress in Indian rice through foliar injury and differential protein profile. Environ Monit Assess 161:205–215. https://doi.org/10.1007/s10661-008-0738-z

Sarkar A, Agrawal SB (2010b) Elevated Ozone and two modern wheat cultivars: an assessment of dose dependent sensitivity with respect to growth, reproductive and yield parameters. Environ Exp Bot 69(3):328–337. https://doi.org/10.1016/j.envexpbot.2010.04.016

Sarkar A, Rakwal R, Agrawal SB, Shibato J, Ogawa Y, Yoshida Y, Agrawal GK, Agrawal M (2010) Investigating the impact of elevated levels of ozone on tropical wheat using integrated phenotypical, physiological, biochemical, and proteomics approaches. J Proteome Res 9(9):4565–4584. https://doi.org/10.1021/pr1002824

Sarkar A, Datta S, Singh P (2017) Tropospheric ozone pollution, agriculture, and food security. In: Singh RP, Singh A, Srivastava V (eds) Environmental issues surrounding human overpopulation. IGI Global, USA, pp 233–252. https://doi.org/10.4018/978-1-5225-1683-5.ch014

Shi G, Yang L, Wang Y, Kobayashi K, Zhu J, Tang H, Pan S, Chen T, Liu G, Wang Y (2009) Impact of elevated ozone concentration on yield of four Chinese rice cultivars under fully open-air field conditions. Agric Ecosys Environ 131(3–4):178–184. https://doi.org/10.1016/j.agee.2009.01.009

Singh P, Agrawal M, Agrawal SB, Singh S, Singh A (2015) Genotypic differences in utilization of nutrients in wheat under ambient Ozone concentrations: growth, biomass and yield. Agric Ecosys Environ 199:26–33. https://doi.org/10.1016/j.agee.2014.07.021

Singh S, Solanki V, Bardhan K, Kansara R, Vyas TK, Gandhi K, Dhakan D, Ali HM, Siddiqui MH (2022) Evaluation of ozonation technique for pesticide residue removal in okra and green chili using GC-ECD and LC-MS/MS. Plants 11:3202. https://doi.org/10.3390/plants11233202

Singh P, Agrawal M, Agrawal SB (2009) Evaluation of physiological, growth and yield responses of a tropical oil crop (*Brassica campestris* L. var. Kranti) under ambient ozone pollution at varying NPK levels. Environ Pollut 157(3):871–880. https://doi.org/10.1016/j.envpol.2008.11.008

Staehelin J (2003) Ozone measurements and trends (Troposphere). In: Meyers RA (eds) Encyclopedia of physical science and technology, 3rd edn. Academic Press, New York, pp 539–561. https://doi.org/10.1016/B0-12-227410-5/00037-5

Tarasick D, Galbally IE, Cooper OR, Schultz MG, Ancellet G, Leblanc T, Wallington TJ, Ziemke J, Liu X, Steinbacher M, Staehelin J (2019) Tropospheric ozone assessment report: tropospheric ozone from 1877 to 2016, observed levels, trends and uncertainties. Elem Sci Anth 7:39. https://doi.org/10.1525/elementa.376

Thwe AA, Vercambre G, Gautier H, Gay F, Phattaralerphong J, Kasemsap P (2015) Effects of acute ozone stress on reproductive traits of tomato, fruit yield and fruit composition. J Sci Food Agric 95(3):614–620. https://doi.org/10.1002/jsfa.6798

# Chapter 9
# Synergistic Effects of Heavy Metal-Induced Stress and High Tropospheric Ozone Concentration on Plant Productivity

**Khadija Mika Dawud, Usman Lawal Usman, Chongtham Allaylay Devi, and Ashutosh K. Pandey**

**Abstract** Heavy metal pollution and high tropospheric ozone concentrations are two major environmental stresses that can significantly affect plant growth and productivity. In recent years, there has been increasing concern about the synergistic effects of these two stresses on plant physiology and biochemistry. This chapter explores the synergistic impacts of heavy metal-induced stress and high tropospheric ozone concentration on plant productivity, physiology, and biochemistry. Heavy metal pollution and tropospheric ozone are two major environmental stresses that affect plants globally. However, their combined effects on plants are less understood. The chapter reviews current literature on the physiological and biochemical responses of plants to heavy metal stress and high ozone concentrations and the interactions between them. It discusses the impacts of these stresses on plant growth, photosynthesis, and the antioxidant defense system. Overall, this chapter provides insights into the complex interplay between heavy metal pollution, high tropospheric ozone concentrations, and plant productivity, which can be used to develop effective strategies for improving plant growth and productivity in polluted environments.

**Keywords** Heavy metal · Tropospheric ozone · Plant productivity · Photosynthesis · Oxidative stress

K. M. Dawud
Sharda School of Agricultural Sciences, Sharda University, Greater Noida, India

U. L. Usman
Department of Biology, Umaru Musa Yar'adua University, Katsina, Nigeria

C. A. Devi
Amity Institute of Horticultural Studies and Research, Amity University, Noida, Uttar Pradesh, India

A. K. Pandey (✉)
Department of Life Science, Sharda School of Bio-Science & Technology, Sharda University, Greater Noida, Uttar Pradesh, India
e-mail: Aashu.p20@gmail.com; Ashutosh.pandey1@sharda.ac.in

 
A. K. Pandey and E. Oksanen (eds.), *Tropospheric Ozone and Food Security*,
https://doi.org/10.1007/978-3-032-03608-7_9

## 9.1 Introduction

Plants are constantly exposed to a variety of environmental stressors that can significantly impact their growth, development, and overall productivity. One such stressor is heavy metal pollution, which has become a growing concern due to industrial activities, mining, and agricultural practices. Heavy metals, such as lead, cadmium, mercury, and arsenic, are non-degradable pollutants that accumulate in the soil and can be taken up by plants (Ackova 2018). Heavy metal-induced stress in plants arises from the ability of these toxic elements to interfere with essential physiological and biochemical processes. The mechanisms of heavy metal uptake and accumulation in plants involve complex interactions between the roots and the surrounding soil environment. Once absorbed, heavy metals can disrupt the normal functioning of various plant systems, leading to impaired growth, reduced photosynthetic efficiency, and nutrient imbalances (Pande et al. 2022). The consequences of heavy metal-induced stress on plants can be wide-ranging. Reduced growth, stunted development, chlorosis, and necrosis are commonly observed symptoms of heavy metal toxicity. Moreover, heavy metals can induce oxidative stress by generating reactive oxygen species (ROS) and impairing antioxidant defense systems. Imbalances in nutrient uptake and mineral nutrition, as well as disturbances in hormonal regulation, further contribute to the overall negative impact on plant growth and productivity (Riyazuddin et al. 2022).

High levels of tropospheric ozone are a serious environmental stressor that can have a major effect on plant physiology and productivity. Precursor pollutants including nitrogen oxides (NOx) and volatile organic compounds (VOCs), which are emitted from anthropogenic activities like industrial emissions, car exhaust, and agricultural practices, are involved in complex photochemical reactions that result in the formation of tropospheric ozone (Pinto et al. 2010). Due to their proximity to the atmosphere, plants are particularly susceptible to tropospheric ozone. The primary stomatal apertures on leaf surfaces are where ozone enters plants. From there, it diffuses through the intercellular gaps and then reaches the mesophyll cells. Ozone can cause several physiological and biochemical changes once it enters the plant, which can have a negative impact on the growth, development, and productivity of the plant (Bhatia et al. 2012). High levels of tropospheric ozone have a wide range of effects on plants. One of the main impacts is the reduction of photosynthesis, primarily because Rubisco, the essential enzyme involved in carbon fixation, is degrading more quickly. Lower energy generation and lower carbon uptake as a result of this photosynthetic impairment ultimately contribute to diminished plant growth and productivity. High ozone levels also disrupt other physiological processes in plants in addition to photosynthesis (Emberson et al. 2018). Plant tissues produce ROS, which leads to oxidative stress brought on by ozone. Cellular malfunction and metabolic pathway impairment can result from ROS damage to cellular components such as membranes, proteins, and DNA. In addition, excessive ozone concentrations can impair stomatal function, altering water relations and lowering plant water usage efficiency. As plants struggle to maintain a proper water balance when exposed to ozone, this can worsen the detrimental impacts of drought stress on plant productivity. High levels of tropospheric ozone have a significant negative effect on plant biochemistry (Xie et al. 2019). As a defence against oxidative stress, ozone can stimulate the creation of antioxidant substances and enzymes. The antioxidant defence mechanisms can be overpowered by continuous exposure to high ozone concentrations, which can cause oxidative damage and changes in cellular homeostasis. Due to the prevalence of high tropospheric ozone concentrations and their ramifications for both natural ecosystems and agriculture, it is crucial to comprehend how these conditions affect plants. Ozone pollution can have an impact on crop productivity and quality, costing farmers much (Tripathy and Oelmüller 2012). Additionally, it can affect the variety and stability of ecosystems by changing the makeup and behaviour of plant communities in natural settings.

In this chapter, we explore how heavy metal-induced stress and high tropospheric ozone concentrations interact to affect plant physiology, biochemistry, and productivity. We can learn more about the intricate relationships and mechanisms underlying plant responses to many stressors by analysing the combined effects of these stressors. The ability to minimise the negative effects of high tropospheric ozone concentrations on plant productivity and improve plant resistance in ozone-polluted environments depends on this understanding. We add to the knowledge of the complex interrelationship between high tropospheric ozone concentrations and plant physiology through this thorough investigation. Furthermore, we underscore the importance of considering multiple stressors in plant research and environmental management practices, providing a foundation for sustainable approaches to mitigate the impacts of ozone pollution on plant productivity and ecosystem health.

## 9.2 Heavy Metals and High Tropospheric Ozone Concentrations in Plant

Studying the synergistic impacts of heavy metal-induced stress and high tropospheric ozone concentrations on plant productivity physiology and biochemistry is of significant importance for several reasons. Firstly, both heavy metal pollution and high tropospheric ozone concentrations are prevalent environmental stressors that can have detrimental effects on plant growth, development, and productivity as depicted in Table 9.1 (Hafeez et al. 2023); (Chauhan and Sharma 2023). Understanding the combined effects of these stressors is crucial because, in many polluted environments, plants are simultaneously exposed to multiple stressors rather than just one. Studying the synergistic impacts provides a more realistic representation of the challenges that plants face in polluted environments, allowing us to better comprehend the overall stress burden on plants and the subsequent implications for ecosystem functioning and agricultural productivity (Watanabe et al. 2017). Secondly, investigating the synergistic effects provides insights into the underlying mechanisms and interactions between heavy metals and ozone. Heavy metals and ozone can influence common physiological and biochemical pathways in plants, and their combined effects may lead to amplified or unique responses compared to individual stressors. By studying these synergistic interactions, we can unravel the complex molecular and cellular processes involved, such as the interplay between oxidative stress, antioxidant defense systems, and metabolic adjustments (Renaut et al. 2009). This knowledge enhances our understanding of plant stress responses and can potentially identify novel targets for developing strategies to mitigate the negative impacts of multiple stressors on plant productivity. Moreover, studying the synergistic impacts has practical implications for crop production and environmental management. Agriculture is often carried out in areas where heavy metal pollution and high tropospheric ozone concentrations coexist. The combined effects of these stressors can lead to significant yield losses, reduced crop quality, and increased vulnerability to other biotic and abiotic stresses. By elucidating the synergistic impacts, we can develop targeted management strategies that consider the combined stress burden and provide effective solutions to minimize crop losses and optimize agricultural productivity. This information can also aid in the development of guidelines and policies for sustainable land use practices and pollution control measures to protect both human health and the environment.

## 9.3 Heavy Metal-Induced Stress and High Tropospheric Ozone Concentrations

The co-occurrence of heavy metal-induced stress and high tropospheric ozone concentrations can have synergistic impacts on plant physiology, biochemistry, and overall productivity. When plants are exposed to both stressors simultaneously, their responses and tolerance mechanisms may be altered compared to individual stress conditions (Shabbir et al. 2022). Understanding the synergistic effects of heavy metals and ozone pollution is crucial for comprehensively assessing the ecological consequences and potential risks to plant health.

### *9.3.1 Physiological Responses*

The combined effects of heavy metal-induced stress and high tropospheric ozone concentrations can lead to complex physiological responses in plants. For instance, heavy metals can interfere with stomatal functioning, reducing the ability of plants to regulate water loss through transpiration. This can exacerbate the impact of ozone stress on stomatal conductance and further impair water use efficiency. Additionally, heavy metals can interact with the photosynthetic apparatus, affecting the efficiency of light absorption, electron transport, and carbon assimilation. When coupled with ozone-induced oxidative stress, these interactions can lead to a cumulative reduction

**Table 9.1** Individual effects of heavy metals and tropospheric ozone in plant

| Effects | Heavy metal-induced stress | High tropospheric ozone concentrations |
|---|---|---|
| Plant damage | Heavy metals can accumulate in plants and cause structural damage and physiological disturbances. This can lead to stunted growth, chlorosis (yellowing of leaves), necrosis (tissue death), and reduced productivity | Ozone damages plant cells by disrupting the cellular membrane and chlorophyll molecules. It affects photosynthesis, reducing the rate of $CO_2$ assimilation and inhibiting the production of ATP and other energy-rich compounds |
| Oxidative stress | Heavy metals generate reactive oxygen species (ROS) through redox reactions. ROS can damage DNA, proteins, and lipids, leading to oxidative stress. Plants respond by producing antioxidants to mitigate oxidative damage | Ozone also generates ROS, leading to oxidative stress in plants. The excess ROS can cause lipid peroxidation, protein degradation, and DNA damage, resulting in increased susceptibility to diseases |
| Nutrient imbalance | Heavy metals can interfere with nutrient uptake and translocation in plants. They can inhibit the absorption of essential nutrients such as iron, manganese, and zinc, causing nutrient imbalances | Ozone exposure affects the uptake and utilization of essential nutrients like nitrogen, potassium, and calcium, leading to nutrient imbalances and deficiencies in plants |
| Ecological impact | Heavy metal pollution can disrupt ecosystems, affecting plant diversity and reducing overall productivity | High tropospheric ozone concentrations contribute to air pollution, which has a negative impact on plant communities, including changes in species composition, reduced biodiversity, and altered ecosystem dynamics |
| Morphological changes | Heavy metal stress can lead to alterations in plant morphology, including reduced root length and biomass, abnormal root branching, and altered leaf shape and size | High ozone concentrations can result in visible changes in plant morphology, such as leaf injury, leaf chlorosis (yellowing), and necrosis (tissue death). These visible changes can impact the overall plant structure and reduce its ability to carry out vital functions |
| Reproductive impairment | Heavy metals can affect plant reproduction by inhibiting pollen germination, reducing pollen viability, and impairing fertilization. It can also cause abnormalities in seed development, leading to reduced seed quality and viability | Ozone exposure can disrupt reproductive processes in plants. It can lead to reduced seed set, decreased pollen viability, and altered floral morphology, ultimately affecting plant reproduction and reproductive success |

(continued)

**Table 9.1** (continued)

| Effects | Heavy metal-induced stress | High tropospheric ozone concentrations |
|---|---|---|
| Accumulation of toxic elements | Plants exposed to heavy metals can accumulate these toxic elements in their tissues, posing potential health risks to humans and animals through the food chain | Ozone can cause the accumulation of toxic compounds in plant tissues. Ozone reacts with volatile organic compounds (VOCs) emitted by plants, forming secondary pollutants that can have detrimental effects on plant health and ecosystem functioning |
| Remediation and mitigation | Various strategies can be employed for heavy metal remediation, such as phytoremediation (using plants to remove or stabilize heavy metals) and soil amendments to improve soil quality | Measures to reduce tropospheric ozone concentrations include reducing emissions of precursor pollutants, implementing air quality regulations, and promoting sustainable practices to reduce ozone-forming emissions. Additionally, plant breeding and genetic engineering approaches can be employed to develop ozone-tolerant plant varieties |
| Interaction with other stresses | Heavy metal stress can interact with other environmental stresses, such as drought, salinity, and high temperature, amplifying their negative effects on plants | The effects of high tropospheric ozone concentrations on plants can be exacerbated when combined with other environmental stresses. For example, ozone-induced oxidative stress can interact with drought stress, leading to synergistic effects and increased plant susceptibility to damage, and reduced resilience |

in photosynthetic rates and chlorophyll content, compromising plant growth and productivity (Fodor 2002).

## 9.3.2 *Biochemical Interactions*

The combined exposure to heavy metals and high tropospheric ozone concentrations can result in intricate biochemical interactions within plant tissues. Heavy metals can enhance the production of reactive oxygen species (ROS), exacerbating oxidative stress when combined with ozone pollution. The accumulation of ROS can lead to lipid peroxidation, protein oxidation, and DNA damage, compromising cellular integrity and functionality. Moreover, heavy metals and ozone can disrupt antioxidant defense mechanisms, impairing the plant's ability to scavenge ROS effectively. The altered balance between ROS production and antioxidant capacity can further exacerbate oxidative damage and disrupt cellular processes (Shahid et al. 2014).

### 9.3.3 Nutrient Imbalances

Both heavy metals and high ozone concentrations can interfere with nutrient uptake, assimilation, and distribution within plants. Heavy metals may disrupt essential nutrient uptake, leading to deficiencies or imbalances. This can exacerbate the negative impacts of ozone stress on nutrient metabolism and compromise plant growth and productivity. Furthermore, heavy metals can interact with nutrient transporters and affect the translocation of essential elements within plants. These interactions can disrupt nutrient homeostasis, leading to impaired physiological functions and altered biochemical processes (Sandeep et al. 2019).

### 9.3.4 Gene Expression and Signaling

The synergistic impacts of heavy metal-induced stress and high tropospheric ozone concentrations can also influence gene expression and signaling pathways in plants. The combined stressors may modulate the expression of stress-responsive genes and the activation of signaling molecules involved in stress perception and adaptive responses. Changes in gene expression patterns can affect the synthesis of stress-related proteins, antioxidants, and secondary metabolites. These alterations in gene expression can have cascading effects on plant defense mechanisms, growth regulation, and overall stress tolerance (Dutta et al. 2018).

## 9.4 Interactive Effects on Plant Physiology

The simultaneous presence of heavy metal-induced stress and high tropospheric ozone concentrations can result in interactive effects on plant physiology. The combined exposure to these stressors can lead to complex alterations in various physiological processes, influencing the overall health and functioning of plants.

### 9.4.1 Photosynthesis and Carbon Assimilation

Heavy metals and high ozone concentrations can independently impair photosynthetic processes in plants. Heavy metals can disrupt chloroplast structure and function, affecting light absorption, electron transport, and carbon assimilation (Muhammad et al. 2021). Ozone, on the other hand, can damage photosynthetic pigments and disrupt the photosynthetic machinery. When plants are exposed to both stressors simultaneously, the effects on photosynthesis can be compounded. The combined stressors can lead to a greater reduction in photosynthetic efficiency, resulting in decreased carbon assimilation and limited energy production for growth and development (Yang et al. 2020).

### *9.4.2 Stomatal Regulation and Water Relations*

Heavy metals can disrupt stomatal functioning, affecting the opening and closure of stomata. This interference can result in altered stomatal conductance, reducing the plant's ability to regulate water loss through transpiration. Reduced stomatal conductance combined with high ozone concentrations can exacerbate water stress in plants, leading to decreased water use efficiency and potential drought susceptibility (Guo et al. 2023).

### *9.4.3 Nutrient Uptake and Transport*

Both heavy metals and ozone pollution can interfere with nutrient uptake and transport mechanisms in plants. Heavy metals can inhibit the uptake of essential nutrients, leading to deficiencies or imbalances. Ozone, on the other hand, can impair nutrient assimilation and translocation within plants. The interactive effects of heavy metals and ozone on nutrient uptake and transport can exacerbate nutrient deficiencies and imbalances. This can have detrimental effects on plant growth, development, and overall physiological functions (Muhammad et al. 2021).

### *9.4.4 Hormonal Regulation*

Heavy metals and high ozone concentrations can disrupt hormonal regulation in plants. These stressors can alter the synthesis, signaling, and response to plant hormones, such as auxins, cytokinins, gibberellins, and abscisic acid. The interactive effects of heavy metals and ozone on hormonal regulation can result in imbalanced hormone levels and disrupted physiological processes. This can influence plant growth, development, and various stress responses (Bücker-Neto et al. 2017).

### *9.4.5 Oxidative Stress and Antioxidant Defense*

Heavy metals and high ozone concentrations can independently induce oxidative stress in plants, leading to the overproduction of reactive oxygen species (ROS). ROS can cause cellular damage and disrupt physiological processes. The interactive effects of these stressors can further enhance ROS production and oxidative damage, overwhelming the plant's antioxidant defense systems. The interactive effects of heavy metals and ozone on oxidative stress can result in cumulative damage to cellular components and compromise plant vitality and productivity (Hasanuzzaman et al. 2020).

## 9.5 Synergistic Effects of Heavy Metal and Tropospheric Ozone on Plant

The synergistic effects of heavy metal-induced stress and high tropospheric ozone concentration on plants have several similarities as shown in Fig. 9.1. It is important to note that the specific effects and mechanisms may vary depending on the type and concentration of heavy metals, as well as the duration and intensity of ozone exposure. Additionally, the combined effects of heavy metals and high ozone concentrations on plants can be complex and may vary depending on the plant species and their physiological and genetic characteristics. Here are some of the commonalities:

(i) Oxidative stress: Both heavy metals and high ozone concentrations can induce oxidative stress in plants. Heavy metals generate reactive oxygen species (ROS) through redox reactions, while high ozone levels lead to the formation of ozone-derived ROS. Both types of ROS can cause damage to cellular components, including proteins, lipids, and DNA (Singh et al. 2016). (ii) Impaired photosynthesis: Both heavy metals and high ozone concentrations can disrupt photosynthesis in plants. Heavy metals can inhibit the activity of enzymes involved in photosynthetic processes, such as chlorophyll biosynthesis, electron transport, and carbon fixation.

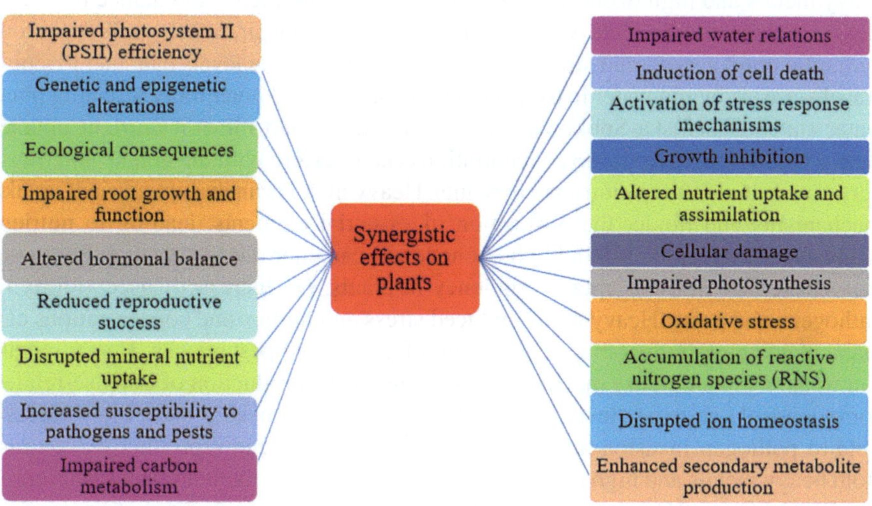


**Fig. 9.1** Synergistic effects of heavy metal and tropospheric ozone on plant

High ozone levels can damage chloroplast membranes and reduce the efficiency of light harvesting and electron transport (Muhammad et al. 2021). (iii) Cellular damage: Heavy metals and high ozone concentrations can both cause damage to plant cell structures. Heavy metals can accumulate in the cell walls, vacuoles, and organelles, disrupting their integrity and function. High ozone levels can lead to the breakdown of cell membranes and the leakage of cellular contents (Hossain et al. 2012). (iv) Altered nutrient uptake and assimilation: Both heavy metals and high ozone concentrations can affect the uptake and assimilation of essential nutrients in plants. Heavy metals can compete with nutrient ions for uptake by roots, leading to nutrient imbalances and deficiencies. High ozone levels can impair the uptake and assimilation of nitrogen, affecting plant growth and development (Watanabe et al. 2017). (v) Growth inhibition: Heavy metals and high ozone concentrations can both inhibit plant growth. Heavy metals can impair cell division and elongation, leading to stunted growth and reduced biomass production. High ozone levels can suppress cell division and expansion, resulting in decreased leaf area and overall plant size (Muhammad et al. 2021). (vi) Activation of stress response mechanisms: Plants respond to both heavy metal-induced stress and high ozone concentrations by activating various stress response mechanisms. These include the synthesis of antioxidant enzymes, such as superoxide dismutase (SOD) and catalase, to scavenge ROS. Plants also produce stress-related proteins and osmolytes to protect cellular structures and maintain osmotic balance (Muhammad et al. 2021). (vii) Induction of cell death: Both heavy metals and high ozone concentrations can induce programmed cell death (PCD) in plants. Heavy metals can trigger PCD through the generation of ROS and the activation of signaling pathways. High ozone levels can directly induce PCD in leaf tissues through the accumulation of ROS and the disruption of cell membrane integrity (Petrov et al. 2015). (viii) Impaired water relations: Both heavy metals and high ozone concentrations can disrupt the water balance in plants. Heavy metals can interfere with water uptake and transpiration by affecting root hydraulic conductivity and stomatal regulation. High ozone levels can lead to stomatal closure, reducing water loss through transpiration and potentially affecting plant water status (Rucińska-Sobkowiak 2016). (iv) Disrupted mineral nutrient uptake: Heavy metals and high ozone concentrations can interfere with the uptake and transport of essential mineral nutrients in plants. Heavy metals can disrupt nutrient uptake mechanisms and hinder the absorption of essential elements, leading to nutrient imbalances (Arif et al. 2016). High ozone levels can alter nutrient availability in the soil and reduce nutrient uptake efficiency in plants. (x) Increased susceptibility to pathogens and pests: Heavy metal-induced stress and high ozone concentrations can make plants more vulnerable to pathogen infections and pest attacks. Heavy metals can weaken plant defense mechanisms and impair the production of defense-related compounds. High ozone levels can damage leaf tissues, making them more susceptible to pathogen invasion and insect feeding. (xi) Reduced reproductive success: Both heavy metals and high ozone concentrations can negatively impact plant reproductive processes. Heavy metals can affect pollen viability and germination, leading to reduced seed set and fertility. High ozone levels can disrupt flower development, pollen tube growth, and fertilization, resulting in decreased fruit and seed production

(Sénéchal et al. 2015). (xii) Genetic and epigenetic alterations: Heavy metal-induced stress and high ozone concentrations can induce genetic and epigenetic changes in plants. Heavy metals can cause DNA damage and mutations, potentially leading to genetic variations. High ozone levels can affect gene expression patterns and induce DNA methylation changes, altering the epigenetic regulation of gene activity (Dutta et al. 2018). (xiii) Ecological consequences: The combined effects of heavy metal-induced stress and high ozone concentrations on plants can have broader ecological consequences. Both stressors can affect plant growth and survival, leading to changes in plant community composition and ecosystem dynamics. They can also impact interactions with other organisms, such as herbivores, pollinators, and soil microbes, influencing ecological processes and biodiversity (Sénéchal et al. 2015). (xiv) Impaired root growth and function: Both heavy metals and high ozone concentrations can adversely affect root growth and function. Heavy metals can inhibit root elongation and reduce root biomass, limiting the plant's ability to absorb water and nutrients from the soil. High ozone levels can lead to the formation of reactive oxygen species in the root system, damaging root cells and impairing nutrient uptake (Sénéchal et al. 2015). (xv) Altered hormonal balance: Heavy metals and high ozone concentrations can disrupt the balance of plant hormones, affecting growth and development. Heavy metals can interfere with hormone synthesis, signaling pathways, and hormone transport (Bücker-Neto et al. 2017). High ozone levels can alter the production and distribution of hormones, influencing processes such as seed germination, root elongation, and flowering. (xvi) Impaired carbon metabolism: Both heavy metals and high ozone concentrations can disrupt carbon metabolism in plants. Heavy metals can inhibit enzymes involved in carbohydrate metabolisms, such as sucrose synthesis and starch degradation. High ozone levels can impair photosynthesis and reduce carbon assimilation, leading to altered carbon allocation and reduced biomass production (Muhammad et al. 2021). (xvii) Enhanced secondary metabolite production: Heavy metal-induced stress and high ozone concentrations can stimulate the production of secondary metabolites in plants. Heavy metals can trigger the synthesis of phytochelatins, metallothioneins, and other detoxification compounds. High ozone levels can induce the production of phenolic compounds, flavonoids, and antioxidants, which play a role in plant defense against oxidative stress (Hossain et al. 2012). (xviii) Disrupted ion homeostasis: Both heavy metals and high ozone concentrations can disrupt ion homeostasis in plants. Heavy metals can interfere with the uptake and translocation of essential ions, such as calcium, potassium, and magnesium. High ozone levels can disrupt ion transport processes across cell membranes, leading to imbalances in ion concentrations and ion toxicity (Arif et al. 2016). (xix) Impaired photosystem II (PSII) efficiency: Heavy metals and high ozone concentrations can both affect the efficiency of photosystem II (PSII), a key component of the photosynthetic machinery. Heavy metals can inhibit PSII activity and damage the D1 protein, reducing the plant's ability to capture light energy. High ozone levels can accelerate the degradation of the D1 protein and disrupt PSII electron flow (Sasi et al. 2018). (xx) Accumulation of reactive nitrogen species (RNS): Both heavy metals and high ozone concentrations can lead to the accumulation of reactive nitrogen species (RNS) in plants. Heavy metals can promote the production of nitric oxide (NO) and other reactive nitrogen intermediates. High ozone levels can enhance the formation of nitric oxide and nitrogen dioxide through reactions involving ozone and nitric oxide (Saddhe et al. 2019).

## 9.6 Combined Impact on Photosynthesis and Carbon Assimilation

Photosynthesis is a fundamental process for plant growth and productivity, and both heavy metal-induced stress and high tropospheric ozone concentrations can individually impair photosynthetic efficiency. However, when these stressors occur simultaneously, their combined impact on photosynthesis and carbon assimilation can be more severe than the sum of their individual effects.

Heavy metal-induced stress can disrupt the photosynthetic apparatus by impairing chloroplast structure and function. It can affect the light-harvesting complexes, electron transport chain, and carbon fixation enzymes, leading to reduced photosynthetic rates. This results in decreased carbon assimilation and limited energy availability for plant growth and development (Arif et al. 2016). High tropospheric ozone concentrations can also negatively affect photosynthesis by damaging photosynthetic pigments and disrupting photosynthetic reactions. Ozone can cause chlorophyll degradation and reduce the efficiency of electron transport, leading to decreased photosynthetic capacity (Emberson et al. 2018). When plants are exposed to both heavy metals and high ozone concentrations, the combined impact on photosynthesis can be amplified. The presence of heavy metals can exacerbate the negative effects of ozone on photosynthetic processes. Heavy metals can impair the repair mechanisms of the photosynthetic apparatus and disrupt the balance between energy production and utilization. These further compromise the plant's ability to capture light, transport electrons, and fix carbon dioxide. The combined impact of heavy metal-induced stress and high tropospheric ozone concentrations on photosynthesis can lead to a significant reduction in photosynthetic rates, limiting the availability of carbohydrates for plant growth and metabolic processes. This reduction in carbon assimilation can have cascading effects on plant development, biomass accumulation, and ultimately, overall productivity.

### *9.6.1 Effects on Stomatal Conductance and Water Relations*

Stomata are small pores on the surface of plant leaves that regulate gas exchange, including the uptake of carbon dioxide ($CO_2$) for photosynthesis and the release of oxygen ($O_2$) and water vapor ($H_2O$). Both heavy metal-induced stress and high tropospheric ozone concentrations can independently disrupt stomatal functioning, impacting stomatal conductance and water relations. When these stressors occur together, their combined effects can further alter stomatal behavior and exacerbate water stress in plants. Heavy metal-induced stress can interfere with stomatal regulation by affecting the guard cells, which control the opening and closing of stomata (Jarvis et al. 1999). This interference can lead to reduced stomatal aperture and limited gas exchange. Consequently, the plants may experience decreased $CO_2$ uptake for photosynthesis, resulting in a reduction in carbon assimilation. High tropospheric ozone concentrations can also influence stomatal conductance by promoting stomatal closure. Ozone pollution triggers the production of reactive oxygen species (ROS) in the leaf tissues, which can lead to oxidative stress and the subsequent closure of stomata as a protective mechanism. This closure reduces gas exchange and transpiration, leading to water conservation but also limiting $CO_2$ uptake (Li et al. 2019). When plants are exposed to both heavy metals and high ozone concentrations, the combined effects on stomatal conductance and water relations can be more pronounced. The presence of heavy metals can amplify the response of stomata to ozone pollution, resulting in increased stomatal closure and reduced water loss through transpiration. This can lead to water stress and limited availability of $CO_2$ for photosynthesis. The combined impact on stomatal conductance and water relations can have detrimental effects on plant growth and productivity. Reduced stomatal conductance limits the availability of $CO_2$ for photosynthesis, impairing carbon assimilation and overall plant energy production (Guo et al. 2023). Moreover, the decreased transpiration resulting from stomatal closure can affect the plant's ability to cool itself, regulate leaf temperature, and transport nutrients through the xylem.

### 9.6.2 Changes in Nutrient Uptake and Distribution

Both heavy metal-induced stress and high tropospheric ozone concentrations can have significant impacts on nutrient uptake and distribution in plants. When these stressors occur concurrently, their combined effects can lead to altered nutrient availability, uptake efficiency, and nutrient distribution within the plant as shown in Table 9.2 (Sachdev et al. 2021); (Hafeez et al. 2023). Heavy metal-induced stress can interfere with nutrient uptake mechanisms in plants. Certain heavy metals, such as cadmium, lead, and mercury, can competitively inhibit the uptake of essential nutrients, including iron (Fe), zinc (Zn), manganese (Mn), and calcium (Ca). This interference can result in nutrient deficiencies and imbalances, affecting various physiological processes. High tropospheric ozone concentrations can also affect nutrient uptake and assimilation. Ozone pollution can impair the absorption and utilization of certain nutrients, such as nitrogen (N) and sulfur (S). This can disrupt nitrogen metabolism and lead to imbalances in other essential nutrients (Ackova 2018). When plants are exposed to both heavy metals and high ozone concentrations, the combined effects on nutrient uptake and distribution can be complex. The presence of heavy metals can enhance the negative impact of ozone on nutrient uptake efficiency. Heavy metals can interfere with nutrient transporters and disrupt the balance between nutrient absorption and exclusion. This can further exacerbate nutrient deficiencies or imbalances within the plant. The altered nutrient availability and distribution resulting from the combined effects of heavy metals and ozone can have significant consequences for plant growth and development. Nutrient deficiencies can impair various physiological processes, such as photosynthesis, enzyme activity, and hormone regulation. Imbalances in nutrient ratios can disrupt nutrient-dependent signaling pathways and metabolic functions (Kumar et al. 2021). Furthermore, the changes in nutrient uptake and distribution can have cascading effects on plant productivity, biomass accumulation, and overall nutritional quality. The nutritional status of plants can also influence their interactions with other organisms, such as herbivores, pathogens, and symbiotic microorganisms, which in turn can affect ecosystem dynamics.

### *9.6.3 Interactive Effects on Plant Biochemistry*

The synergistic impacts of heavy metal-induced stress and high tropospheric ozone concentrations can have profound effects on plant biochemistry. These combined stressors can elicit complex biochemical changes that affect various metabolic pathways and biochemical reactions within plants.

a. *Oxidative Stress and Reactive Oxygen Species (ROS) Production*
   Both heavy metals and high ozone concentrations can independently induce oxidative stress in plants, leading to the overproduction of reactive oxygen species (ROS). ROS, such as superoxide radicals ($O_2^{\bullet-}$), hydrogen peroxide ($H_2O_2$), and hydroxyl radicals (•OH), can cause oxidative damage to cellular components, including lipids, proteins, and DNA. When plants are exposed to both stressors simultaneously, the interactive effects can lead to an intensified generation of ROS and oxidative stress. Heavy metals can disrupt antioxidant defense systems and exacerbate ROS production triggered by high ozone concentrations. This cumulative oxidative stress can result in significant damage to cellular structures and impair various biochemical processes (Sachdev et al. 2021).
b. *Activation of Antioxidant Defense Systems*
   Plants possess an array of antioxidant defense mechanisms to counteract the detrimental effects of ROS and maintain redox homeostasis. These defense systems include enzymatic antioxidants such as superoxide dismutase (SOD), catalase (CAT), peroxidases (POX), and non-enzymatic antioxidants like ascorbate (vitamin C), glutathione (GSH), and tocopherols (vitamin E). The interactive effects of heavy metal-induced stress and high ozone concentrations can stimulate the activation of antioxidant defense systems. Plants exposed to both stressors may exhibit enhanced activity of antioxidant enzymes and increased synthesis of non-enzymatic antioxidants. This upregulation of antioxidant defenses aims to counterbalance the excessive ROS production and minimize oxidative damage (Sachdev et al. 2021).
c. *Alterations in Primary and Secondary Metabolite Production*

**Table 9.2** Combined effects of heavy metals and tropospheric ozone on plant productivity

| Synergistic effects | Implications on plants |
|---|---|
| Increased oxidative stress | Heavy metals and tropospheric ozone both contribute to the production of reactive oxygen species (ROS), leading to oxidative stress in plant tissues |
| Reduced photosynthetic rate | Heavy metals and tropospheric ozone can independently impair photosynthesis in plants. When combined, they have a cumulative negative impact on photosynthetic activity |
| Altered nutrient uptake | Heavy metals and tropospheric ozone can disrupt nutrient uptake mechanisms in plants, leading to imbalances and deficiencies in essential nutrients |
| Stunted growth | The combined exposure to heavy metals and tropospheric ozone can hinder plant growth and development, resulting in reduced biomass accumulation |
| Increased cell damage | Heavy metals and tropospheric ozone can damage plant cell membranes, impairing their integrity and function. This can lead to cell death and tissue necrosis |
| Impaired antioxidant defense | Both heavy metals and tropospheric ozone can interfere with the plant's antioxidant defense system, reducing its ability to neutralize ROS and protect against oxidative damage |
| Decreased reproductive yield | Plants exposed to heavy metals and tropospheric ozone may experience reduced flower production, pollen viability, and seed/fruit yield, impacting their reproductive capacity |
| Increased root damage | Heavy metals and tropospheric ozone can both cause damage to plant roots. When combined, they can have a synergistic effect, leading to reduced root growth and impaired nutrient absorption |
| Disruption of stomatal function | Heavy metals and tropospheric ozone can independently affect stomatal function, but their combined presence can exacerbate the negative impact. This can lead to increased water loss and reduced plant water-use efficiency |
| Altered gene expression | The interaction between heavy metals and tropospheric ozone can result in altered gene expression patterns in plants. This can influence various physiological processes, such as defense mechanisms and growth regulation |
| Increased susceptibility to pathogens | Plants exposed to both heavy metals and tropospheric ozone may become more susceptible to pathogen infections. The pollutants can weaken the plant's defense mechanisms, making them more vulnerable to diseases |
| Impaired reproductive success | The combined exposure to heavy metals and tropospheric ozone can lead to reduced pollination success and hinder the development of viable seeds or fruits. This can result in diminished plant reproductive success and population decline |
| Altered secondary metabolism | Heavy metals and tropospheric ozone can affect the production and accumulation of secondary metabolites in plants, such as phenolics and flavonoids. This can disrupt their roles in defense against herbivores and pathogens |
| Impaired nutrient cycling | Heavy metals and tropospheric ozone can disrupt nutrient cycling processes in soil. This can affect the availability and uptake of essential nutrients by plants, leading to nutrient imbalances and reduced growth |
| Reduced carbon assimilation | The combination of heavy metals and tropospheric ozone can result in decreased carbon assimilation in plants. This can lead to reduced energy production and hinder the plant's overall growth and productivity |

(continued)

**Table 9.2** (continued)

| Synergistic effects | Implications on plants |
|---|---|
| Altered root-shoot ratio | Exposure to heavy metals and tropospheric ozone can influence the allocation of resources between plant roots and shoots. This can lead to changes in the root-to-shoot ratio, affecting the overall plant architecture and performance |
| Increased susceptibility to abiotic stress | Heavy metals and tropospheric ozone can weaken plants' ability to tolerate and recover from other environmental stressors, such as drought or high temperatures. This can further compromise their resilience and survival |
| Disrupted symbiotic relationships | The presence of heavy metals and tropospheric ozone can interfere with beneficial symbiotic relationships, such as mycorrhizal associations. This can disrupt nutrient uptake and impair the plant's ability to form beneficial partnerships |

The combined effects of heavy metal-induced stress and high tropospheric ozone concentrations can lead to alterations in primary and secondary metabolite production in plants. Primary metabolites include carbohydrates, proteins, and nucleic acids, which are essential for plant growth and development. Secondary metabolites, such as phenolics, flavonoids, and alkaloids, play diverse roles in plant defense, UV protection, and signaling. Under the interactive stress of heavy metals and ozone, plants may exhibit changes in primary metabolite allocation and composition. These alterations can affect energy metabolism, carbon allocation, and growth processes. Additionally, the production of secondary metabolites may be modulated as a response to oxidative stress and environmental cues (Hafeez et al. 2023). The synthesis and accumulation of specific secondary metabolites may increase, serving as defense compounds against the combined stressors.

d. *Synergistic Oxidative Stress and Reactive Oxygen Species (ROS) Production*

One of the key biochemical changes observed when plants are exposed to the synergistic impacts of heavy metal-induced stress and high tropospheric ozone concentrations is the generation of oxidative stress and the overproduction of reactive oxygen species (ROS). Oxidative stress occurs when there is an imbalance between the production of ROS and the ability of plants to detoxify them, leading to cellular damage and disruption of various physiological processes. Heavy metal-induced stress and high tropospheric ozone concentrations can independently trigger the production of ROS in plant cells. Heavy metals can generate ROS through redox cycling and the Fenton reaction, resulting in the formation of highly reactive hydroxyl radicals (•OH) and other ROS. High ozone concentrations, on the other hand, can stimulate the production of ROS directly through the breakdown of ozone molecules within plant tissues. When these stressors occur together, their synergistic effects can lead to amplified oxidative stress and ROS production. Heavy metals can enhance the generation of ROS induced by high ozone concentrations, and vice versa. The presence of heavy metals can disrupt antioxidant defense systems and compromise the ability of plants to scavenge and neutralize ROS, resulting in the accumulation of ROS within cells. The overproduction of ROS can have detrimental effects on cellular components and biochemical processes. ROS can oxidize lipids, proteins, and nucleic acids, leading to membrane damage, protein degradation, and DNA mutations (Hasanuzzaman et al. 2020). This oxidative damage can disrupt cellular integrity and impair the functioning of organelles and enzymes. Moreover, the accumulation of ROS can trigger a cascade of secondary reactions and signaling pathways. ROS serve as signaling molecules that regulate various physiological and biochemical processes, including cell proliferation, programmed cell death (apoptosis), defense responses, and stress adaptation. The synergistic effects of heavy metal-induced stress and high tropospheric ozone concentrations can alter ROS-mediated signaling, leading to modifications in gene expression, enzyme activities, and metabolic pathways (Saddhe et al. 2019).

### *9.6.4 Combined Activation of Antioxidant Defense Systems*

The synergistic impacts of heavy metal-induced stress and high tropospheric ozone concentrations can result in the combined activation of antioxidant defense systems in plants. Antioxidant defense systems play a crucial role in neutralizing and detoxifying reactive oxygen species (ROS) generated during stress conditions. The simultaneous exposure to heavy metals and high ozone concentrations can trigger a coordinated response of these defense systems, aiming to alleviate oxidative stress and maintain cellular homeostasis. Heavy metal-induced stress can disrupt the normal functioning of antioxidant defense systems in plants. Heavy metals can impair the activities of antioxidant enzymes, such as superoxide dismutase (SOD), catalase (CAT), peroxidases (POX), and glutathione peroxidase (GPX). They can also deplete non-enzymatic antioxidants like ascorbate (vitamin C), glutathione (GSH), and tocopherols (vitamin E), which act as scavengers of ROS. Similarly, high tropospheric ozone concentrations can directly induce oxidative stress in plants and trigger the activation of antioxidant defense systems. Ozone pollution leads to the production of ROS within plant tissues, stimulating the synthesis and activation of antioxidant enzymes and non-enzymatic antioxidants. This response aims to counterbalance the oxidative damage caused by ozone and restore redox homeostasis. When plants are exposed to both stressors simultaneously, the combined effects can lead to enhanced activation of antioxidant defense systems (Krishnamurthy and Rathinasabapathi 2013). The presence of heavy metals can amplify the response of plants to high ozone concentrations, resulting in increased synthesis and activity of antioxidant enzymes and accumulation of non-enzymatic antioxidants. This coordinated response is crucial for maintaining cellular integrity and mitigating the detrimental effects of oxidative stress. The combined activation of antioxidant defense systems helps to scavenge and neutralize the excess ROS generated during heavy metal-induced stress and high ozone concentrations. Antioxidant enzymes, such as SOD, CAT, POX, and GPX, catalyze the breakdown of ROS, converting them into less harmful compounds. Non-enzymatic antioxidants, such as ascorbate and glutathione, directly interact with ROS, acting as sacrificial molecules to protect cellular components from oxidative damage. The combined activation of antioxidant defense systems not only helps in detoxifying ROS but also influences other physiological processes. Antioxidant enzymes and non-enzymatic antioxidants can interact with signaling molecules and transcription factors, modulating gene expression and cellular responses to stress. This coordination ensures the maintenance of cellular redox balance and supports plant survival under challenging environmental conditions (Rajput et al. 2021).

### *9.6.5 Alterations in Primary and Secondary Metabolite Synthesis*

The synergistic impacts of heavy metal-induced stress and high tropospheric ozone concentrations can lead to significant alterations in the synthesis and accumulation of both primary and secondary metabolites in plants. These changes in metabolite profiles play critical roles in plant adaptation, defense responses, and overall biochemical functioning. Primary metabolites are essential compounds involved in fundamental cellular processes, such as energy metabolism, growth, and development. They include carbohydrates, amino acids, organic acids, and nucleotides. Secondary metabolites, on the other hand, are not directly involved in primary metabolic pathways but contribute to plant adaptation, defense against environmental stresses, and interactions with other organisms (Isah 2019). Under the influence of heavy metals and high ozone concentrations, plants may exhibit modifications in primary metabolite synthesis and allocation. Heavy metal-induced stress can disrupt metabolic pathways involved in carbohydrate metabolism, leading to alterations in sugar accumulation, starch synthesis, and glycolytic activities. High tropospheric ozone concentrations can also affect primary metabolism by impairing photosynthetic carbon fixation and altering energy allocation. The combined effects of heavy metals and high ozone concentrations on primary metabolism can result in shifts in carbon allocation, affecting plant growth, development, and overall productivity. Reduced photosynthesis and impaired carbon assimilation can lead to decreased carbohydrate levels, limiting energy availability for various physiological processes.

Additionally, the synthesis and accumulation of secondary metabolites can be modulated in response to the synergistic impacts of heavy metal-induced stress and high ozone concentrations. Both heavy metals and ozone are known to stimulate the production of secondary metabolites, such as phenolics, flavonoids, alkaloids, and terpenoids. These compounds are involved in plant defense mechanisms, antioxidant activities, and interactions with herbivores, pathogens, and symbiotic microorganisms. The interactive effects of heavy metals and ozone on secondary metabolite synthesis can lead to changes in the composition and abundance of specific compounds. The presence of heavy metals can enhance the biosynthesis of secondary metabolites induced by high ozone concentrations, resulting in increased accumulation of defense-related compounds. These secondary metabolites can act as chemical barriers against oxidative stress, UV radiation, and other environmental challenges (Chen et al. 2022).

### 9.6.6 Mechanistic Insights into the Synergistic Interactions

The synergistic impacts of heavy metal-induced stress and high tropospheric ozone concentrations on plant productivity, physiology, and biochemistry arise from complex and intertwined mechanisms. Unraveling these mechanistic insights is crucial for understanding the underlying processes and interactions that drive the synergistic effects observed in plants.

i. *Oxidative Stress and Cellular Signaling*

   The synergistic interactions between heavy metals and high ozone concentrations can result in oxidative stress, characterized by the overproduction of reactive oxygen species (ROS) and oxidative damage to cellular components. ROS not only cause cellular damage but also act as signaling molecules, triggering various cellular responses. The interplay between ROS production, antioxidant defense systems, and signaling pathways plays a vital role in mediating the synergistic effects observed in plants (Tripathy and Oelmüller 2012).

ii. *Disruption of Cellular Homeostasis*

   Both heavy metals and ozone can disrupt cellular homeostasis by affecting key processes such as photosynthesis, nutrient uptake, and hormone signaling. Heavy metals can interfere with photosynthetic machinery, leading to reduced carbon assimilation and energy production. Ozone can damage chloroplasts and impair photosynthetic pigments, affecting the efficiency of light capture and utilization. Together, these stressors disrupt cellular energy balance and compromise plant growth and development (Kosová et al. 2018).

iii. *Altered Gene Expression and Transcriptional Regulation*

   The synergistic interactions between heavy metals and high ozone concentrations can modulate gene expression and transcriptional regulation in plants. Both stressors can trigger the activation or repression of specific genes involved in stress responses, defense mechanisms, and metabolic pathways. The coordinated regulation of gene expression plays a crucial role in the adaptive responses of plants to combined stresses (Ludwikow and Sadowski 2008).

iv. *Crosstalk between Stress Signaling Pathways*

   The presence of heavy metals and high ozone concentrations can activate common stress signaling pathways in plants. These signaling pathways involve the activation of various protein kinases, transcription factors, and secondary messengers that coordinate stress responses. The crosstalk between different signaling pathways allows plants to integrate and prioritize responses to multiple stressors, leading to synergistic effects on plant physiology and biochemistry (Sharma et al. 2012).

v. *Nutrient Imbalances and Feedback Mechanisms*

Heavy metals and high ozone concentrations can disrupt nutrient uptake and assimilation processes in plants. Heavy metals may inhibit the uptake and translocation of essential nutrients, leading to nutrient imbalances and deficiencies (Sperdouli 2022). High ozone concentrations can impair nutrient uptake and reduce nutrient use efficiency. The altered nutrient status can further exacerbate oxidative stress and impact plant metabolism and growth.

## 9.7 Reactive Oxygen Species (ROS) in Mediating the Synergistic Effects

Reactive oxygen species (ROS) play a pivotal role in mediating the synergistic effects of heavy metal-induced stress and high tropospheric ozone concentrations on plant productivity, physiology, and biochemistry. ROS are highly reactive molecules that include superoxide radicals ($O_2^{\bullet-}$), hydrogen peroxide ($H_2O_2$), and hydroxyl radicals (•OH). While excessive ROS can cause cellular damage, they also act as signaling molecules involved in stress responses and regulatory processes. The presence of heavy metals and high ozone concentrations can lead to the overproduction of ROS within plant cells. Heavy metals can directly generate ROS through redox reactions and indirectly promote ROS production by disrupting cellular redox balance and impairing antioxidant defense systems (Sachdev et al. 2021). High ozone concentrations, on the other hand, can directly induce ROS formation through the interaction of ozone with cellular components, such as lipids, proteins, and photosynthetic pigments. The synergistic interactions between heavy metals and high ozone concentrations result in an amplified production of ROS in plants. This elevated ROS production can have detrimental effects on cellular components, including lipids, proteins, nucleic acids, and membranes. ROS can oxidize these cellular components, leading to oxidative damage and impairing cellular integrity and function. Furthermore, ROS act as signaling molecules that initiate a cascade of cellular responses to cope with stress conditions. ROS can modulate gene expression, activate specific signaling pathways, and regulate the activity of key enzymes involved in stress responses. They can also function as secondary messengers in various signaling pathways, including those related to antioxidant defense, hormonal regulation, and cellular redox homeostasis (Krishnamurthy and Rathinasabapathi 2013). The synergistic effects of heavy metals and high ozone concentrations on ROS production have implications for plant physiology and biochemistry. Elevated ROS levels can lead to oxidative stress, disrupting cellular processes such as photosynthesis, respiration, and nutrient uptake. Additionally, ROS can induce DNA damage, protein degradation, and lipid peroxidation, further compromising plant cellular function and overall performance.

### *9.7.1 Interaction Between Heavy Metals and Ozone at the Molecular Level*

The interaction between heavy metals and ozone at the molecular level plays a crucial role in the synergistic effects observed on plant productivity, physiology, and biochemistry. Understanding the molecular interactions between these stressors provides insights into the underlying mechanisms driving their combined effects.

a. *Oxidative Stress Amplification*
   Heavy metals and ozone individually induce oxidative stress in plants by promoting the production of reactive oxygen species (ROS). However, when present simultaneously, they can interact synergistically and amplify the oxidative stress response. Heavy metals can enhance the ozone-induced generation of ROS through various mechanisms, including the disruption of antioxidant defense systems and the promotion of ROS-generating processes. This amplification of oxidative stress leads to more severe damage to cellular components and greater perturbations in plant physiology. ROS-Mediated Signaling Pathways: The interaction between heavy metals and ozone influences ROS-mediated signaling pathways in plants. Both stressors can activate common signaling cascades, such as mitogen-activated protein kinase (MAPK) pathways, calcium signaling, and transcription factor activation (Hasanuzzaman et al. 2020). These signaling pathways regulate the expression of stress-responsive genes and coordinate cellular responses to multiple stressors. The interaction between heavy metals and ozone can modulate the intensity and duration of these signaling pathways, altering the plant's ability to adapt and cope with stress.

b. *Disruption of Cellular Redox Homeostasis*
   Heavy metals and ozone can disrupt cellular redox homeostasis by altering the balance between ROS production and antioxidant defense systems. Heavy metals interfere with antioxidant enzyme activities and deplete cellular antioxidants, leading to an imbalance in the cellular redox state. Ozone, on the other hand, directly interacts with cellular components, generating ROS and impairing antioxidant capacity. The combined presence of heavy metals and ozone exacerbates the disruption of cellular redox homeostasis, leading to sustained oxidative stress and cellular damage (Isah 2019).

c. *DNA Damage and Repair*
   Heavy metals and ozone can induce DNA damage in plants through different mechanisms. Heavy metals can directly interact with DNA, causing DNA strand breaks and cross-linking. Ozone exposure leads to the formation of DNA adducts and oxidative damage to DNA bases. When heavy metals and ozone act synergistically, the extent of DNA damage can be increased, overwhelming the plant's DNA repair mechanisms. The accumulation of DNA damage compromises genomic stability and can lead to long-term effects on plant growth and development (Das and Roychoudhury 2014).

d. *Alterations in Gene Expression and Metabolic Pathways*

The interaction between heavy metals and ozone can result in alterations in gene expression and metabolic pathways in plants. Heavy metals and ozone individually influence the expression of stress-responsive genes, metabolic enzymes, and signaling molecules. However, their combined presence can induce more extensive changes in gene expression, affecting various metabolic pathways such as photosynthesis, respiration, and secondary metabolite synthesis. These alterations in gene expression and metabolic pathways contribute to the synergistic effects observed at the physiological and biochemical levels (Mareri et al. 2022).

### 9.7.2 Impact of Synergistic Interactions on Cellular Processes and Signaling Pathways

The synergistic interactions between heavy metal-induced stress and high tropospheric ozone concentrations have profound effects on cellular processes and signaling pathways in plants. These interactions disrupt normal cellular functions and initiate complex responses that alter plant physiology and biochemical pathways as depicted in Fig. 9.2.

a. *Cellular Oxidative Stress*

The combined exposure to heavy metals and ozone leads to increased oxidative stress within plant cells. The synergistic interaction enhances the production of reactive oxygen species (ROS), including superoxide radicals ($O_2^{\bullet-}$), hydrogen peroxide ($H_2O_2$), and hydroxyl radicals (•OH). Elevated ROS levels result in oxidative damage to cellular components, including lipids, proteins, nucleic acids, and membranes. This oxidative stress disrupts cellular homeostasis and affects key cellular processes (Sachdev et al. 2021).

b. *Disruption of Photosynthesis*

Synergistic interactions between heavy metals and ozone impair the process of photosynthesis. Heavy metals can directly inhibit photosynthetic machinery, such as photosystem II (PSII), by disrupting the electron transport chain and reducing the efficiency of light capture and utilization. Ozone, on the other hand, damages chloroplasts and impairs photosynthetic pigments, leading to reduced carbon assimilation and decreased photosynthetic efficiency. The combined effects of heavy metals and ozone result in significant reductions in photosynthesis, leading to decreased plant productivity (Aggarwal and Sharma 2012).

c. *Altered Nutrient Uptake and Transport*

The synergistic interactions between heavy metals and ozone influence nutrient uptake and transport processes in plants. Heavy metals can compete with essential nutrients for uptake and binding sites, leading to nutrient imbalances and deficiencies (Hossain et al. 2012). Ozone exposure affects nutrient uptake by altering root architecture and reducing nutrient assimilation efficiency. These combined stressors disrupt the nutrient acquisition and distribution systems, impacting plant growth, development, and overall nutrient status.

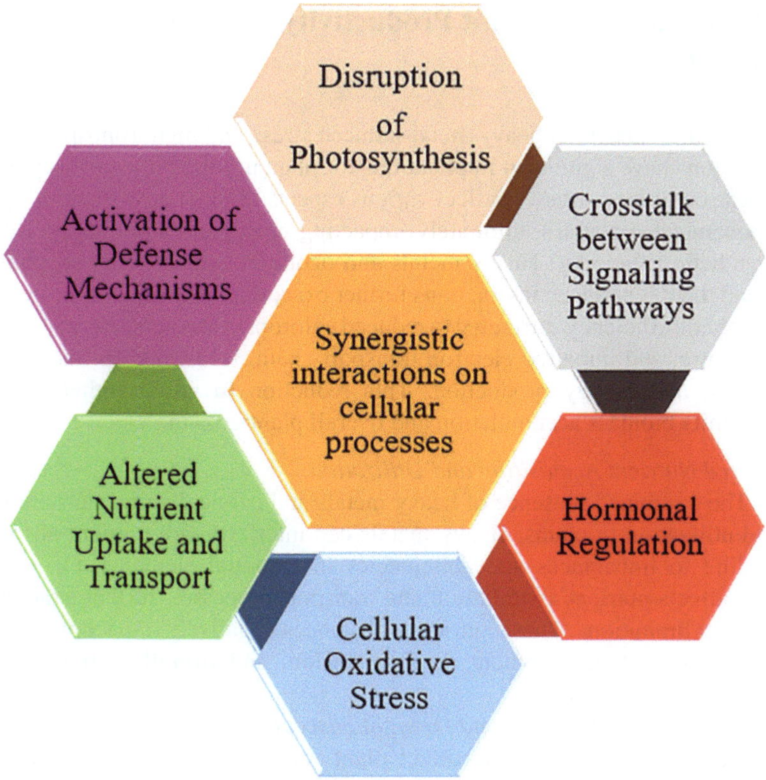


**Fig. 9.2** Effect of synergistic interactions on cellular processes

d. *Hormonal Regulation*

Synergistic interactions between heavy metals and ozone affect hormonal regulation in plants. Heavy metals can interfere with hormone synthesis, transport, and signaling pathways, leading to imbalances in hormone levels (Bücker-Neto et al. 2017). Ozone exposure also disrupts hormonal regulation by affecting hormone synthesis and metabolism. The combined presence of heavy metals and ozone alters hormonal signaling, affecting various plant processes, including growth, development, and stress responses.

e. *Activation of Defense Mechanisms*

The synergistic interactions between heavy metals and ozone trigger the activation of defense mechanisms in plants. Both stressors induce the expression of stress-responsive genes and the synthesis of defense-related compounds. This includes the activation of antioxidant enzymes, such as superoxide dismutase (SOD), catalase (CAT), and peroxidases, to counteract ROS-mediated damage (Tripathy & Oelmüller 2012). The synergistic effects enhance the activation of defense mechanisms, enabling plants to cope with the combined stressors.

## 9.8 Implications for Plant Productivity and Crop Production

The synergistic impacts of heavy metal-induced stress and high tropospheric ozone concentrations have significant implications for plant productivity and crop production. These combined stressors affect various aspects of plant growth, development, and biochemical processes, ultimately impacting crop yield and quality. Reduced Photosynthetic Efficiency: Heavy metals and ozone individually impair photosynthesis, and their synergistic interactions further exacerbate this effect. The decreased photosynthetic efficiency, attributed to reduced chlorophyll content, disrupted chloroplast structure, and impaired electron transport chain, leads to a decline in carbon assimilation and energy production. This reduction in photosynthetic capacity directly limits biomass accumulation and overall plant growth.

i. *Altered Nutrient Acquisition and Utilization*

   The combined presence of heavy metals and ozone disrupts nutrient uptake and utilization in plants. Heavy metals can interfere with nutrient absorption, leading to imbalances and deficiencies in essential elements. Ozone exposure affects nutrient assimilation and transport processes, further exacerbating nutrient limitations. These nutrient imbalances negatively impact plant growth, impairing root development, leaf expansion, and overall crop productivity (Nawab et al. 2022).

ii. *Impaired Water Relations and Drought Susceptibility*

   Heavy metals and ozone affect plant water relations and increase water stress susceptibility. Heavy metals can reduce water uptake and transpiration rates, affecting stomatal conductance and water use efficiency. Ozone exposure induces stomatal closure, limiting water loss through transpiration. The combination of these stressors disrupts water balance in plants, making them more susceptible to drought stress. Reduced water availability compromises crop productivity and yield (Rucińska-Sobkowiak 2016).

iii. *Compromised Biochemical Pathways*

   The synergistic interactions between heavy metals and ozone disrupt biochemical pathways essential for crop productivity. Heavy metals and ozone individually influence primary and secondary metabolic pathways, including those involved in carbohydrate metabolism, amino acid synthesis, and secondary metabolite production. Their combined effects alter metabolic processes, resulting in reduced energy availability, imbalanced nutrient partitioning, and altered production of bioactive compounds. These disruptions negatively impact crop quality and nutritional value (Molina and Segura 2021).

iv. *Increased Oxidative Stress and Cellular Damage*

   Heavy metals and ozone induce oxidative stress by promoting the production of reactive oxygen species (ROS). The synergistic interactions between these stressors further enhance ROS production, leading to elevated oxidative stress levels. This increased oxidative stress damages cellular components, including lipids, proteins, DNA, and membranes, impairing cellular integrity and function. The cumulative cellular damage negatively affects plant health, leading to reduced crop productivity and compromised yield (Saddhe et al. 2019).

v. *Challenges for Food Security:* The combined impacts of heavy metal-induced stress and high tropospheric ozone concentrations pose challenges for food security. As these stressors affect multiple aspects of plant physiology and biochemistry, crop productivity and yield potential are compromised. Reduced crop yields, impaired crop quality, and increased susceptibility to environmental stresses can have severe implications for food production and availability. Furthermore, regions with high heavy metal contamination and elevated ozone concentrations may face difficulties in sustaining agricultural practices and ensuring a stable food supply.

### *9.8.1 Strategies for Mitigating the Synergistic Impacts*

Mitigating the synergistic impacts of heavy metal-induced stress and high tropospheric ozone concentrations on plant productivity, physiology, and biochemistry is essential for sustainable crop production and environmental health. Implementing appropriate strategies can help minimize the negative effects of these stressors and enhance crop resilience. Here are some strategies for mitigating the synergistic impacts:

i. *Soil Remediation:* Implementing soil remediation techniques is crucial for reducing heavy metal contamination in agricultural soils. Soil amendments, such as organic matter, lime, and certain minerals, can help immobilize heavy metals, reducing their bioavailability to plants. Phytoremediation, which involves using specific plant species to extract and accumulate heavy metals from the soil, can also be employed. By improving soil quality and reducing heavy metal concentrations, the negative impacts of heavy metal-induced stress can be minimized (Azhar et al. 2022).

ii. Ozone Pollution Control Measures: To mitigate the impacts of high tropospheric ozone concentrations, it is important to implement measures to reduce ozone pollution. This can be achieved through the implementation of emission control strategies, such as reducing industrial emissions, promoting cleaner energy sources, and adopting strict air quality regulations. Additionally, public awareness campaigns and community initiatives can educate individuals about the importance of reducing ozone-forming pollutants and promoting sustainable practices.

iii. Crop Management Techniques: Implementing appropriate crop management techniques can help mitigate the effects of synergistic stressors. Practices such as crop rotation, intercropping, and cover cropping can improve soil health, enhance nutrient cycling, and reduce heavy metal uptake by plants. Adequate irrigation and water management practices can minimize the impact of water stress caused by heavy metals and ozone exposure. Furthermore, precision farming techniques, such as site-specific nutrient management and controlled-release fertilizer application, can optimize nutrient use efficiency and minimize nutrient imbalances.

iv. Breeding for Stress Tolerance: Developing crop varieties with enhanced stress tolerance is crucial for mitigating the synergistic impacts. Plant breeding programs can focus on selecting and developing cultivars that are resistant to heavy metal toxicity and ozone damage. Breeding for traits such as efficient heavy metal exclusion, tolerance to oxidative stress, and enhanced antioxidant defense mechanisms can improve plant resilience to combined stressors. Modern biotechnological tools, such as genetic engineering and genome editing, can also be utilized to enhance stress tolerance in crops.

v. Chemical Amendments and Biostimulants: The application of chemical amendments and biostimulants can aid in mitigating the impacts of synergistic stressors. Chemical chelating agents, such as EDTA (ethylenediaminetetraacetic acid), can help mobilize heavy metals in the soil, making them less available for plant uptake (Hasan et al. 2019). Biostimulants, including plant growth-promoting substances and biofertilizers, can enhance plant growth, stimulate root development, and improve nutrient uptake efficiency. These amendments and biostimulants can help mitigate the negative effects of combined stressors on plant physiology and biochemistry.

vi. Integrated Pest and Disease Management: Given the increased susceptibility of plants to pests and diseases under synergistic stress, implementing integrated pest and disease management practices is essential. This approach involves combining various strategies, including cultural, biological, and chemical control methods, to manage pest and disease pressures. By implementing integrated pest management techniques, crop health can be maintained, reducing the additional stress imposed by pest and disease attacks.

vii. Climate-Resilient Agricultural Practices: Adopting climate-resilient agricultural practices is crucial for mitigating the impacts of synergistic stressors in a changing climate. This includes practices such as conservation agriculture, agroforestry, and precision farming, which promote sustainable land management, soil conservation, and resource-use efficiency. Implementing climate-smart agricultural techniques can enhance crop resilience, reduce environmental risks, and contribute to long-term sustainability.

By implementing these strategies, farmers, researchers, and policymakers can work together to mitigate the synergistic impacts of heavy metal-induced stress and high tropospheric ozone concentrations. Through a combination of soil management, pollution control, crop breeding, and sustainable agricultural practices, the negative effects on plant productivity, physiology, and biochemistry can be minimized, ensuring a more resilient and sustainable agricultural system.

### 9.8.2 Genetic Approaches

Genetic approaches play a crucial role in mitigating the synergistic impacts of heavy metal-induced stress and high tropospheric ozone concentrations on plant productivity, physiology, and biochemistry. These approaches involve the identification, selection, and development of crop varieties with enhanced stress tolerance and resilience. Here are some genetic strategies for mitigating the synergistic impacts:

a. *Breeding for Stress Tolerance:* Plant breeding programs focus on selecting and developing crop varieties with improved tolerance to heavy metals and ozone stress. This involves screening germplasm collections for natural genetic variations that confer stress tolerance traits. Traits such as efficient heavy metal exclusion, enhanced antioxidant defense mechanisms, and improved stomatal regulation under ozone exposure are targeted. Through traditional breeding techniques, such as cross-breeding and selection, crop varieties with improved stress tolerance can be developed.
b. *Marker-Assisted Selection (MAS):* Marker-assisted selection is a genetic approach that utilizes DNA markers linked to specific stress tolerance genes. By identifying and utilizing molecular markers associated with stress tolerance traits, breeders can expedite the selection process for desirable traits. MAS enables the early identification of stress-tolerant individuals in breeding populations, allowing breeders to develop improved varieties with enhanced tolerance to both heavy metals and ozone stress (Mareri et al. 2022).
c. *Genetic Engineering and Genome Editing:* Modern biotechnological tools, such as genetic engineering and genome editing, offer precise and targeted approaches for improving stress tolerance in crops. Genes are responsible for stress tolerance can be transferred or modified in crop genomes to confer enhanced resistance to heavy metal-induced stress and ozone damage. For example, genes encoding metal transporters, antioxidant enzymes, or regulatory proteins involved in stomatal regulation can be introduced or manipulated to enhance stress tolerance (Mareri et al. 2022). Genome editing technologies, such as CRISPR-Cas9, provide efficient and precise tools for making targeted modifications in plant genomes, facilitating the development of stress-tolerant crop varieties.
d. *Transgenic Approaches:* Transgenic approaches involve the introduction of genes from unrelated organisms into crop genomes to confer specific stress tolerance traits. For instance, the introduction of genes encoding metal-binding peptides or enzymes involved in detoxification pathways can enhance heavy metal tolerance in crops. Similarly, the incorporation of genes that enhance antioxidant defense mechanisms or regulate stomatal responses can improve plant tolerance to high tropospheric ozone concentrations (Parmar et al. 2017). Transgenic approaches offer opportunities to engineer crops with multiple stress tolerance traits, providing robust protection against the combined impacts of heavy metals and ozone stress.
e. *Genomic Studies and Molecular Mechanisms:* Understanding the molecular mechanisms underlying plant responses to heavy metal-induced stress and ozone

exposure is crucial for developing effective genetic strategies. Genomic studies, such as transcriptomics, proteomics, and metabolomics, enable the identification of genes, proteins, and metabolites associated with stress responses. By deciphering the regulatory networks and molecular pathways involved in stress tolerance, researchers can identify key targets for genetic manipulation and develop crop varieties with improved resilience to synergistic stressors (Isah 2019). Genetic approaches offer promising solutions for mitigating the synergistic impacts of heavy metal-induced stress and high tropospheric ozone concentrations on plant productivity, physiology, and biochemistry. Through breeding for stress tolerance, marker-assisted selection, genetic engineering, and genome editing, crop varieties with enhanced tolerance to combined stressors can be developed. These genetically improved crops contribute to sustainable agriculture, ensuring food security and environmental stewardship in the face of increasing stressors in our changing climate.

### 9.8.3 *Agronomic Practices*

In addition to genetic approaches, implementing appropriate agronomic practices is vital for mitigating the synergistic impacts of heavy metal-induced stress and high tropospheric ozone concentrations on plant productivity, physiology, and biochemistry. Agronomic practices focus on optimizing crop management techniques and environmental conditions to reduce the negative effects of combined stressors. Here are some agronomic strategies for mitigating the synergistic impacts as shown in Fig. 9.3.

a. *Soil Management:* Implementing effective soil management practices is essential for minimizing the impacts of heavy metal-induced stress. Soil amendments, such as organic matter, compost, and biochar, can improve soil structure, nutrient availability, and microbial activity, leading to enhanced plant health and reduced metal uptake (Alkharabsheh et al. 2021). Additionally, adjusting soil pH and applying amendments like lime can help alleviate heavy metal toxicity. Proper soil drainage and erosion control measures should be implemented to prevent heavy metal runoff and contamination of adjacent areas.
b. *Nutrient Management:* Optimizing nutrient management practices is crucial for reducing the negative impacts of combined stressors on plant physiology. Balanced nutrient application, based on soil testing and crop nutrient requirements, ensures that plants have sufficient nutrients for growth and development. Adequate nutrient levels can help mitigate the negative effects of heavy metals and ozone on nutrient uptake and distribution in plants. Using slow-release fertilizers or organic fertilizers can enhance nutrient use efficiency and minimize nutrient imbalances.
c. *Irrigation Management:* Proper irrigation management plays a critical role in mitigating the impacts of combined stressors on plant water relations. Water stress

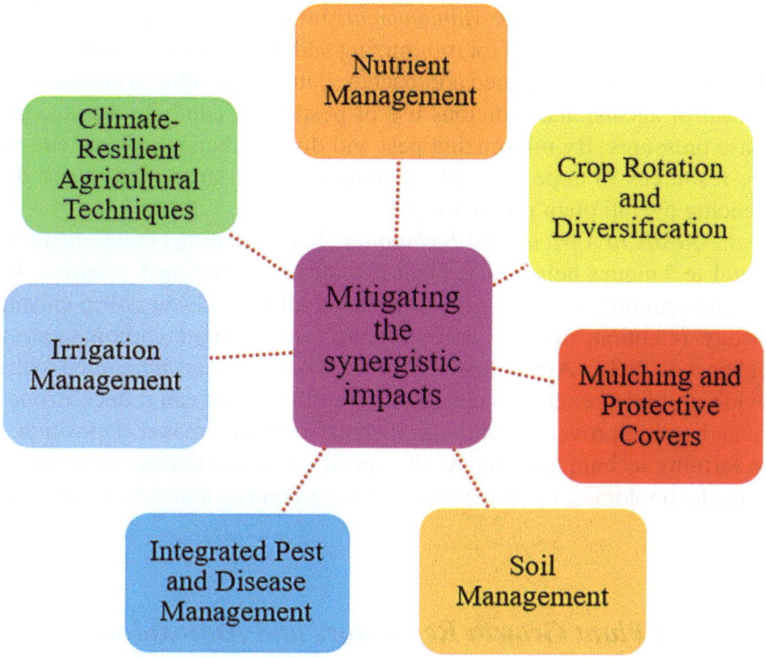


**Fig. 9.3** Agronomic strategies for mitigating the synergistic impacts

caused by heavy metals and high ozone concentrations can exacerbate the negative effects on plant physiology. Implementing efficient irrigation techniques, such as drip irrigation or precision irrigation, helps optimize water use efficiency and minimize water stress. Monitoring soil moisture levels and applying irrigation based on plant needs can prevent excessive water loss and reduce the risk of drought-induced stress.

d. *Crop Rotation and Diversification:* Implementing crop rotation and diversification practices can help mitigate the impacts of combined stressors. Rotating crops with different nutrient requirements and stress tolerances helps break pest and disease cycles, reduce heavy metal accumulation, and enhance soil health. Intercropping or mixed cropping can provide a diverse plant community, which can improve nutrient cycling, enhance pest and disease management, and reduce the overall impact of stressors on individual crops (Norris and Congreves 2018).
e. *Mulching and Protective Covers:* Using mulching materials, such as straw, plastic films, or cover crops, can provide multiple benefits in mitigating stress impacts. Mulching helps conserve soil moisture, regulate soil temperature, suppress weed growth, and reduce heavy metal uptake by plants (Iqbal et al. 2020). Protective covers, such as shade nets or screen houses, can offer protection against high ozone concentrations, reducing the direct exposure of plants to ozone and mitigating its negative effects on plant physiology and biochemistry.

f. *Integrated Pest and Disease Management:* Integrated pest and disease management practices are essential for minimizing additional stress on plants exposed to combined stressors. Integrated approaches, combining cultural practices, biological control agents, and judicious use of pesticides, can help manage pest and disease pressures. By minimizing pest and disease damage, plants can allocate more resources to cope with heavy metal-induced stress and ozone damage, enhancing overall plant productivity.
g. *Climate-Resilient Agricultural Techniques:* Implementing climate-resilient agricultural techniques helps build plant resilience to combined stressors. Conservation agriculture practices, such as reduced tillage and soil cover, enhance soil moisture retention, improve soil structure, and promote carbon sequestration (Sekhar et al. 2022). Agroforestry systems, where trees are integrated with crops, provide multiple benefits, including shade effects that can reduce ozone exposure and soil improvement through increased organic matter deposition. Precision farming techniques, such as site-specific nutrient management and remote sensing technologies, optimize resource use efficiency and reduce environmental impacts.

### *9.8.4 Use of Plant Growth Regulators and Antioxidants*

The application of plant growth regulators and antioxidants offers potential solutions for mitigating the synergistic impacts of heavy metal-induced stress and high tropospheric ozone concentrations on plant productivity, physiology, and biochemistry. These compounds can modulate plant responses and alleviate the negative effects of combined stressors. Here are some strategies involving the use of plant growth regulators and antioxidants:

a. *Plant Growth Regulators:* Plant growth regulators, such as phytohormones, play a crucial role in regulating plant growth, development, and stress responses. Exogenous application of certain phytohormones, such as auxins, cytokinins, and abscisic acid, can enhance plant tolerance to heavy metal-induced stress and ozone damage. These regulators can improve root growth, nutrient uptake, and photosynthetic efficiency, thereby mitigating the negative impacts of stressors on plant physiology (Asif et al. 2022). Additionally, plant growth regulators can modulate stomatal behavior, reducing ozone uptake and minimizing ozone-induced oxidative stress.
b. *Antioxidants:* Antioxidants are compounds that protect plants against oxidative stress caused by reactive oxygen species (ROS). ROS production is enhanced under heavy metal-induced stress and high tropospheric ozone concentrations. Exogenous applications of antioxidants, such as ascorbic acid, glutathione, or polyphenols, can enhance the antioxidant defense system in plants (Kasote et al. 2015). Antioxidants scavenge ROS, minimizing oxidative damage to cellular components and reducing the negative impacts on plant physiology and

biochemistry. Additionally, antioxidants can regulate gene expression and signal transduction pathways, enhancing stress tolerance in plants.

c. *Biostimulants:* Biostimulants are substances that promote plant growth, development, and stress tolerance. They can enhance plant physiological processes, nutrient uptake, and antioxidant defense mechanisms. Biostimulants derived from natural sources, such as seaweed extracts, humic acids, or microbial-based products, have shown the potential in mitigating the negative impacts of combined stressors (Drobek et al. 2019). These biostimulants improve plant resilience, increase nutrient availability, and enhance the plant's capacity to withstand heavy metal-induced stress and high ozone concentrations.
d. *Nanoparticles:* Nanotechnology-based approaches involving the use of nanoparticles have emerged as potential tools for mitigating stress impacts. Nanoparticles, such as titanium dioxide or cerium oxide, can scavenge ROS and reduce oxidative stress in plants (Huang et al. 2021). They can also help regulate stomatal conductance, reducing ozone uptake and its negative effects on plant physiology. Additionally, nanoparticles can enhance nutrient availability, improve water use efficiency, and facilitate the delivery of growth regulators or antioxidants to targeted plant tissues, optimizing their efficacy.
e. *Foliar Sprays and Seed Priming:* Applying growth regulators and antioxidants as foliar sprays or during seed priming can provide direct and targeted delivery to plants. Foliar sprays ensure rapid uptake and utilization of these compounds by the plant, particularly during critical growth stages or under stress conditions. Seed priming involves the pre-soaking or coating of seeds with growth regulators or antioxidants before planting. This enhances seed germination, early seedling growth, and stress tolerance in emerging plants.
f. *Application Timing and Dose Optimization:* The timing and dose of plant growth regulators and antioxidants are crucial for their effectiveness. Optimal application timing ensures that plants receive treatments during periods of high-stress susceptibility, maximizing their protective effects. Similarly, optimizing the dosage of these compounds based on plant species, growth stage, and stress severity is important to avoid phytotoxicity while providing adequate stress mitigation benefits.

## 9.9 Conclusion

This chapter has explored the synergistic impacts of heavy metal-induced stress and high tropospheric ozone concentrations on plant productivity, physiology, and biochemistry. Through an examination of the mechanisms, physiological responses, and biochemical changes, as well as the interactive effects and mechanistic insights, several key findings have emerged:

i. *Heavy Metal-Induced Stress:* Plants exposed to heavy metals experience alterations in nutrient uptake and distribution, leading to reduced photosynthesis and chlorophyll content. The oxidative stress caused by heavy metals triggers the production of reactive oxygen species (ROS), which can damage cellular components. However, plants also activate antioxidant defense systems to counteract oxidative damage.

ii. *High Tropospheric Ozone concentrations:* Elevated ozone concentrations have detrimental effects on plant productivity by disrupting photosynthesis and carbon assimilation. Stomatal conductance and water use efficiency are negatively affected, leading to water stress. Ozone exposure also induces ROS generation and oxidative stress in plants, which can impair cellular processes and trigger antioxidant defense mechanisms.

iii. *Synergistic Interactions:* The combined effects of heavy metal-induced stress and high tropospheric ozone concentrations on plants are greater than the sum of their individual effects. The interactive impacts are manifested through alterations in photosynthesis, stomatal conductance, nutrient uptake, and distribution. Oxidative stress is intensified, leading to higher ROS production and subsequent activation of antioxidant defense systems. Primary and secondary metabolite synthesis is also influenced by synergistic interactions, affecting plant growth, development, and overall productivity.

### 9.9.1 Future Research Directions

While significant progress has been made in understanding the synergistic impacts of heavy metal-induced stress and high tropospheric ozone concentrations on plant productivity, physiology, and biochemistry, there are still several areas that warrant further investigation. Future research efforts should focus on the following key directions:

i. *Mechanistic Insights:* Elucidating the underlying mechanisms of the synergistic interactions between heavy metal-induced stress and high tropospheric ozone concentrations is crucial. Further research is needed to investigate the molecular pathways, gene expression patterns, and signaling networks involved in the combined stress response. This will provide a deeper understanding of the interactive effects and help identify key targets for developing effective mitigation strategies.

ii. *Long-Term Effects:* Most studies conducted thus far have focused on short-term responses to combined stressors. However, long-term effects and their implications for plant growth, development, and ecosystem dynamics remain relatively unexplored. Investigating the cumulative effects of prolonged exposure to heavy metals and high ozone concentrations will provide insights into the chronic impacts on plant productivity and ecosystem functioning.

iii. *Crop-Specific Responses:* Different plant species and cultivars exhibit varying sensitivities and responses to combined stressors. It is essential to investigate crop-specific responses and identify genotypes or traits that confer tolerance to heavy metal-induced stress and high ozone concentrations. Crop breeding programs can utilize this knowledge to develop cultivars with enhanced stress tolerance and productivity.

iv. *Interactions with Other Environmental Factors:* Plants are often subjected to multiple environmental stressors simultaneously. Understanding the interactions between heavy metal-induced stress, high tropospheric ozone concentrations, and other stress factors, such as drought, heat, or nutrient deficiency, is crucial for comprehensive stress management strategies. Future research should explore the combined effects of these stressors and their synergistic or antagonistic interactions.

v. *Ecological Consequences:* The impacts of combined stressors on plant productivity and biochemistry extend beyond individual plants to ecosystem dynamics. Investigating the ecological consequences of heavy metal-induced stress and high ozone concentrations will help elucidate the cascading effects on plant communities, biodiversity, nutrient cycling, and ecosystem functioning. Such studies will provide a broader perspective on the implications of synergistic stress interactions in natural ecosystems.

vi. *Mitigation Strategies:* While various strategies, including genetic approaches, agronomic practices, and the use of plant growth regulators and antioxidants, have shown promise in mitigating the synergistic impacts, further research is needed to optimize their efficacy and sustainability. Developing practical and cost-effective mitigation strategies that can be implemented at field scales is crucial for their real-world application and adoption by farmers.

## References

Ackova DG (2018) Heavy metals and their general toxicity on plants. In Plant Sci Today. https://doi.org/10.14719/pst.2018.5.1.355

Aggarwal A, Sharma I (2012) Metal Toxicity and Photosynthesis. Photosynthesis: …

Alkharabsheh HM, Seleiman MF, Battaglia ML, Shami A, Jalal RS, Alhammad BA, Almutairi KF, Al-Saif AM (2021) Biochar and its broad impacts in soil quality and fertility, nutrient leaching and crop productivity: a review. In Agronomy. https://doi.org/10.3390/agronomy11050993

Arif N, Yadav V, Singh S, Singh S, Ahmad P, Mishra RK, Sharma S, Tripathi DK, Dubey NK, Chauhan DK (2016) Influence of high and low levels of plant-beneficial heavy metal ions on plant growth and development. In Front Environ Sci. https://doi.org/10.3389/fenvs.2016.00069

Asif R, Yasmin R, Mustafa M, Ambreen A, Mazhar M, Rehman A, Umbreen S, Ahmad M (2022) Phytohormones as plant growth regulators and safe protectors against biotic and abiotic stress. In: Plant hormones-recent advances, new perspectives and applications. https://doi.org/10.5772/intechopen.102832

Azhar U, Ahmad H, Shafqat H, Babar M, Shahzad Munir HM, Sagir M, Arif M, Hassan A, Rachmadona N, Rajendran S, Mubashir M, Khoo KS (2022) Remediation techniques for elimination of heavy metal pollutants from soil: a review. Environ Res. https://doi.org/10.1016/j.envres.2022.113918

Bhatia A, Tomer R, Kumar V, Singh SD, Pathak H (2012) Impact of tropospheric ozone on crop growth and productivity—a review. In J Sci Ind Res

Bücker-Neto L, Paiva ALS, Machado RD, Arenhart RA, Margis-Pinheiro M (2017) Interactions between plant hormones and heavy metals responses. Genet Mol Biol. https://doi.org/10.1590/1678-4685-gmb-2016-0087

Chauhan P, Sharma N (2023) Effect of elevated $O_3$ on plants growth, active constituents, and production. In: Plants and their interaction to environmental pollution. https://doi.org/10.1016/b978-0-323-99978-6.00020-0

Chen D, Mubeen B, Hasnain A, Rizwan M, Adrees M, Naqvi SAH, Iqbal S, Kamran M, El-Sabrout AM, Elansary HO, Mahmoud EA, Alaklabi A, Sathish M, Din GMU (2022) Role of promising secondary metabolites to confer resistance against environmental stresses in crop plants: current scenario and future perspectives. In Front Plant Sci. https://doi.org/10.3389/fpls.2022.881032

Das K, Roychoudhury A (2014) Reactive oxygen species (ROS) and response of antioxidants as ROS-scavengers during environmental stress in plants. In Front Environ Sci. https://doi.org/10.3389/fenvs.2014.00053

Drobek M, Frąc M, Cybulska J (2019) Plant biostimulants: importance of the quality and yield of horticultural crops and the improvement of plant tolerance to abiotic stress—a review. Agronomy. https://doi.org/10.3390/agronomy9060335

Dutta S, Mitra M, Agarwal P, Mahapatra K, De S, Sett U, Roy S (2018) Oxidative and genotoxic damages in plants in response to heavy metal stress and maintenance of genome stability. In Plant Signal Behav. https://doi.org/10.1080/15592324.2018.1460048

Emberson LD, Pleijel H, Ainsworth EA, van den Berg M, Ren W, Osborne S, Mills G, Pandey D, Dentener F, Büker P, Ewert F, Koeble R, Van Dingenen R (2018) Ozone effects on crops and consideration in crop models. Eur J Agron. https://doi.org/10.1016/j.eja.2018.06.002

Fodor F (2002) Physiological responses of vascular plants to heavy metals. In: Physiology and biochemistry of metal toxicity and tolerance in plants. https://doi.org/10.1007/978-94-017-2660-3_6

Guo Z, Gao Y, Yuan X, Yuan M, Huang L, Wang S, Liu C (2023). Effects of heavy metals on stomata in plants: a review

Hafeez A, Rasheed R, Ashraf MA, Qureshi FF, Hussain I, Iqbal M (2023) Effect of heavy metals on growth, physiological and biochemical responses of plants. In: Plants and their interaction to environmental pollution. https://doi.org/10.1016/b978-0-323-99978-6.00006-6

Hasan MM, Uddin MN, Ara-Sharmeen I, Alharby HF, Alzahrani Y, Hakeem KR, Zhang L (2019) Assisting phytoremediation of heavy metals using chemical amendments. Plants. https://doi.org/10.3390/plants8090295

Hasanuzzaman M, Bhuyan MHMB, Parvin K, Bhuiyan TF, Anee TI, Nahar K, Hossen MS, Zulfiqar F, Alam MM, Fujita M (2020) Regulation of ros metabolism in plants under environmental stress: a review of recent experimental evidence. In Int J Mol Sci. https://doi.org/10.3390/ijms21228695

Hossain MA, Piyatida P, da Silva JAT, Fujita M (2012) Molecular mechanism of heavy metal toxicity and tolerance in plants: central role of glutathione in detoxification of reactive oxygen species and methylglyoxal and in heavy metal chelation. J Bot. https://doi.org/10.1155/2012/872875

Huang X, He D, Pan Z, Luo G, Deng J (2021) Reactive-oxygen-species-scavenging nanomaterials for resolving inflammation. In Mater Today Bio. https://doi.org/10.1016/j.mtbio.2021.100124

Iqbal R, Raza MAS, Valipour M, Saleem MF, Zaheer MS, Ahmad S, Toleikiene M, Haider I, Aslam MU, Nazar MA (2020) Potential agricultural and environmental benefits of mulches—a review. Bull Natl Res Cent. https://doi.org/10.1186/s42269-020-00290-3

Isah T (2019) Stress and defense responses in plant secondary metabolites production. In Biol Res. https://doi.org/10.1186/s40659-019-0246-3

Jarvis AJ, Mansfield TA, Davies WJ (1999) Stomatal behaviour, photosynthesis and transpiration under rising $CO_2$. Plant Cell Environ. https://doi.org/10.1046/j.1365-3040.1999.00407.x

Kasote DM, Katyare SS, Hegde MV, Bae H (2015) Significance of antioxidant potential of plants and its relevance to therapeutic applications. In Int J Biol Sci. https://doi.org/10.7150/ijbs.12096

Kosová K, Vítámvás P, Urban MO, Prášil IT, Renaut J (2018) Plant abiotic stress proteomics: the major factors determining alterations in cellular proteome. In Front Plant Sci. https://doi.org/10.3389/fpls.2018.00122

Krishnamurthy A, Rathinasabapathi B (2013) Oxidative stress tolerance in plants. Plant Signal Behav. https://doi.org/10.4161/psb.25761

Kumar S, Kumar S, Mohapatra T (2021) Interaction between macro- and micro-nutrients in plants. In Front Plant Sci. https://doi.org/10.3389/fpls.2021.665583

Li S, Courbet G, Ourry A, Ainsworth EA (2019) Elevated ozone concentration reduces photosynthetic carbon gain but does not alter leaf structural traits, nutrient composition or biomass in switchgrass. Plants. https://doi.org/10.3390/plants8040085

Ludwikow A, Sadowski J (2008) Gene networks in plant ozone stress response and tolerance. In J Integr Plant Biol. https://doi.org/10.1111/j.1744-7909.2008.00738.x

Mareri L, Parrotta L, Cai G (2022) Environmental Stress and Plants. In Int J Mol Sci. https://doi.org/10.3390/ijms23105416

Molina L, Segura A (2021) Biochemical and metabolic plant responses toward polycyclic aromatic hydrocarbons and heavy metals present in atmospheric pollution. In Plants. https://doi.org/10.3390/plants10112305

Muhammad I, Shalmani A, Ali M, Yang QH, Ahmad H, Li FB (2021) Mechanisms regulating the dynamics of photosynthesis under abiotic stresses. Front Plant Sci 11(January):1–25. https://doi.org/10.3389/fpls.2020.615942

Nawab J, Ghani J, Khan S, Khan MA, Ali A, Rahman Z, Alam M, Hesham A E-L, Lei M (2022). Nutrient uptake and plant growth under the influence of toxic elements doi:https://doi.org/10.1007/978-3-030-91499-8_5

Norris CE, Congreves KA (2018) Alternative management practices improve soil health indices in intensive vegetable cropping systems: a review. In Front Environ Sci. https://doi.org/10.3389/fenvs.2018.00050

Pande A, Mun BG, Methela NJ, Rahim W, Lee DS, Lee GM, Hong JK, Hussain A, Loake G, Yun BW (2022) Heavy metal toxicity in plants and the potential NO-releasing novel techniques as the impending mitigation alternatives. In Front Environ Sci. https://doi.org/10.3389/fpls.2022.1019647

Parmar N, Singh KH, Sharma D, Singh L, Kumar P, Nanjundan J, Khan YJ, Chauhan DK, Thakur AK (2017) Genetic engineering strategies for biotic and abiotic stress tolerance and quality enhancement in horticultural crops: a comprehensive review. In 3 Biotech. https://doi.org/10.1007/s13205-017-0870-y

Petrov V, Hille J, Mueller-Roeber B, Gechev TS (2015) ROS-mediated abiotic stress-induced programmed cell death in plants. In Front Plant Sci. https://doi.org/10.3389/fpls.2015.00069

Pinto DM, Blande JD, Souza SR, Nerg AM, Holopainen JK (2010) Plant volatile organic compounds (vocs) in ozone ($O_3$) polluted atmospheres: The ecological effects. J Chem Ecol. https://doi.org/10.1007/s10886-009-9732-3

Rajput VD, Harish Singh RK, Verma KK, Sharma L, Quiroz-Figueroa FR, Meena M, Gour VS, Minkina T, Sushkova S, Mandzhieva S (2021) Recent developments in enzymatic antioxidant defence mechanism in plants with special reference to abiotic stress. In Biology. https://doi.org/10.3390/biology10040267

Renaut J, Bohler S, Hausman JF, Hoffmann L, Sergeant K, Ahsan N, Jolivet Y, Dizengremel P (2009) The impact of atmospheric composition on plants: a case study of ozone and poplar. Mass Spectrom Rev. https://doi.org/10.1002/mas.20202

Riyazuddin R, Nisha N, Ejaz B, Khan MIR, Kumar M, Ramteke PW, Gupta R (2022) A comprehensive review on the heavy metal toxicity and sequestration in plants. In Biomolecules. https://doi.org/10.3390/biom12010043

Rucińska-Sobkowiak R (2016) Water relations in plants subjected to heavy metal stresses. In Acta Physiol Plant. https://doi.org/10.1007/s11738-016-2277-5

Sachdev S, Ansari SA, Ansari MI, Fujita M, Hasanuzzaman M (2021) Abiotic stress and reactive oxygen species: generation, signaling, and defense mechanisms. In Antioxidants. https://doi.org/10.3390/antiox10020277

Saddhe AA, Malvankar MR, Karle SB, Kumar K (2019) Reactive nitrogen species: paradigms of cellular signaling and regulation of salt stress in plants. Environ Exp Bot. https://doi.org/10.1016/j.envexpbot.2018.11.010

Sandeep G, Vijayalatha KR, Anitha T (2019) Heavy metals and its impact in vegetable crops. Int J Chem Stud:1612

Sasi S, Venkatesh J, Daneshi RF, Gururani MA (2018) Photosystem ii extrinsic proteins and their putative role in abiotic stress tolerance in higher plants. Plants. https://doi.org/10.3390/plants7040100

Sekhar G, Goswami J, Naveen J, Priyanka E (2022). Climate resilient agronomic techniques for sustainable agriculture. *July*

Sénéchal H, Visez N, Charpin D, Shahali Y, Peltre G, Biolley JP, Lhuissier F, Couderc R, Yamada O, Malrat-Domenge A, Pham-Thi N, Poncet P, Sutra JP (2015) A review of the effects of major atmospheric pollutants on pollen grains, pollen content, and allergenicity. In Sci World J. https://doi.org/10.1155/2015/940243

Shabbir R, Singhal RK, Mishra UN, Chauhan J, Javed T, Hussain S, Kumar S, Anuragi H, Lal D, Chen P (2022) Combined abiotic stresses: challenges and potential for crop improvement. In Agronomy. https://doi.org/10.3390/agronomy12112795

Shahid M, Pourrut B, Dumat C, Nadeem M, Aslam M, Pinelli E (2014) Heavy-metal-induced reactive oxygen species: phytotoxicity and physicochemical changes in plants. In Rev Environ Contam Toxicol. https://doi.org/10.1007/978-3-319-06746-9_1

Sharma P, Jha AB, Dubey RS, Pessarakli M (2012) Reactive oxygen species, oxidative damage, and antioxidative defense mechanism in plants under stressful conditions. J Bot. https://doi.org/10.1155/2012/217037

Singh S, Parihar P, Singh R, Singh VP, Prasad SM (2016) Heavy metal tolerance in plants: role of transcriptomics, proteomics, metabolomics, and ionomics. *Front Plant Sci 6*(FEB 2016):1–36. https://doi.org/10.3389/fpls.2015.01143

Sperdouli I (2022) Heavy metal toxicity effects on plants. In Toxics. https://doi.org/10.3390/toxics10120715

Tripathy BC, Oelmüller R (2012) Reactive oxygen species generation and signaling in plants. Plant Signal Behav. https://doi.org/10.4161/psb.22455

Watanabe M, Hoshika Y, Koike T, Izuta T (2017) Combined effects of ozone and other environmental factors on Japanese trees. In: Air pollution impacts on plants in East Asia. https://doi.org/10.1007/978-4-431-56438-6_6

Xie X, He Z, Chen N, Tang Z, Wang Q, Cai Y (2019) The roles of environmental factors in regulation of oxidative stress in plant. In Bio Med Res Int. https://doi.org/10.1155/2019/9732325

Yang Y, Zhang L, Huang X, Zhou Y, Quan Q, Li Y, Zhu X (2020) Response of photosynthesis to different concentrations of heavy metals in Davidia involucrata. PLoS One. https://doi.org/10.1371/journal.pone.0228563

[illegible]

[illegible]

Springer Nature Customer Service Center GmbH
[illegible]platz 3, 69115 Heidelberg, Germany

[illegible] by Libri Plureos GmbH
[illegible], Germany

MIX
Papier aus verantwortungsvollen Quellen
Paper from responsible sources
FSC® C105338

If you have any concerns about our products,
you can contact us on
**ProductSafety@springernature.com**

In case Publisher is established outside the EU,
the EU authorized representative is:
**Springer Nature Customer Service Center GmbH**
**Europaplatz 3, 69115 Heidelberg, Germany**

Printed by Libri Plureos GmbH
in Hamburg, Germany